Vitamins and Hormones

VOLUME 73

Editorial Board

TADHG P. BEGLEY

ANTHONY R. MEANS

BERT W. O'MALLEY

LYNN RIDDIFORD

ARMEN H. TASHJIAN, JR.

INSECT HORMONES

VITAMINS AND HORMONES
ADVANCES IN RESEARCH AND APPLICATIONS

Editor-in-Chief

GERALD LITWACK

Professor and Chair Emeritus
Department of Biochemistry and Molecular Pharmacology
Thomas Jefferson University Medical College
Philadelphia, Pennsylvania
Visiting Scholar
Department of Biological Chemistry
David Geffen School of Medicine at UCLA
Toluca Lake, California

VOLUME 73

AMSTERDAM • BOSTON • HEIDELBERG • LONDON
NEW YORK • OXFORD • PARIS • SAN DIEGO
SAN FRANCISCO • SINGAPORE • SYDNEY • TOKYO
Academic Press is an imprint of Elsevier

Elsevier Academic Press
525 B Street, Suite 1900, San Diego, California 92101-4495, USA
84 Theobald's Road, London WC1X 8RR, UK

This book is printed on acid-free paper. ∞

Copyright © 2005, Elsevier Inc. All Rights Reserved.

No part of this publication may be reproduced or transmitted in any form or by any means, electronic or mechanical, including photocopy, recording, or any information storage and retrieval system, without permission in writing from the Publisher.

The appearance of the code at the bottom of the first page of a chapter in this book indicates the Publisher's consent that copies of the chapter may be made for personal or internal use of specific clients. This consent is given on the condition, however, that the copier pay the stated per copy fee through the Copyright Clearance Center, Inc. (www.copyright.com), for copying beyond that permitted by Sections 107 or 108 of the U.S. Copyright Law. This consent does not extend to other kinds of copying, such as copying for general distribution, for advertising or promotional purposes, for creating new collective works, or for resale. Copy fees for pre-2005 chapters are as shown on the title pages. If no fee code appears on the title page, the copy fee is the same as for current chapters.
0083-6729/2005 $35.00

Permissions may be sought directly from Elsevier's Science & Technology Rights Department in Oxford, UK: phone: (+44) 1865 843830, fax: (+44) 1865 853333, E-mail: permissions@elsevier.com. You may also complete your request on-line via the Elsevier homepage (http://elsevier.com), by selecting "Support & Contact" then "Copyright and Permission" and then "Obtaining Permissions."

For all information on all Elsevier Academic Press publications
visit our Web site at www.books.elsevier.com

ISBN-13: 978-0-12-709873-9
ISBN-10: 0-12-709873-9

PRINTED IN THE UNITED STATES OF AMERICA
05 06 07 08 09   9 8 7 6 5 4 3 2 1

**Working together to grow
libraries in developing countries**

www.elsevier.com | www.bookaid.org | www.sabre.org

ELSEVIER   BOOK AID International   Sabre Foundation

Former Editors

ROBERT S. HARRIS
Newton, Massachusetts

JOHN A. LORRAINE
*University of Edinburgh
Edinburgh, Scotland*

PAUL L. MUNSON
*University of North Carolina
Chapel Hill, North Carolina*

JOHN GLOVER
*University of Liverpool
Liverpool, England*

GERALD D. AURBACH
*Metabolic Diseases Branch
National Institute of Diabetes
and Digestive and Kidney Diseases
National Institutes of Health
Bethesda, Maryland*

KENNETH V. THIMANN
*University of California
Santa Cruz, California*

IRA G. WOOL
*University of Chicago
Chicago, Illinois*

EGON DICZFALUSY
*Karolinska Sjukhuset
Stockholm, Sweden*

ROBERT OLSEN
*School of Medicine
State University of New York
at Stony Brook
Stony Brook, New York*

DONALD B. McCORMICK
*Department of Biochemistry
Emory University School of Medicine
Atlanta, Georgia*

Contents

Contributors xi
Preface xiii

1

Hormonal Control of Insect Ecdysis: Endocrine Cascades for Coordinating Behavior with Physiology

James W. Truman

I. Introduction 2
II. The Physiology and Behaviors of the Ecdysis Sequence 3
III. Endocrine and Neuroendocrine Factors Involved in Ecdysis Control 6
IV. Interaction of Endocrine Signals in Controlling the Behavioral Phases 16
V. Conclusions 25
    References 26

2

A MOLECULAR GENETIC APPROACH TO THE BIOSYNTHESIS OF THE INSECT STEROID MOLTING HORMONE

LAWRENCE I. GILBERT AND JAMES T. WARREN

I. Introduction   32
II. Ecdysteroids   33
III. Neuropeptide Control of Ecdysteroidogenesis   36
IV. Subcellular Translocation of Ecdysteroid Intermediates   38
V. The "Black Box"   39
VI. Terminal Hydroxylations (From the Diketol to 20-Hydroxyecdysone)   41
VII. Epilogue   54
References   54

3

ECDYSTEROID RECEPTORS AND THEIR APPLICATIONS IN AGRICULTURE AND MEDICINE

SUBBA R. PALLI, ROBERT E. HORMANN, UWE SCHLATTNER, AND MARKUS LEZZI

I. Introduction   60
II. Ecdysteroid Receptors   61
III. Applications   78
IV. Conclusions and Perspectives   91
References   91

4

LIGAND-BINDING POCKET OF THE ECDYSONE RECEPTOR

ISABELLE M. L. BILLAS AND DINO MORAS

I. Introduction   102
II. Sequence Comparison of Invertebrate EcR Ligand-Binding Domains   104

III. Crystal Structures of the EcR Ligand-Binding Domain   105
IV. Conclusions   124
    References   126

5

Nonsteroidal Ecdysone Agonists

Yoshiaki Nakagawa

  I. Introduction   132
 II. Diacylhydrazines   134
III. Other Nonsteroidal Ecdysone Agonists   160
IV. Conclusions   161
    References   162

6

Juvenile Hormone Molecular Actions and Interactions During Development of Drosophila melanogaster

Edward M. Berger and Edward B. Dubrovsky

  I. Introduction   176
 II. Hormones and Development   177
III. Hormones and Reproduction   184
IV. Molecular Biology of Hormone Action   187
 V. Hormonal Cross-Talk   197
    References   201

7

INSECT NEUROPEPTIDE AND PEPTIDE HORMONE RECEPTORS: CURRENT KNOWLEDGE AND FUTURE DIRECTIONS

ILSE CLAEYS, JEROEN POELS, GERT SIMONET, VANESSA FRANSSENS,
TOM VAN LOY, MATTHIAS B. VAN HIEL, BERT BREUGELMANS, AND
JOZEF VANDEN BROECK

I. Introduction   218
II. Members of the Rhodopsin (GPCR Class A) Family   222
III. Members of Other 7TM Families   242
IV. Single Transmembrane Receptors   245
V. Emerging Concepts and Future Developments   260
References   265

INDEX   283

Contributors

Numbers in parenthesis indicate the pages on which the authors' contributions begin.

Edward M. Berger (175) Department of Biology, Dartmouth College, Hanover, New Hampshire 03755.

Isabelle M. L. Billas (101) IGBMC, Laboratoire de génomique et Biologie structurales, CNRS/INSERM/Université Louis Pasteur, Parc d'Innovation BP10142, 67404 Illkirch cedex, France.

Bert Breugelmans (217) Laboratory for Developmental Physiology, Genomics and Proteomics, Department of Animal Physiology and Neurobiology, Zoological Institute, K.U.Leuven, Naamsestraat 59, B-3000 Leuven, Belgium.

Ilse Claeys (217) Laboratory for Developmental Physiology, Genomics and Proteomics, Department of Animal Physiology and Neurobiology, Zoological Institute, K.U.Leuven, Naamsestraat 59, B-3000 Leuven, Belgium.

Edward B. Dubrovsky (175) Department of Biology, Dartmouth College, Hanover, New Hampshire 03755.

Vanessa Franssens (217) Laboratory for Developmental Physiology, Genomics and Proteomics, Department of Animal Physiology and Neurobiology, Zoological Institute, K.U.Leuven, Naamsestraat 59, B-3000 Leuven, Belgium.

Lawrence I. Gilbert (31) Department of Biology, University of North Carolina, Chapel Hill, North Carolina 27599.

Robert E. Hormann (59) RheoGene Inc., Norristown, Pennsylvania 19403.

Markus Lezzi (59) Institute of Cell Biology, Swiss Federal Institute of Technology (ETH), CH-8093 Zurich, Switzerland.

Dino Moras (101) IGBMC, Laboratoire de génomique et biologie structurales, CNRS/INSERM/Université Louis Pasteur, Parc d'Innovation BP10142, 67404 Illkirch cedex, France.

Yoshiaki Nakagawa (131) Division of Applied Life Sciences, Graduate School of Agriculture Kyoto University, Kyoto 606–8502, Japan.

Subba R. Palli (59) Department of Entomology, College of Agriculture, University of Kentucky, Lexington, Kentucky 40546.

Jeroen Poels (217) Laboratory for Developmental Physiology, Genomics and Proteomics, Department of Animal Physiology and Neurobiology, Zoological Institute, K.U.Leuven, Naamsestraat 59, B-3000 Leuven, Belgium.

Uwe Schlattner (59) Institute of Cell Biology, Swiss Federal Institute of Technology (ETH) CH-8093 Zurich, Switzerland.

Gert Simonet (217) Laboratory for Developmental Physiology, Genomics and Proteomics, Department of Animal Physiology and Neurobiology, Zoological Institute, K.U.Leuven, Naamsestraat 59, B-3000 Leuven, Belgium.

James W. Truman (1) Department of Biology, University of Washington, Seattle, Washington 98195.

Matthias B. Van Hiel (217) Laboratory for Developmental Physiology, Genomics and Proteomics, Department of Animal Physiology and Neurobiology, Zoological Institute, K.U.Leuven, Naamsestraat 59, B-3000 Leuven, Belgium.

Tom Van Loy (217) Laboratory for Developmental Physiology, Genomics and Proteomics, Department of Animal Physiology and Neurobiology, Zoological Institute, K.U.Leuven, Naamsestraat 59, B-3000 Leuven, Belgium.

Jozef Vanden Broeck (217) Laboratory for Developmental Physiology, Genomics and Proteomics, Department of Animal Physiology and Neurobiology, Zoological Institute, K.U.Leuven, Naamsestraat 59, B-3000 Leuven, Belgium.

James T. Warren (31) Department of Biology, University of North Carolina, Chapel Hill, North Carolina 27599.

PREFACE

The discovery of hormones and receptors in insects has provided important tools for the molecular biologist in directing the expression of a specific gene in a mammalian system by using an insect marker that is foreign to the mammalian cell. An often used strategy is to locate the ecdysone receptor in front of a specific gene and turn on the expression of that gene with ecdysone. More importantly, the field of insect biochemistry and molecular biology has made stunning progress. This volume collects manuscripts on insect hormones and receptors.

The first review is by J. W. Truman on "Hormonal Control of Insect Ecdysis: Endocrine Cascades for Coordinating Behavior with Physiology." "Biosynthesis of Molting Hormone" is offered by L. I. Gilbert and J. T. Warren. S. R. Palli, R. E. Hormann, U. Schlattner and M. Lezzi report on "Ecdysteroid Receptors and their Applications in Agriculture and Medicine." In terms of X-ray determined structure, I. M. L. Billas and D. Moras present: "Ligand-Binding Pocket of the Ecdysone Receptor." Y. Nakagawa describes: "Non steroidal Ecdysone Agonists." "Juvenile Hormone Molecular Actions and Interactions during Development of *Drosophila melanogaster*" is contributed by E. M. Berger and E. B. Dubrovsky. Finally, I. Claeys, J. Poels, G. Simonet, V. Franssens, T. Van Loy, M. B. Van Hiel, B. Breugelmans and J. Vanden Broeck offer: "Insect Neuropeptide and Peptide Hormone Receptors: Current Knowledge and Future Directions."

Preparation of thematic volumes, although infrequent, allows for the collection of manuscripts covering the recent progress in exciting areas of vitamin and hormone research. Availability of this Serial on-line should increase access.

Once again, the Editor-in-Chief is grateful to Academic Press/Elsevier Science for excellent cooperation in mounting this continuing source of up-to-date review information.

Gerald Litwack
Toluca Lake, California
April, 2005

1

HORMONAL CONTROL OF INSECT ECDYSIS: ENDOCRINE CASCADES FOR COORDINATING BEHAVIOR WITH PHYSIOLOGY

JAMES W. TRUMAN

Department of Biology, University of Washington, Seattle, Washington 98195

    I. Introduction
   II. The Physiology and Behaviors of the Ecdysis Sequence
  III. Endocrine and Neuroendocrine Factors involved in Ecdysis Control
      A. Eclosion Hormone
      B. Ecdysis Triggering Hormone and Pre-ecdysis Triggering Hormone
      C. Crustacean Cardioactive Peptide
      D. Corazonin
      E. Bursicon
      F. 20-Hydroxyecdysone
  IV. Interaction of Endocrine Signals in Controlling the Behavioral Phases
      A. Pre-ecdysis Behavior
      B. Activation of the Ecdysis Phase—Role for Delay Circuits
      C. Transition to Post-ecdysis Behavior: Shift of Phases Out of the Sequence
   V. Conclusions
      References

I. INTRODUCTION

The adaptive value of behavior lies in its ability to be adjusted for accommodating the changing physiological or developmental state of the animal. Information about this internal state is typically conveyed through the endocrine and neuroendocrine systems, and the hormonal signals that alter physiology often act in parallel on the CNS to adjust behavior accordingly. Invertebrate systems have played a major role in our understanding of how hormones and other neuromodulators adjust the functioning of the nervous system. Studies on the stomatogastric ganglion of decapod crustaceans led the way by showing how modulators sculpt out functional circuits by altering synaptic strengths and membrane properties of selected neurons (Harris-Warrick et al., 1997; Marder and Thirumalai, 2002). Thus, as modulators ebb and flow, CNS circuits dynamically remodel themselves to keep pace with changing brain chemistry. In this way, hormones and other neuromodulators serve as the key architects for building different behavioral states. In the context of complex behavior the challenge is to establish the modulator code for each behavioral state and to understand how the sign stimuli that shift the animal from one state to the other impact these modulatory systems.

Behaviors that are goal directed are often organized into sequences of discrete phases with each phase having its own characteristic components. Such organization is most obvious for reproductive behaviors that can progress through discrete courtship, copulation, and mate-guarding phases. But unlike reproductive behaviors that are widespread throughout animal groups, insects and other arthropods display a goal-directed behavior, their ecdysis behavior, that is the outcome of their periodic need to molt. With the making of a new exoskeleton, the old one must then be shed, a process termed ecdysis. Like other goal-directed behaviors, ecdysis involves a complex sequence of behaviors, called the ecdysis sequence, which is embedded in a particular physiological context, the ongoing endocrinology and physiology of the molting process. The ecdysis sequence provides certain advantages as a model for analyzing how modulators can organize complex behavioral programs: (1) There has been intense selection pressure on this behavior because the failure to do it properly on the first attempt results in crippling deformities or death. Hence, the sequence is robust and stereotyped. (2) Unlike many behavioral sequences that involve an interplay with another organism, ecdysis is a solitary undertaking and the insect interacts with an environment of its own making—its old skin. This simplicity of context reduces the types of sensory information that need to be considered. (3) In some cases, the onset of the behavior can be predicted down to a matter of minutes, although proximate stimuli may then modify the timing of the sequence or the progression through it. (4) Some of the modulators are associated only with a particular phase of the ecdysis sequence and so the manipulation of that modulator produces very selective effects on behavior.

Ecdysis control in insects has been reviewed in depth by Ewer and Reynolds (2002), and the reader is referred there for more details. The goal of this study is to provide an overview of the ecdysis control system, emphasizing lessons from ecdysis control that may help in understanding behavioral organization in animals in general.

II. THE PHYSIOLOGY AND BEHAVIORS OF THE ECDYSIS SEQUENCE

In insects, as in other arthropods, growth and changes in form require the periodic production of a new cuticular exoskeleton, a process termed molting (e.g., Gilbert, 1989). The molt is caused by the steroid hormone 20-hydroxyecdysone (20E) and begins with apolysis, the detachment of the epidermis from the overlying cuticle. The epidermis secretes a molting fluid into the space between it and the old cuticle and then proceeds to secrete the layers of new cuticle. As the molt progresses, some of the new cuticular structures begin to pigment and harden. This process is most evident for new hairs and scales but also occurs in selected regions of the new cuticle that require rigidity during ecdysis or will be used to rupture the covering of old cuticle. As the molt enters its terminal phases, preparations begin for the insect's escape. Proenzymes in the molting fluid are activated and degrade the old cuticle from the inside out. The protein and chitin of the internal layers are readily digested, but the outer exocuticle is highly cross-linked and resistant to degradation. Certain regions of this outer cuticle, though, are not cross-linked and provide points of weakness, the "ecdysial seams," that can be ruptured to allow the insect to emerge. The molting fluid is then resorbed when the digestion is complete. The cuticular linings of the tracheal system are also replaced, and the new trachea fills with gas during the final phases of molting fluid resorption. After the insect sheds its old cuticle, it then inflates its new cuticle to its final form. Inflation is typically associated with the insect swallowing air or water to expand to a larger size. The proteins in this expanded cuticle are then cross-linked by a quinone-based tanning, and the new size and shape of the insect are then fixed until its next molt.

The physiological processes that occur during the terminal phases of the molt are coordinated with the behaviors of the ecdysis sequence (Ewer and Reynolds, 2002). As illustrated for the cricket in Fig. 1, the insect goes through discrete behavioral phases during shedding of the old cuticle and expansion of the new one (Carlson, 1977). Each phase is characterized by motor patterns that are unique to that phase and recruited in a stereotyped order. There is plasticity though, and depending on sensory feedback, a phase may be extended or shortened and the ordering of motor patterns can be adjusted. The major phases, however, occur in a strict order, and there are typically abrupt transitions from one behavioral phase to the next.

The ecdysis sequence is usually divided into three phases. During the pre-ecdysis phase the insect secures itself to the substrate, followed by movements that break the remaining muscle connections between the insect and its old cuticle. Caterpillars of the tobacco hornworm, *Manduca sexta*, secure themselves to the substrate early in the molting process, but their movements to loosen the old cuticle begin about an hour and a half before ecdysis (Fig. 2; Copenhaver and Truman, 1982; Zitnan *et al.*, 1999). The larva shows two main motor patterns during this phase (Novicki and Weeks, 1993, 1995; Zitnan *et al.*, 1999). Pre-ecdysis I involves rhythmic dorsoventral compressions along the length of the caterpillar and is driven by ascending interneurons from the terminal ganglion (Novicki and Weeks, 1995). Pre-ecdysis II is recruited later and includes rhythmic retraction movements of the abdominal prolegs. During pre-ecdysis II, the insect finishes resorption of the molting fluid and fills its new trachea with air.

The ecdysis phase includes the motor patterns used for shedding the old cuticle. In most insects, including *Manduca*, the major motor pattern involves rhythmic peristaltic contractions that move anteriorly along the body (Fig. 2; Weeks and Truman, 1984; Zitnan *et al.*, 1996). In insects that have long appendages, such as crickets, the peristaltic movements are coordinated

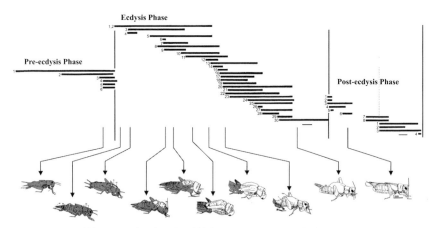

FIGURE 1. Ethogram showing the shift in motor programs during the various phases of the adult ecdysis sequence in a cricket. Each bar shows the onset and duration of a distinct motor pattern. The shift from one phase to the next involves the abrupt recruitment of a new set of motor patterns. Modified from Carlson (1977).

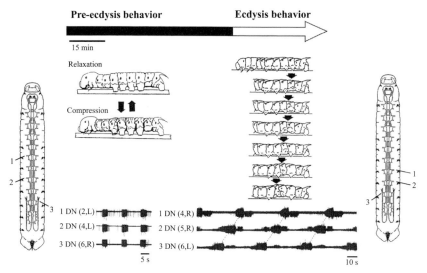

FIGURE 2. Temporal organization of the pre-ecdysis and ecdysis phases in larvae of *Manduca sexta*. Drawings show the main abdominal movements during the pre-ecdysis and ecdysis phases. The electrophysiological records under each show the phase of motor bursts that produce the respective movements. Electrode placement is indicated on the accompanying diagram. Pre-ecdysis data based on Miles and Weeks (1991); ecdysis data on Weeks and Truman (1984).

with motor subroutines that withdraw the appendages from their sheaths of old cuticle (Carlson, 1977). Prior to ecdysis, much of the new cuticle is permeable to water. As the old cuticle is being shed, dermal glands secrete their products into the ecdysial space between the two cuticles and the old cuticle helps spread this waterproofing material over the surface of the new cuticle as it is being withdrawn.

The post-ecdysis phase is devoted to the expansion and hardening of the new cuticle. In many instances, the cuticle expansion is coincident with ecdysis and the expanding new cuticle helps to rupture the old one. The expansion involves a change in cardiac activity and the swallowing of air or water to expand the body cavity. For the adult ecdysis of some moths and flies, the pupa may be hidden in a protected site, and the newly emerged adult often needs to escape from that site before it expands its delicate wings. In these cases, the process of cuticular expansion is delayed for a variable period after ecdysis until the insect finds an appropriate wing inflation site. Early in expansion, the cuticle becomes plasticized so that it can be inflated readily. The cuticle proteins then are cross-linked, giving the strength and rigidity to the new exoskeleton.

III. ENDOCRINE AND NEUROENDOCRINE FACTORS INVOLVED IN ECDYSIS CONTROL

Research on the control over the ecdysis sequence began with studies on the ecdysis of adult giant silk moths 35 years ago (Truman and Riddiford, 1970). Research soon shifted to the tobacco hornworm moth, *M. sexta*, which has provided a growing list of peptide hormones that orchestrate the ecdysis sequence. *Drosophila* has been used to test the effects of removal of specific peptides and to identify new components in this regulatory pathway.

A. ECLOSION HORMONE

The first indication of a hormonal control over the ecdysis sequence came from studies on the giant silk moths, *Hyalophora cecropia* and *Antheraea pernyi*. Adult ecdysis (also termed eclosion) of these moths is under a strong circadian control, with each species emerging at a characteristic time of the day: *H. cecropia* early in the day and *A. pernyi* in the late evening. Brain extirpation and implantation experiments within and between species showed that the brain controlled the timing of the behavior and could exert this control even if transplanted to the abdomen (Truman and Riddiford, 1970). The removal of the brain did not block the ability of moths to shed the pupal cuticle, but the behavior now occurred randomly with respect to time of day. Moreover, there was no longer behavioral coordination, and behavioral components were often seen in isolation, rather than part of a strict sequence (Truman, 1971). The fact that a transplanted brain could restore the timing and coordination of the ecdysis sequence in a brainless host clearly showed that a hormone must be involved. The search for the hormone shifted to *M. sexta* and culminated with the isolation and sequencing of eclosion hormone (EH) by two groups in 1987 (Kataoka *et al.*, 1987; Marti *et al.*, 1987). *M. sexta* EH (MasEH; Fig. 3) was a novel 62 amino acid peptide constrained by three disulphide bridges.

The EH gene was first sequenced from *Manduca* (Horodyski *et al.*, 1989) and encodes a conceptual precursor with a 26 amino acid signal sequence followed by one copy of EH. In *Manduca*, there is no posttranslational modification, and the active peptide results from removal of the leader sequence. The precursor from the *Drosophila* EH gene encodes a slightly larger peptide that shows a 10 amino acid N-terminal extension and an amino acid C-terminal addition as compared to MasEH (Horodyski *et al.*, 1993). The extended N-terminus has a potential processing site that would yield a 62 aa peptide, but there is no direct evidence that this cleavage actually occurs in *Drosophila*. Partial EH sequences have been found from a variety of insects (Horodyski, 1996), and it is clear that EH is widely distributed within the Insecta, but it has yet to be found outside of this group.

Ecdysis-triggering hormone & pre-ecdysis-triggering hormone
  Manduca sexta (MasETH)
    SNEAISPFDQGMMGYVIKTNKNIPRM-NH$_2$
  (MasPETH)
    SFIKPNNVPRV-NH$_2$
Eclosion hormone (EH)
  Manduca sexta (MasEH)
    NPAIATGYDPMEICIENCAQCKKMLGAWFEGPLCAESCIKFKGKLIPECEDFASIAPFLNKL
Crustacean cardioactive peptide (CCAP)
  PFCNAFTGC-NH$_2$
Bursicon & partner of bursicon
  Drosophila melanogaster (DrmBurs)
    QPDSSVAATDNDITHLGDDCQVTPVIHVLQYPGCVPKPIPSFACVGRCASYIQVSGSKIW
    QMERCMCCQESGEREAAVSLFCPKVKPGERKFKKVLTKAPLECMCRPCTSIEEGIIPQEI
    AGYSDEGPLNNHFRRIALQ[139]

  (DrmPburs)
    RYSQGTGDENCETLKSEIHLIKEEFDELGRMQRTCNADVIVNKCEGLCNSQVQPSVITPT
    GFLKECYCCRESFLKEKVITLTHCYDPDGTRLTSPEMGSMDIRLREPTECKCFKCGDFTR[120]
Corazonin
  pETFQYSRGWTN-NH$_2$

FIGURE 3. The primary structure of the peptide hormones that are involved in the various phases of the ecdysis sequence.

The eclosion hormone and its mRNA are confined to one or two pairs of neurons in the ventromedial region of the brain (Fig. 4). These are large neurosecretory neurons, which project up to the length of the CNS and also into the periphery at cephalic and/or terminal abdominal neurohemal sites (Truman and Copenhaver, 1989). The eclosion hormone is released both into the blood and locally within the CNS (Hewes and Truman, 1991).

Both *in vivo* and *in vitro* studies show that EH is a potent activator of guanylate cyclase, causing a marked increase in cGMP in its target cells. The stimulation of the cyclase, however, is indirect and appears to be through activation of phospholipase C via an as yet unidentified receptor (Morton and Simpson, 2002). The steps between phospholipase C activation and guanylate cyclase stimulation still need to be resolved, but the latter is an atypical guanylate cyclase, MsGC-β3, which is cytoplasmic but insensitive to nitric oxide (Morton and Anderson, 2003). MsGC-β3 is expressed in at least some of the known EH target cells, including the Inka cells, described further (Morton and Simpson, 2002). Studies on the Inka cells show that EH exposure results in enhanced intracellular Ca^{2+} as well the cGMP increase and that these two pathways may work together in driving ecdysis triggering hormone (ETH) secretion (Kingan *et al.*, 2001).

The eclosion hormone acts on both peripheral and central targets. Peripherally, EH is required for the resorption of fluid from the lumen of the

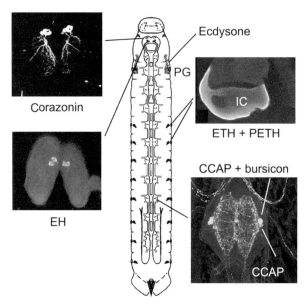

FIGURE 4. Anatomy of the endocrine and neuroendocrine components involved in ecdysis control in larvae of the tobacco hornworm, *Manduca sexta*. IC, Inka cell; PG, prothoracic gland. Image of brain corazonin cells based on data from Shiga *et al.* (2003).

new trachea, thereby allowing them to fill with air just prior to ecdysis. *Drosophila* that lack their EH neurons are also defective in tracheal airfilling, and many die soon after ecdysis with their trachea remaining filled with fluid (Baker *et al.*, 1999; McNabb *et al.*, 1997). The tracheal filling defects can be cured by injecting these insects with membrane-permeable analogs of cGMP just prior to ecdysis (Baker *et al.*, 1999). Circulating EH is also necessary for the dermal glands to secrete their waterproofing products onto the surface of the new cuticle during the ecdysis phase (Hewes and Truman, 1991). The Inka cells, described further, are key endocrine cells involved in ecdysis control, and EH is a potent stimulator of peptide release from these cells (Ewer *et al.*, 1997; Kingan *et al.*, 1997; Zitnan *et al.*, 1996). At pupal ecdysis in *Manduca*, a compact collection of cells at the anterior margin of each abdominal ganglion shows a prominent increase in cGMP in response to EH (Hesterlee and Morton, 2000). This tissue has proved very convenient for working out some of the biochemical steps in EH action (Morton and Simpson, 2002), although its function at pupal ecdysis is unkown.

The presence of the blood–brain barrier complicated the interpretation of many of the early studies of the central action of EH. Despite its large size, it was thought that EH could somehow pass the blood–brain barrier and that circulating peptide had access to targets within the CNS. This conclusion was based on the observation that injection of EH into the animal induced the complete behavioral sequence (e.g., Truman, 1971) and that EH could evoke the corresponding motor programs when added to the bath around isolated CNS preparations (Truman, 1978). For EH to work *in vitro* though, the tracheal system supplying the CNS had to be included. It was assumed that the tracheal requirement reflected a need for proper oxygen levels to generate complex neural programs, but we now know that the CNS tracheal trunks harbor an endocrine cell, the Inka cell, which secretes small peptides that can pass the blood–brain barrier to start the ecdysis sequence (Zitnan *et al.*, 1996). The issue is further confused because one of their central targets is the EH cells themselves (Gammie and Truman, 1997b, 1999) and because the EH neurons release their peptide into the CNS (Hewes and Truman, 1991) as well as into the circulation.

At the time of EH release, a small network of central neurons shows a striking spike of cGMP production. These include the abdominal neurons that contain CCAP and bursicon (cells 27(A) and 704(A)), as well as their homologs in the subesophageal and thoracic ganglia (Ewer *et al.*, 1994). Besides this network of 50 neurons, there is a segmental set of two neurosecretory cells whose peptide content is unknown (Fuse and Truman, 2002). In isolated CNS preparations, this set of 50 neurons does not show a cGMP response if the CNS is bathed in EH. They do show a cGMP increase, however, if some of the ganglia are desheathed prior to the addition of the peptide (Gammie and Truman, 1999). Thus, it appears that the EH that normally stimulates these cells comes from local release within the CNS rather than from the circulation.

B. ECDYSIS TRIGGERING HORMONE AND PRE-ECDYSIS TRIGGERING HORMONE

Over 25 years elapsed between the discoveries of EH and a second peptide involved in ecdysis control. Ecdysis triggering hormone (ETH) was also first isolated from *Manduca*. It is a 26 amino acid amidated peptide that can also induce precocious ecdysis (Fig. 3; Zitnan *et al.*, 1996). The subsequent isolation of the ETH gene from *Manduca* showed that proETH is cleaved into three peptides, MasETH, pre-ecdysis triggering hormone (MasPETH), and ETH-associated peptide, the last product having no known biological activity. Both MasPETH and MasETH have a C-terminal PRXamide motif, but the behavioral actions for the two *Manduca* peptides are qualitatively distinct (Fig. 3; Zitnan *et al.*, 1999) and will be described further. The *Drosophila* ETH gene also encodes for two active peptides (DrmETH1 and DrmETH2), but as suggested by their name, these show more sequence similarity to ETH rather than PETH (Park *et al.*, 1999). Two ETHs are similarly encoded by the *Anopheles* precursor (Zitnan *et al.*, 2003). However, *Bombyx* has a ETH–PETH pair, as is likely also the case for the cockroach *Nauphoeta cinera* and the bug *Pyrrhocoris apterus* (Zitnan *et al.*, 2003). At this time, it is not clear if the divergence or convergence in the structure of these peptides is related to the complexity of the pre-ecdysis behaviors shown by the respective insects.

The source of the ETH peptides is endocrine, coming from single-celled glands, the Inka cell (Fig. 4), associated with an exocrine epitracheal gland on specific sites on the trachea. In moths the Inka cells reach a diameter of over 100 μm, with a single cell located on the ventral tracheal trunk leaving each of the segmental spiracles (Zitnan *et al.*, 1996). The presence of a single giant cell per hemisegment is a feature of many higher insects, whereas more basal groups show many smaller cells distributed over the trachea of each segment (O'Brien and Taghert, 1998; Zitnan *et al.*, 2003).

The processing of pro-ETH has been studied in *Manduca*. The accumulation of ETH transcripts begins as rising ecdysteroid titers initiate the molting process. The processing of pro-ETH, however, is incomplete by the time the glands begin to secrete their peptides prior to ecdysis. Consequently, there is more PETH released than ETH (Zitnan *et al.*, 1999). The behavioral significance of this imbalance will be discussed in a later section.

Ecdysis triggering hormone action is mediated through G-protein–coupled receptors related to the thyrotropin-releasing hormone/neuromedin U receptor family in vertebrates (Park *et al.*, 2003). The gene encodes two major splice variants CG5911a and CG5911b. The b form has almost equal sensitivity to DrmETH1 and DrmETH2 and almost 500-fold greater affinity than does CG5911a. The latter also shows a 10-fold higher affinity for DrmETH1 versus DrmETH2. When tested with the corresponding peptides from *Manduca*, the two receptors showed an inversion of their sensitivity. In other words, MasETH was more active than MasPETH on CG5911a but the opposite is the case for CG5911b.

Most studies on ETH and PETH have focused on their central actions in evoking behaviors of the ecdysis sequence (Zitnan *et al.*, 1996, 1999). At this time their only suggested peripheral target are tracheae, which do not show proper air-filling in *Drosophila* that are mutant for ETH (Park *et al.*, 2002a). The stimulation of tracheal air-filling has also been claimed to be a function of EH (Baker *et al.*, 1999), so it is not clear whether the role of ETH in tracheal filling is direct or indirect via its stimulation of EH release (Gammie and Truman, 1999).

In terms of their central actions, the ETH peptides are necessary and sufficient for evoking the pre-ecdysis motor programs. Ecdysis triggering hormone mutants in *Drosophila* omits the pre-ecdysis program, but these behaviors are completely rescued by injecting mutant larvae with DrmETH (Park *et al.*, 2002a). In *Drosophila* ETH1 and ETH2 have qualitatively similar effects on adult ecdysis (Park *et al.*, 1999), with DrmETH1 being about 10-fold more potent than DrmETH2 though the situation is different for larval ecdysis. In both wild type and mutant individuals, DrmETH1 induces the normal range of pre-ecdysis behaviors followed by ecdysis, whereas DrmETH2 does not induce the pre-ecdysis components, but ecdysis occurs after the appropriate latency (Park et al., 2002a). The behavioral differences between the two products of the ETH gene are especially marked in *Manduca* larvae. PETH evokes the pre-ecdysis I behavior (Zitnan et al., 1999) through action on the terminal abdominal ganglion. This effect of PETH can be seen in the intact animal, one that has been neck - ligated or in the isolated chain of segmental ganglia. Ecdysis triggering hormone, by contrast, evokes pre-ecdysis-II, through action on a set of pattern generators that are distributed through the segmental ganglia. The motor pattern begins a few minutes after treatment with ETH and is seen in whole animals and in isolated abdomens. Likewise, challenging the isolated CNS (Zitnan et al., 1996) or the isolated abdominal ganglia (Gammie and Truman, 1999) with ETH results in the pre-ecdysis II motor program.

Ecdysis behavior typically follows the pre-ecdysis behavior evoked by ETH treatment. There is controversy, though, as to whether the action of ETH on ecdysis is direct (Zitnan et al., 1996, 1999) or mediated through stimulation of EH release (Gammie and Truman, 1997b, 1999). In both intact larvae and entire isolated CNSs, treatment with ETH results in the rapid onset of the pre-ecdysis movements followed later with an abrupt transition into the ecdysis phase (Zitnan et al., 1996). The transition to ecdysis, however, requires the participation of the head, so decapitated larvae or isolated CNSs, lacking the cephalic ganglia, show a good pre-ecdysis motor pattern after ETH treatment, but the transition to ecdysis fails to occur (Gammie and Truman, 1997b, 1999). In *Drosophila*, the genetic lesioning of the EH neurons also results in flies that no longer show precocious ecdysis in response to ETH injection (Baker et al., 1999; McNabb et al., 1997). In *Manduca*, timed decapitation experiments show that by 10–15 min after ETH injection, insects no longer need the influence from the head in order to subsequently show ecdysis behavior (Fuse and Truman, 2002). This time coincides with that of EH release (Ewer et al., 1997). Also, acutely isolated EH neurons respond to ETH by spike broadening and tonic firing (Gammie and Truman, 1999). These results suggest that although ETH directly triggers pre-ecdysis behavior, its action in evoking ecdysis motor pattern is indirect and occurs through EH release.

C. CRUSTACEAN CARDIOACTIVE PEPTIDE

Crustacean cardioactive peptide is a nine amino acid peptide, which is amidated on its C-terminus and has a three to nine disulphide bridge (Fig. 3). It was first sequenced from shore crabs based on its ability to stimulate heart rate (Stangier et al., 1998), but later also found to regulate cardiac function in insects, such as *Manduca* (Loi et al., 2001). The CCAP gene in *Drosophila* encodes a 155 amino acid precursor, which is cleaved to yield one copy of CCAP and three other peptides of unknown function (Ewer et al., 2001).

A potential CCAP receptor, GC6111, has been identified (Park et al., 2002b). This G-protein–coupled receptor has high relatedness to the vasopressin receptors of vertebrates. The relatively high concentration of ligand needed to cause activation ($EC_{50} = 150$ nM) of this receptor, through, is in marked contrast to other *Drosophila* peptide receptors that bind in the low nM range.

Crustacean cardioactive peptide is expressed by a number of central neurons. The ones most relevant to ecdysis are two pairs of cells (cells 27(A) and 704(A)) located in most abdominal ganglia (Fig. 4). Cell 27(A) is a segmental neurosecretory cells that projects to peripheral neurohemal sites, whereas cell 704(A) is an interneuron that ramifies through the segmental neuropils (Davis et al., 1993). These cells became relevant to ecdysis when it was shown that they produce a spike of cGMP at the time of ecdysis, suggesting that they were EH target cells (Ewer et al., 1994). Intracellular recordings from the cell 27(A) showed that the increase in intracellular cGMP markedly decreases the spike threshold of these cells and allows them to generate trains of action potentials (Gammie and Truman, 1997a). Application of CCAP to the desheathed CNS results in the rapid activation of the ecdysis motor program and this program persists for as long as the peptide is present (Gammie and Truman, 1997b). Crustacean cardioactive peptide, though, apparently cannot pass the blood–brain barrier, suggesting that its normal avenue to the ecdysis circuitry is through local release within the CNS, which is most likely from cell 704(A).

Park et al. (2003) used targeted cell killing of the CCAP neurons in *Drosophila* to examine their role in ecdysis. The lack of these neurons, however, did not necessarily result in the lack of ecdysis behavior. Larvae showed ecdysis behavior, albeit somewhat disorganized, and the adult ecdysis movements were present but weaker than normal. However, the CCAP cell knockouts had severe problems with head eversion, the *Drosophila* equivalent of pupal ecdysis, and most failed to complete this behavior (Park et al., 2003).

D. CORAZONIN

This blocked undecapeptide (Fig. 3) was discovered based on its physiological actions on the heart of the cockroach *Periplaneta americana* (Veenstra, 1989), but it was implicated in ecdysis control (Kim et al., 2004). For larval ecdysis in *Manduca*, corazonin appears in the blood prior to EH or the ETH peptides. Its involvement in the ecdysis triggering cascade became evident with the identification of the corazonin receptor in *Drosophila* (Park et al., 2002) and finding that the Inka cells contained high levels of this receptor. Injections of corazonin evoke the premature pre-ecdysis and ecdysis by stimulating ETH and PETH release. The peptide acts on the Inka cells through a Ca^{2+} mediated pathway (Kim et al., 2004). The corazonin receptor appears to desensitize rapidly, and so this peptide does not evoke the massive levels of ETH release observed after EH exposure. Although this peptide is involved in larval ecdysis in *Manduca*, its role in other insects needs to be confirmed.

E. BURSICON

Bursicon, the insect tanning hormone, was identified in the early 1960s by Cottrell (1962) and Frankel and Hsio (1962). It causes the tanning and the hardening of the new cuticle that occur after ecdysis. The determination of its structure was difficult because of its large size of about 30,000 Da. Bursicon was eventually purified from the cockroach, *P. americana*, and the sequences of peptide fragments were used to search for similar proteins encoded in the *Drosophila* genome. A significant match was found for CG 13419 (now called *bursicon* [*burs*]) that encoded a secreted cystine-knot protein of 139 amino acids (Fig. 3). Mutations in this gene resulted in flies that failed to show post-ecdysial tanning and lacked bursicon bioactivity in their CNS (Dewey et al., 2004). It was originally thought that the active hormone was a bursicon homodimer, but the expressed peptide was devoid of bioactivity. Subsequent work revealed that other fragments from the cockroach hormone matched sequences found in a related cystine knot peptide, named *partner of bursicon* (*pbur*). The products of neither *burs* nor *pbur* are active by themselves, but the heterodimer has full tanning activity, both *in vivo* and *in vitro* (Luo et al., 2005). This heterodimeric cystine-knot protein is the first to be found in invertebrates.

The mutation, *rickets*, was proposed to be in the receptor for bursicon because these mutants fail to tan after ecdysis and do not respond to injected bursicon, although they show strong tanning when treated with cAMP, the second messenger that mediates bursicon action (Baker and Truman, 2002). The *rickets* mutation disrupts the gene encoding a G-protein coupled receptor with a large extracellular domain, bearing numerous leucine-rich repeats (Baker and Truman, 2002), a gene that had been previously identified and named *Drosophila* glycoprotein hormone receptor 2 (DGHR2) (Eriksen et al., 2000). With the discovery of *pbur*, it was shown that the bur/pbur dimer does bind to DGHR2 and stimulates the production of cAMP (Luo et al., 2005).

A close association of bursicon and CCAP was suspected because one of the CCAP neurons, cell 27(A) (Fig. 4), was originally identified as containing bursicon bioactivity (Taghert and Truman, 1982) as were the homologous neurons in crickets (Kostron et al., 1996). Subsequent studies showed that these cells contain bursicon immunoreactivity (Honegger et al., 2002), and in *Drosophila*, the messages for both *bursicon* (Dewey et al., 2004) and *partner of bursicon* (Luo *et al.*, 2005) are found in a subset of the CCAP neurons. As will be discussed in a later section, this colocalization brings up some interesting issues because the two peptides are associated with different phases of the ecdysis sequence.

F. 20-HYDROXYECDYSONE

The *raison d'etre* for ecdysis is the production of a new cuticle, the latter being driven by ecdysone and its metabolite, 20E. Ecdysone is secreted from the endocrine prothoracic glands (Fig. 4) and converted in the periphery to 20E. These steroids are not simply the initiators of the molting process, they also micromanage the progression through the molting process, especially those events occurring at the end of the molt. Sláma (1980) first showed that the decline in ecdysteroids towards the end of the molt was essential for ecdysis and that ecdysis could be delayed in a dosage-dependent fashion by the injections of 20E. This requirement for withdrawal of ecdysteroids prior to ecdysis is also seen in *Manduca* (Truman *et al.*, 1983) and *Drosophila* (Robinow *et al.*, 1993).

One way by which 20E affects the ecdysis sequence is through actions on the peptide systems described in an earlier section. The steroid decline, late in the molt, increases the excitability of the EH neurons (Hewes and Truman, 1994) and also renders the insect sensitive to injected EH (Truman *et al.*, 1983). Originally, it was thought that this sensitivity was at the level of CNS targets for EH, but with the discovery of ETH, it became evident that this behavioral sensitivity resides at the level of the Inka cells (Zitnan *et al.*, 1999).

For the Inka cells, both peptide synthesis and release are linked to the ecdysteroid titer (Zitnan *et al.*, 1999). The cells become completely depleted of both ETH and PETH at ecdysis and remain so during the subsequent feeding period. The rising titers of 20E that initiate the next molt also stimulate the Inka cells to transcribe and translate the *ETH* gene. This gene has an ecdysone receptor response element upstream of the promoter, so the steroid induction of transcription may be direct. The ability of the cells to subsequently secrete ETH requires that the 20E then be withdrawn. The cells become responsive to EH in two steps during this steroid withdrawal. Relatively early in the decline the cells become capable of activating guanylate cyclase in response to EH, but this does not result in ETH secretion (Ewer *et al.*, 1997; Kingan *et al.*, 1997). Similarly, exposure of these cells to cGMP analogs does not provoke peptide secretion although hours later, exposure to either EH or cGMP results in the massive secretion of ETH peptides. The ability of EH to drive secretion by the glands corresponds to when they begin to show a low basal level of secretion (Kingan *et al.*, 1997). Although the EH neurons and the Inka cells clearly need steroid priming for normal functioning at ecdysis, it is not known if CCAP and bursicon neurons also require steroid exposure.

20-Hydroxyecdysone also has effects on the central circuits, which respond to some of these peptide hormones. Molting larvae of *Manduca* become responsive to injections of PETH and ETH about 30 h before their expected ecdysis, a time that corresponds to peak levels of 20E (Zitnan *et al.*, 1999). Initially they respond with just the pre-ecdysis behaviors, but later they show both pre-ecdysis followed by ecdysis. This shift to ecdysis occurs when the EH neurons first become competent to release their peptide in response to ETH (Ewer *et al.*, 1997). *In vitro* studies showed that the competence of the CNS to respond to ETH is due to the direct action of 20E on the CNS. Incubation of the isolated CNS with 20E for 24 h was sufficient to render it completely responsive to ETH (Zitnan *et al.*, 1999; Zitnanova *et al.*, 2001). Unlike the situation for ETH, there appears to be no steroid requirement for the CNS to respond to CCAP. Nervous systems from intermolt larvae respond to CCAP with a robust ecdysis motor pattern (Gammie and Truman, 1997b). Consequently, this peptide can evoke the ecdysis motor program at any time regardless of the history of steroid exposure of the nervous system. Hence, there is a complex relationship between 20E and the sensitivity of neurons to the various peptides involved with ecdysis control. With the identification of some of the receptors for the relevant peptides (Park *et al.*, 2002b, 2003), this will likely be fertile ground for future analysis of steroid–peptide interactions in the CNS.

20-Hydroxyecdysone has effects on ecdysis beyond its action in the assembly of the ETH–EH–CCAP axis. Mutation and selected cell killing experiments in *Drosophila* show that the insect can still perform some components of the ecdysis movements despite the lack of ETH (Park *et al.*, 2002a), EH (McNabb *et al.*, 1997), or CCAP (Park *et al.*, 2003). In the absence of these peptides, ecdysis behavior is typically weaker than normal but it occurs at approximately its normal time. In *Drosophila*, a circadian clock restricts the time of adult ecdysis to a temporal gate that occurs early in the day. This gating shows a circadian periodicity in constant darkness, but if a lights-on signal is given during a circadian gate, there is an abrupt burst of eclosion. In flies that lack either their EH neurons (McNabb *et al.*, 1997) or their CCAP cells (Park *et al.*, 2003), the circadian gating of adult ecdysis persists, but the flies fail to show a lights-on response (Fig. 5). This phenotype will be discussed in later section, but the cell knockouts clearly show that the circadian influence over the timing of adult ecdysis in *Drosophila* is not mediated through either EH or CCAP. Subsequent work by Myers *et al.* (2003) showed that the circadian timing of adult ecdysis is disrupted if *timeless*, one of the circadian clock components, is tonically expressed in the prothoracic glands. Thus, the circadian timing of ecdysis may be linked directly to fluctuations in the ecdysteroid titer itself. The circadian pattern of ecdysone secretion is best understood in the bug, *Rhodnius prolixus* (Vafopoulou and Steel, 1991), and a circadian modulation of the ecdysone decline in *Manduca* synchronizes some of the late events that

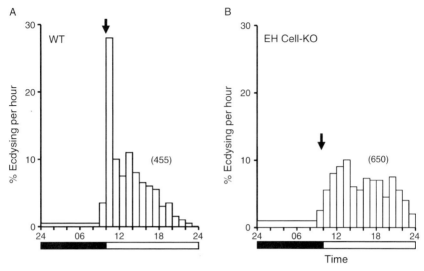

FIGURE 5. The timing of adult ecdysis in populations of *Drosophila melanogaster* that have been maintained in a 14L:10D photoperiod regimen. (A) Wild-type flies show a circadian peak of ecdyses with a spike of emergence following the lights-on signal (arrow). (B) Flies lacking their EH neurons show the circadian peak but lack the lights-on response. Modified from McNabb *et al.* (1997).

are preparatory to adult ecdysis (Schwartz and Truman, 1983). Whether the affect of the declining ecdysteroid titer on ecdysis behavior is direct or through other intermediary peptides is unclear. It is interesting, though, that both *Manduca* (Fahrbach and Truman, 1989) and *Drosophila* (Robinow *et al.*, 1993) have a set of neurons that shows exceptionally high levels of the ecdysone receptor prior to adult ecdysis. These cells die immediately after adult ecdysis, suggesting that they play a role in the ecdysis process, and since the adult ecdysis is the terminal molt in higher insects, these neurons are then eliminated. The high levels of receptor may be related to this elimination, but they may also provide an avenue for a direct action of steroid on these cells for behavior control.

IV. INTERACTION OF ENDOCRINE SIGNALS IN CONTROLLING THE BEHAVIORAL PHASES

The adaptive value of a behavioral sequence lies in its ability to adjust to physiological changes and external factors. Among insects, the wealth of variation in the factors that control ecdysis is evident at both the species and the stage level. How this variation is reflected in the organization of the endocrine control system is only starting to be explored. *Manduca* and *Drosophila* have received the most attention in this regard, and there is only scattered information available for other species. Nevertheless, a comparison of what is known for various species will knit together common features of behavioral control and the points to be addressed.

A. PRE-ECDYSIS BEHAVIOR

The relationship of the ETH-related peptides to the pre-ecdysis behavior is clear. In *Manduca*, ETH and PETH appear in the blood coincident with the onset of pre-ecdysis movements (Zitnan *et al.*, 1999). Experiments with isolated CNS preparations in *Manduca* show the onset of pre-ecdysis motor bursts a few minutes after addition of the peptides (Zitnan *et al.*, 1996) and the pre-ecdysis behaviors fail to appear in mutant *Drosophila* that lack a functional ETH gene (Park *et al.*, 2002a). Hence, the ETH peptides are both necessary and sufficient for evoking the pre-ecdysis phase.

In larval *Manduca*, the pre-ecdysis behavior first appears as isolated movements (Copenhaver and Truman, 1982; Zitnan *et al.*, 1999) that gradually increase in frequency and complexity. The incomplete processing of the pro-ETH results in more PETH being released than ETH (Zitnan *et al.*, 1999) and the initial low levels of PETH cause the appearance of sporadic Pre-ecdysis I movements. Pre-ecdysis I movements become more regular as levels of PETH rise. Although ETH levels are always lower than PETH, they eventually raise high enough for this peptide to take control and the behavior shifts to Pre-ecdysis II. But even higher levels of ETH are needed to initiate the ecdysis phase (Zitnan *et al.*, 1999). While ETH acts on the ventral CNS to cause pre-ecdysis, its action in causing ecdysis is mediated through a target in the head. At present the EH neurons are the only demonstrated ETH target cells in the brain (Gammie and Truman, 1997b, 1999), and when these cells are removed in *Drosophila*, ETH can no longer stimulate ecdysis behavior (Baker *et al.*, 1999; McNabb *et al.*, 1997). The positive feedback relationship between EH and ETH results in a massive surge of both peptides and the stores of both are completely exhausted. This provides an all-or-nothing signal that irreversibly commits the insect to an ecdysis attempt.

The factors that lead to the release of the ETH peptides are complex and vary depending on which species or stage is considered (Fig. 6). In *Manduca* (Truman *et al.*, 1983) and other insects (Slama, 1980) the decline of the high molting surge of 20E is an absolute prerequisite for ecdysis to occur. Insects vary, though, as to whether they have requirements beyond the steroid decline. In some instances, the appendages that are withdrawn from the old cuticle are initially soft and unable to effectively support the insect, so it is crucial that the insect has a perch from which to hang during ecdysis as its appendages harden. These insects typically become "restless" as the time for ecdysis approaches and start to search for an ecdysis perch (Hughes, 1980), but they can delay the onset of the sequence for hours until an acceptable site is found. In other insects, the onset of the ecdysis sequence

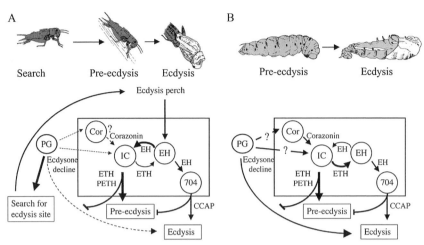

FIGURE 6. Models for the interaction of the various endocrine components involved in ecdysis control for insects that require an ecdysis perch to initiate ecdysis (A) versus ones that do not (B). (A) In this case the declining ecdysone titer likely initiates a search for an ecdysis site and also primes the various peptidergic components of the system. The attainment of a perch works through the eclosion hormone (EH) neurons to activate the Inka cells (IC) and Cell 704 (A) thereby initiating the cascade leading to the initiation of pre-ecdysis and ecdysis motor patterns. (B) In this case the start of the behavior is determined by the ecdysone decline, which either directly or indirectly through corazonin (Cor) release starts the secretion from the Inka cells. Positive feedback between these cells and the EH neurons results in the EH–ETH surge that causes the downstream behaviors of pre-ecdysis and ecdysis. See text for more details. PG, prothoracic glands.

is tightly linked to the timing of the steroid decline and proximate stimuli have little impact, if any.

The ecdyses of both *Manduca* and *Drosophila* fall into the latter category, with behaviors being tightly linked to the declining titer of 20E. For *Manduca* ecdysis, the declining steroid titer makes the EH neurons more excitable (Hewes and Truman, 1994), but it has not been reported to actually cause them to begin firing. The declining 20E titer also acts on the Inka cells, making them competent to secrete (Kingan and Adams, 2000) and it may also start the slow ramping up of PETH release that eventually starts the pre-ecdysis sequence. The 20E action may be direct on the Inka cells or it may be through intermediates, such as the release of the corazonin from the brain (Kim *et al.*, 2004). Although corazonin can definitely induce the release of ETH peptides to initiate pre-ecdysis, it may be just one of the factors that initiate secretion by the Inka cells. This conclusion is based on the observation that molting larvae that lack their heads, the source of circulating corazonin, still show a normal onset of pre-ecdysis (Novicki and Weeks, 1996). Likely, the Inka cells may be responding directly to the declining steroid titers as well as to corazonin. This is suggested by studies with cultured Inka cells that become competent to secrete in response to EH simply by removing 20E from the culture medium (Kingan and Adams, 2000).

We have only fragmentary information about insects that require behavioral cues to trigger the ecdysis sequence. Presumably, in such cases the decline in 20E sensitizes the ETH–EH axis but is not sufficient to initiate secretion of any of the peptides. Corazonin appears not to be necessary for ecdysis in grasshoppers since albino locusts that lack corazonin have no difficulty with ecdysis (Kim *et al.*, 2004). This may be significant since grasshoppers require a perch for ecdysis (Hughes, 1980). A lack of involvement of corazonin may bias the system towards a stronger requirement for EH to initiate ETH release. Thus, the involvement of corazonin in ecdysis control may be facultative and employed when developmental factors are required for initiating the behavior, but dispensable when proximate behavioral stimuli, such as an ecdysis perch, play a prominent role. Finally, the gate of adult ecdysis of the silk moth *Hyalophora cecropia* is uncoupled from the developmental progression of the molt, and the start of pre-ecdysis is not a gradual affair as in *Manduca* larvae, but rather pre-ecdysis shows an abrupt onset and EH is invariably found in the blood of animals that have started the behavior. Figure 6A proposes how the various components may interact when behavioral stimuli are paramount. The declining steroid induces the restlessness in the insect, and it begins its search for an ecdysis site. The steroid withdrawal also enhances sensitivity of ETH and EH sources but does not induce secretion. The EH cells serve as the pathway through which the proper behavioral stimuli can initiate the behavior sequence. The late-appearing sensitivity of the Inka cells, by contrast, monitors the developmental competence of the insect and the positive feedback between the Inka cells and the EH neurons provides the hormone surge needed for an irreversible commitment to an ecdysis attempt. The ETH peptides that are part of the surge initiate a strong pre-ecdysis behavior, whereas EH release sets up the system for the subsequent ecdysis phase as detailed in the next section.

Support for the hypothesis that there are at least two pathways involved in ecdysis control, one that deals with developmental competence and the other with proximate behavioral stimuli, comes from adult ecdysis in *Drosophila*. As described in an earlier section, adult ecdysis is confined to a temporal gate during the day and the timing of the gate is controlled by the prothoracic glands, presumably through modulation of ecdysone secretion (Meyers *et al.*, 2003). During the circadian gate, though, a lights-on signal causes an abrupt "spike" of ecdysis on top of the circadian distribution (Fig. 5). The lights-on signal has two effects, causing a premature release of EH and also shortening the latency between EH release and ecdysis (McNabb and Truman, unpublished). Removal of the EH cells (McNabb *et al.*, 1997) results in flies that show a normal gating of ecdysis but no longer respond to the lights-on signal. Therefore, the EH cells are the major or sole pathway by which proximate stimuli gain access to the ecdysis machinery.

In *Drosophila*, the proximate stimuli have a facultative effect and insects will ecdyse whether or not these stimuli are present. For other insects, though, proximate stimuli, such as an ecdysis perch, are essential. Dragonflies are a group with such a requirement for the adult ecdysis. For these insects, the cautery of the EH neurons completely blocks ecdysis behavior. Consequently, the importance of the EH neurons appears to depend on the relative importance of proximate cues for initiating the ecdysis sequence.

B. ACTIVATION OF THE ECDYSIS PHASE—ROLE FOR DELAY CIRCUITS

Studies on the isolated CNS of *Manduca* first showed that CCAP was a potent activator of the ecdysis motor program (Gammie and Truman, 1997b). The behavior begins a few minutes after peptide addition and persists for as long as CCAP is present. The selective lesioning of the CCAP cells in *Drosophila*, however, showed that ecdysis still occurred without CCAP although in a somewhat weakened form (Park et al., 2003). However, the flies lacked the ability to respond to lights-on signals, placing these cells in the pathway leading from proximate behavioral stimuli through the EH neurons to ecdysis behavior.

The core ecdysis program is still seen in *Drosophila* no matter which of the regulatory peptides are removed (e.g., McNabb et al., 1997; Park et al., 2002a), indicating a second pathway to the behavior that is more closely linked to the developmental progression caused by the 20E decline. The relationship between these two pathways is indicated by the phenotype of the ETH mutants (Park et al., 2002a). Mutant larvae lack the pre-ecdysis behaviors but show a weak ecdysis motor program. Interestingly, their ecdysis movements begin prematurely with respect to the terminal phases of the molt. It appears that although the decline of 20E can initiate the ecdysis program, this activation is normally overridden by the appearance of the ETH peptides. Their appearance causes the ecdysis program to become embedded as part of a behavioral sequence, which is coordinated with ongoing physiological changes and orchestrated by the peptide cascade.

The positive feedback relationship between EH and ETH means that any of these can be either upstream or downstream. Based on simultaneous cGMP spikes, the cells that release ETH, CCAP, and bursicon are all targets of EH (Fig. 7) (Ewer et al., 1994), but the behaviors controlled by each hormone can be separated in time by many minutes or hours. In the Inka cells, peptide release occurs concurrent with the cGMP rise (Ewer et al., 1997; Kingan et al., 1997). The release of CCAP, however, does not occur coincident with the cGMP spike in Cell 704(A) but is delayed for 30 min or more. The cGMP rise in the CCAP neurons enhances their excitability and reduces their spike threshold (Gammie and Truman, 1997a), but the cells remain silent because of descending inhibition. This inhibition is inferred in

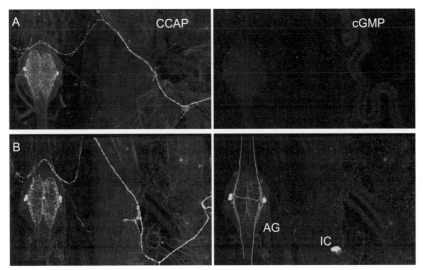

FIGURE 7. Paired con-focal images of a ventral abdominal hemisegment from a larval *Manduca* immunostained for CCAP and cyclic GMP (cGMP) showing the abrupt increase in cGMP that follows EH release. (A) Early in pre-ecdysis, showing the cell bodies and central arbor of Cell 27(A) and 704(A) in the abdominal ganglion (AG) and CCAP-positive axons in the periphery. No cGMP is evident. (B) Late pre-ecdysis, after EH release, the central CCAP neurons show a striking cGMP immunoreactivity as does the peripheral Inka cell (IC). Modified from Ewer *et al.* (1997).

both adult *Manduca* (Fuse and Truman, 2002; Reynolds, 1980; Zitnan and Adams, 2000) and *Drosophila* (Baker *et al.*, 1999), by the results of decapitation experiments at various times late in the molt. Animals decapitated prior to the EH–ETH surge may subsequently show pre-ecdysis but they do not progress to ecdysis. Animals decapitated after the surge subsequently show ecdysis but the behavior now begins a few minutes after decapitation rather than at its normal latency (Fuse and Truman, 2002). In flies that lack their EH neurons, however, decapitation does not lead to immediate ecdysis, no matter how late in the molt it is done (Baker *et al.*, 1999). These observations suggest that EH activates both the CCAP neurons and descending interneurons that inhibit these cells. Transection experiments in *Manduca* show that these cells are not located in the brain but occur in the subesophageal and thoracic ganglia (Fuse and Truman, 2002; Zitnan and Adams, 2000).

The best candidates for the descending inhibitors are the thoracic and subesophageal homologs of Cell 704(A). In *Manduca* these are also EH targets but they do not contain CCAP and they have descending axons that contact the CCAP neurons in the abdominal ganglia. These cells show a dramatic increase in cGMP in concert with the abdominal neurons, but they then show a more rapid loss of the second messenger (Ewer *et al.*, 1994).

Since the EH-induced cGMP enhances the excitability of target neurons, it was proposed that the CCAP neurons and their putative inhibitors are co-excited at the time of the EH–ETH surge (Fuse and Truman, 2002). Consequently, CCAP would not be released at that time because of the strong descending inhibition. The excitability of the descending inhibitors, though, wanes faster than in the CCAP neurons so eventually the inhibition is too weak to suppress the CCAP cells and so peptide release and ecdysis can occur (Fig. 8). A test of the involvement of cGMP degradation in the timing of the delay was done by injecting different levels of zaprinast, an inhibitor of the cGMP-specific phosphodiesterase, prior to challenge with ETH. The presence of the inhibitor resulted in a persistence of cGMP in both the CCAP neurons and the putative descending inhibitors and also caused a dose-dependent delay in the start of the ecdysis motor program (Fuse and Truman, 2002).

Besides establishing a delay between the pre-ecdysis and ecdysis phases of the ecdysis sequence, this inhibitory cassette provides an avenue by which sensory feedback can impact the timing of ecdysis. As described in an earlier section, in *Drosophila* a lights-on signal can markedly shorten the latency to ecdysis for insects that have already released EH and are "waiting" to ecdyse. With respect to their rapid ecdysis, these flies behave similarly to ones that are decapitated after EH release, suggesting that key sensory stimuli block the descending inhibition, thereby permitting the rapid release of CCAP and ecdysis. In the flies lacking their EH neurons, the rapid

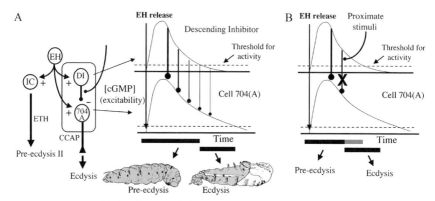

FIGURE 8. Model showing the relationship of descending inhibition to the timing of onset of the ecdysis phase of the ecdysis sequence. On the left, EH is proposed to activate the Inka cells (IC) as well as Cell 704(A) and a descending inhibitor (DI). Ecdysis triggering hormone release occurs immediately to support pre-ecdysis II, but CCAP release is delayed because of the inhibition. The right represents the excitability of the two neurons in terms of their cGMP levels. As the excitability of the DI wanes, Cell 704(A) eventually escapes and releases the CCAP needed for ecdysis. (A) Normal sequence of events. (B) Relevant proximate stimuli that occur after EH release is thought to suppress the inhibition and allow precocious CCAP release and ecdysis (for more details, see text). Based on Fuse and Truman (2002).

response to decapitation as well as the response to a lights-on signal are both missing (Baker *et al.*, 1999).

The properties of the EH target cells then appear to establish the timing for the onset of the next behavioral phase, but they also provide a pathway through which key stimuli can change this timing. The rapid release of ETH initiates the first phase but the co-activation of the CCAP neurons and their inhibitors primes the animal for the second phase. Appropriate stimuli can then advance the entry into the second phase by removing this inhibition (Fig. 8). This design strategy for behavioral sequencing is likely applicable to behaviors beyond insect ecdysis.

The release of EH is phasic, serving as an all-or-nothing signal to propel the insect into an ecdysis attempt. Crustacean cardioactive peptide, by contrast, has a more tonic action and, at least *in vitro* (Gammie and Truman, 1997b), maintains the motor program for as long as the peptide is present. During a normal larval ecdysis in *Manduca* the insect releases about half of its store of CCAP by the time the cuticle is shed. If a molting larva has its old cuticle manipulated so that it cannot be shed, the larva will prolong its ecdysis attempt for about an hour before it finally gives up. By that time, the CCAP supply in cell 704(A) has been essentially exhausted (Gammie and Truman, 1997b) and the termination of the ecdysis attempt may be due to running out of the required neuromodulator.

C. TRANSITION TO POST-ECDYSIS BEHAVIOR: SHIFT OF PHASES OUT OF THE SEQUENCE

The after ecdysis phase is associated with the expansion and hardening of the new cuticle. Many insects are like cockroaches and start expanding their new cuticle early in the ecdysis process, and this expansion aids in the rupturing of the old cuticle. In cockroaches bursicon is released early during ecdysis (Mills *et al.*, 1965) and it acts on the new cuticle to make it more plastic, thereby aiding in its expansion. However, shortly thereafter, the cuticle proteins begin to be cross-linked, and the expanded cuticle becomes rigid. Since bursicon's action of cuticle cross-linking occurs a fixed time after its release, the insect has a limited timeline for shedding the old cuticle. If it encounters difficulty, it runs the risk of the new cuticle tanning when parts are still enclosed in the old skin, locking it in an unexpanded state.

For the cockroach, having CCAP and bursicon present in the same cell (the homolog of cell 27(A)) does not present a problem because the expansion of the new cuticle begins early in the ecdysis phase. Consequently, the corelease of CCAP and bursicon would be appropriate. Things become more problematic for adult ecdysis in some flies and moths because the adults emerge in a protected pupation site but may then delay bursicon release and the expansion and hardening of their new cuticle until they have escaped from that site and found a suitable expansion site (Cottrell, 1962; Fraenkel and Hsiao, 1962, 1965). Assays of bursicon activity in the blood of *Manduca* show that the release of the hormone is delayed until the start of the wing-expansion behavior (Reynolds *et al.*, 1979). The release of cardioacceleratory peptides, including CCAP, into the blood is likewise delayed (Tublitz *et al.*, 1986) and coincident with bursicon release. So what about the involvement of CCAP with ecdysis? It is likely that a dissociation can occur because CCAP is found in two cells, along with bursicon in one case (the neurosecretory cell 27(A)) and by itself in the other (cell 704(A)). The latter cells ramify extensively through central neuropils and are depleted of CCAP during ecdysis (Gammie and Truman, 1977b). It is highly likely that they are the main drivers of the ecdysis motor program. Cell 27(A) by contrast has only sparse central arbor and but extensive peripheral neurohemal release sites. Peptide from the latter sites could act on peripheral targets but cannot pass the blood–brain barrier to act on the CNS. Consequently, during ecdysis of cockroaches or larval *Manduca*, when expansion occurs as part of the ecdysis behavior, one expects that cells 27(A) and 704(A) are coactivated. In cases where cuticle expansion is delayed until ecdysis is completed, one expects that cell 704(A) maintains its normal time of release to induce ecdysis, whereas the activity of cell 27(A) is delayed to be associated with the post-ecdysis expansion of the cuticle. However, this prediction needs to be confirmed.

Genetic studies in *Drosophila* show that bursicon release is not only correlated with expansion behaviors, it is essential for these behaviors to occur. Flies mutant for *bursicon* show essentially normal ecdysis behavior but fail to show wing expansion (Dewey *et al.*, 2004). Similarly, flies that are mutant for the gene *rickets*, the bursicon receptor, progress through ecdysis normally but fail to show wing expansion (Baker and Truman, 2002). Unlike the case with ETH, the behavioral effects of the lack of bursicon cannot be rescued by injection of hormone, although the cuticular tanning deficit is rescued. The failure to rescue the behavior is likely because this large hormone cannot cross the blood–brain barrier. Bursicon released from cell 27(A) within the CNS is likely involved in initiating the last behavioral phase.

Is bursicon directly responsible for evoking the wing expansion motor pattern? One piece of evidence suggests that its action may be indirect (Kimura and Truman, 1990). When flies are neck - ligatured immediately after ecdysis, they neither release bursicon (assessed by cuticular tanning) or show wing expansion. If the ligature is delayed until 15 min, then flies will both tan and show wing expansion. However, flies ligatured in the transition period will sometimes show tanning without wing expansion, but never the other way around. One interpretation of this difference is that there is a head-related signal that is downstream of bursicon release that is necessary for subsequent wing expansion. Consequently, there may be still other peptides that are involved in the expansion process to allow the independent display of the post-ecdysis phase.

V. CONCLUSIONS

Insect ecdysis provides a rich system in which to examine how endocrine and neuroendocrine systems interact to orchestrate complex programs of behavior. During the behavioral progression, each phase has a chemical signature, with ETH and PETH promoting the pre-ecdysis motor patterns, CCAP promoting the ecdysis phase, and bursicon required for the post-ecdysis behaviors. The normal ordering of the behavior patterns in the sequence is reflected in a response hierarchy within the CNS. PETH starts the first motor pattern, pre-ecdysis I, but when ETH levels are high enough, it initiates pre-ecdysis II, despite the presence of higher levels of PETH (Zitnan et al., 1999). The subsequent appearance of CCAP then shifts behavior to the ecdysis phase, even if high levels of ETH are present (Gammie and Truman, 1997b, 1999). A general organizational principle seems to be that the peptide responsible for phase "N" is also capable of suppressing the behavior of phase "N-1."

Temporal ordering is also evident in the interaction of the cells that produce and release the various modulators. A crucial feature of the behavioral program is that it occurs at the appropriate time relative to the developmental processes of the molt. This coordination occurs via the steroid titers that drive the molting process. The appearance of 20E prepares the system by inducing the production of the ETH peptides and rendering the central circuits sensitive to these hormones. The waning steroid level, however, is responsible of the EH sensitivity of the Inka cells, the last link in a positive feedback system that ensures that the ecdysis attempt is an all-or-nothing event.

The positive feedback between EH and ETH is one of only a few positive feedback systems in endocrinology. In its simplest form, it provides a mechanism by which developmental information from the steroid decline, via ETH and the Inka cells, and proximate behavioral stimuli, via the EH neurons, can interact to make the decision to proceed to ecdysis. This feedback relationship has provided both complexity and confusion because of the wide variation seen among the insects in their need for behavioral conditions for ecdysis.

Although eclosion hormone was the first hormone to be found to be associated with ecdysis, it is somewhat ironic that it may not be directly responsible for evoking any of the behavioral programs of this system. It provides, however, a useful model for the logic of how modulatory circuits might be constructed and interact with key sign stimuli. The release of EH activates the cells releasing the modulators that will drive more than one behavioral phase—strong pre-ecdysis occurs in response to the ETH surge and ecdysis via CCAP release. The release of the latter modulator, though, is suppressed by the co-activation of descending inhibition. Consequently EH starts one phase immediately and also primes the modulatory system that will drive the next phase, but only after a delay encoded by the properties of the inhibition. Appropriate stimuli can then adjust the timing of the second phase by suppressing the inhibition. In the case of ecdysis, the relationship of the activation of the inhibitor and the modulatory neuron is such that the second phase will eventually occur. With a stronger level of descending inhibition, the entry into phase 2 could be completely dependent on the right sensory cues. The latter relationship might be seen for sequences in which sign stimuli are essential for behavioral progression, such as the transition from courtship to copulation.

The study of ecdysis has undergone a resurgence over the past 10 years and has shown an unexpected richness of hormonal signals that orchestrate this behavioral sequence. An emphasis on *Drosophila* will continue to be important in testing the role of key components in the control system and the discovery of new players. A comparative expansion to other insects, though, will finally reveal the true complexity of this behavioral control system.

ACKNOWLEDGMENTS

Unpublished experiments were supported by grant IBO 0452009 from NSF.

REFERENCES

Baker, J. D., and Truman, J. W. (2002). Mutations in the *Drosophila* glycoprotein hormone receptor, *rickets*, eliminate neuropeptide-induced tanning and selectively block a stereotyped behavioral program. *J. Exp. Biol.* **205,** 2555–2565.

Baker, J. D., McNabb, S. L., and Truman, J. W. (1999). The hormonal coordination of behavior and physiology at adult ecdysis in *Drosophila melanogaster*. *J. Exp. Biol.* **202,** 3037–3048.

Carlson, J. R. (1977). The imaginal ecdysis of the cricket (*Teleogryllus oceanicus*). I. Temporal structure and organization into motor programmes. *J. Comp. Physiol.* **115,** 299–317.

Copenhaver, P. F., and Truman, J. W. (1982). The role of eclosion hormone in the larval ecdysis of *Manduca sexta*. *J. Insect Physiol.* **28,** 695–701.

Cottrell, C. B. (1962). The imaginal ecdysis of blowflies. Detection of the blood-borne darkening factor and the determination of some of its properties. *J. Exp. Biol.* **39,** 413–430.

Davis, N. T., Homberg, U., Dircksen, H., Levine, R. B., and Hildebrand, J. G. (1993). Crustacean cardioactive peptide-immunoreactive neurons in the hawkmoth *Manduca sexta* and changes in their immunoreactivity during postembryonic development. *J. Comp. Neurol.* **338,** 612–627.

Dewey, E. M., McNabb, S. L., Ewer, J., Kuo, G. R., Takanishi, C. L., Truman, J. W., and Honegger, H. W. (2004). Identification of the gene encoding bursicon, an insect neuropeptide responsible for cuticle sclerotization and wing spreading. *Curr. Biol.* **14,** 1208–1213.

Eriksen, K. K., Hauser, F., Schiott, M., Pedersen, K. M., Sondergaard, L., and Grimmelikhuijzen, C. J. (2000). Molecular cloning, genomic organization, developmental regulation, and a knockout mutant of a novel leu-rich repeats-containing G protein-coupled receptor (DLGR-2) from *Drosophila melanogaster*. *Genome Res.* **10,** 924–938.

Ewer, J., and Reynolds, S. E. (2002). Neuropeptide control of molting in insects. *In* "Hormones, Brain and Behavior" (D. W. Pfaff, A. P. Arnold, A. M. Etgen, S. E. Fahrbach, and R. T. Rubin, Eds.), vol. 3, pp. 1–92. Academic Press, San Diego.

Ewer, J., de Vente, J., and Truman, J. W. (1994). Neuropeptide induction of cyclic GMP increases in the insect CNS: Resolution at the level of single identifiable neurons. *J. Neurosci.* **14**, 7704–7712.

Ewer, J., Gammie, S. C., and Truman, J. W. (1997). Control of insect ecdysis by a positive feedback endocrine system: Roles of eclosion hormone and ecdysis triggering hormone. *J. Exp. Biol.* **200**, 869–881.

Ewer, J., del Campo, M., and Park, J. (2001). Neuroendocrine control of ecdysis. *J. Neurogenet.* **15**, 19.

Fahrbach, S. E., and Truman, J. W. (1989). Autoradiographic identification of ecdysteroid-binding cells in the nervous system of the moth *Manduca sexta. J. Neurobiol.* **20**, 681–702.

Frankel, G., and Hsio, C. (1962). Hormonal and nervous control of tanning in the fly. *Science* **138**, 27–29.

Frankel, G., and Hsio, C. (1965). Bursicon, a hormone which mediates tanning of the cuticle in the adult fly and other insects. *J. Insect Physiol.* **11**, 513–556.

Fuse, M., and Truman, J. W. (2002). Modulation of ecdysis in the moth, *Manduca sexta*: The roles of the suboesophageal and thoracic ganglia. *J. Exp. Biol.* **205**, 1047–1058.

Gammie, S. C., and Truman, J. W. (1997a). An endogenous elevation of cGMP increases the excitability of identified insect neurosecretory cells. *J. Comp. Physiol. A* **180**, 329–338.

Gammie, S. C., and Truman, J. W. (1997b). Neuropeptide hierarchies and the activation of sequential motor behaviors in the hawkmoth, *Manduca sexta. J. Neurosci.* **17**, 4389–4397.

Gammie, S. C., and Truman, J. W. (1999). Eclosion hormone provides a link between ecdysis triggering hormone and crustacean cardioactive peptide in the neuroendocrine cascade that controls ecdysis behaviour. *J. Exp. Biol.* **202**, 343–352.

Gilbert, L. I. (1989). The endocrine control of molting: The tobacco hornworm, *Manduca sexta*, as a model system. *In* "Ecdysone: From Chemistry to Mode of Action" (J. A. Koolman, Ed.), pp. 448–471. Georg Thieme Verlag, Stuttgart.

Harris-Warrick, R. M., Baro, D. J., Coniglio, L. M., Johnson, B. R., Levini, R. M., Peck, J. H., and Zhang, B. (1997). Chemical modulation of crustacean stomatogastric pattern generator networks. *In* "Neurons, Networks and Motor Behavior" (P. S. G. Stein, S. Grillner, A. I. Selverston, and D. G. Stuart, Eds.), pp. 209–215. MIT Press, Cambridge, MA.

Hewes, R. S., and Truman, J. W. (1991). The roles of central and peripheral eclosion hormone release in the control of ecdysis behavior in *Manduca sexta. J. Comp. Physiol. A* **168**, 697–707.

Hewes, R. S., and Truman, J. W. (1994). Steroid regulation of excitability in identified insect neurosecretory cells. *J. Neurosci.* **14**, 1812–1819.

Hesterlee, S., and Morton, D. B. (2000). Identification of the cellular target for eclosion hormone in the abdominal transverse nerves of the tobacco hornworm, *Manduca sexta. J. Comp. Neurol.* **424**, 339–355.

Honegger, H. W., Market, D., Pierce, L. A., Dewey, E. M., Kostron, B., Wilson, M., Choi, D., Klukas, K. A., and Mesce, K. A. (2002). Cellular localization of bursicon using antisera against partial peptide sequences of this insect cuticle-sclerotizing neurohormone. *J. Comp. Neurol.* **452**, 163–177.

Horodyski, F. M. (1996). Neuroendocrine control of insect ecdysis by eclosion hormone. *J. Insect Physiol.* **42**, 917–924.

Horodyski, F. M., Riddiford, L. M., and Truman, J. W. (1989). Isolation and expression of the eclosion hormone gene from the tobacco hornworm, *Manduca sexta. Proc. Natl. Acad. Sci. USA* **86**, 8123–8127.

Horodyski, F. M., Ewer, J., Riddiford, L. M., and Truman, J. W. (1993). Isolation, characterization, and expression of the eclosion hormone gene of *Drosophila melanogaster*. *Eur. J. Biochem.* **215**, 221–228.

Hughes, T. D. (1980). The imaginal ecdysis of the desert locust, *Schistocerca gregaria*. I. Description of the behaviour. *Physiol. Entomol.* **5**, 47–54.

Kataoka, H., Troetschler, R. G., Kramer, S. J., Cesarin, B. J., and Schooley, D. A. (1987). Isolation and primary structure of the eclosion hormone of the tobacco hornworm, *Manduca sexta*. *Biochem. Biophys. Res. Commun.* **146**, 746–750.

Kim, Y. J., Spalovska-Valachova, I., Cho, K. H., Zitnanova, I., Park, Y., Adams, M. E., and Zitnan, D. (2004). Corazonin receptor signaling in ecdysis initiation. *Proc. Natl. Acad. Sci. USA* **101**, 6704–6709.

Kimura, K.-I., and Truman, J. W. (1990). Postmetamorphic cell death in the nervous and muscular systems of *Drosophila melanogaster*. *J. Neurosci.* **10**, 403–411.

Kingan, T. G., and Adams, M. E. (2000). Ecdysteroids regulate secretory competence in Inka cells. *J. Exp. Biol.* **203**, 3011–3018.

Kingan, T. G., Gray, W., Zitnan, D., and Adams, M. E. (1997). Regulation of ecdysis-triggering hormone release by eclosion hormone. *J. Exp. Biol.* **200**, 3245–3256.

Kingan, T. G., Cardullo, R. A., and Adams, M. E. (2001). Signal transduction in eclosion hormone-induced secretion of ecdysis-triggering hormone. *J. Biol. Chem.* **276**, 25136–25142.

Kostron, B., Kaltenhauser, U., Seibel, B., Bräunig, P., and Honegger, H. W. (1996). Localization of bursicon in CCAP-immunoreactive cells in the thoracic ganglia of the cricket *Gryllus bimaculatus*. *J. Exp. Biol.* **199**, 367–377.

Luo, C. W., Dewey, E. M., Sudo, S., Ewer, J., Hsu, S. Y., Honegger, H. W., and Hsueh, A. J. (2005). Bursicon, the insect cuticle-hardening hormone, is a heterodimeric cystine knot protein that activates G protein-coupled receptor LGR2. *Proc. Natl. Acad. Sci. USA* **102**, 2820–2825.

Loi, P. K., Emmal, S. A., Park, Y., and Tublitz, N. J. (2001). Identification, sequence and expression of a crustacean cardioactive peptide (CCAP) gene in the moth *Manduca sexta*. *J. Exp. Biol.* **204**, 2803–2816.

Marder, E., and Thirumalai, V. (2002). Cellular, synaptic and network effects of neuromodulation. *Neural Netw.* **15**, 479–493.

Marti, T., Takio, K., Walsh, K. A., Terzi, G., and Truman, J. W. (1987). Microanalysis of the amino acid sequence of the eclosion hormone from the tobacco hornworm *Manduca sexta*. *FEBS Lett.* **219**, 415–418.

McNabb, S. L., Baker, J. D., Agapite, J., Steller, H., Riddiford, L. M., and Truman, J. W. (1997). Disruption of a behavioral sequence by targeted death of peptidergic neurons in *Drosophila*. *Neuron* **19**, 813–823.

Miles, C. I., and Weeks, J. C. (1991). Developmental attenuation of the pre-ecdysis motor pattern in the tobacco hornworm, *Manduca sexta*. *J. Comp. Physiol.* **168**, 179–190.

Mills, R. R., Mather, R. B., and Guerra, A. A. (1965). Studies on the hormonal control of tanning in the American cockroach. I. Release of an activation factor from the terminal abdominal ganglion. *J. Insect Physiol.* **11**, 1047–1053.

Morton, D. B., and Anderson, E. J. (2003). MsGC-beta3 forms active homodimers and inactive heterodimers with NO-sensitive soluble guanylyl cyclase subunits. *J. Exp. Biol.* **206**, 937–947.

Morton, D. B., and Simpson, P. J. (2002). Cellular signaling in eclosion hormone action. *J. Insect Physiol.* **48**, 1–13.

Myers, E. M., Yu, J., and Sehgal, A. (2003). Circadian control of eclosion: Interaction between a central and peripheral clock in *Drosophila melanogaster*. *Curr. Biol.* **13**, 526–533.

Novicki, A., and Weeks, J. C. (1993). Organization of the larval pre-ecdysis motor pattern in the tobacco hornworm, *Manduca sexta*. *J. Comp. Physiol. A* **173**, 151–162.

Novicki, A., and Weeks, J. C. (1995). A single pair of interneurons controls motor neuron activity during pre-ecdysis compression behavior in larval *Manduca sexta*. *J. Comp. Physiol. A* **176**, 45–54.

Novicki, A., and Weeks, J. C. (1996). The initiation of pre-ecdysis and ecdysis behaviors in larval *Manduca sexta*: The roles of the brain, terminal ganglion and eclosion hormone. *J. Exp. Biol.* **199**, 1757–1769.

O' Brien, M. A., and Taghert, P. H. (1998). A peritracheal neuropeptide system in insects: Release of myomodulin-like peptides at ecdysis. *J. Exp. Biol.* **201**, 193–209.

Park, J. H., Schroeder, A. J., Helfrich-Forster, C., Jackson, F. R., and Ewer, J. (2003). Targeted ablation of CCAP neuropeptide-containing neurons of *Drosophila* causes specific defects in execution and circadian timing of ecdysis behavior. *Development* **130**, 2645–2656.

Park, Y., Zitnan, D., Gill, S. S., and Adams, M. E. (1999). Molecular cloning and biological activity of ecdysis-triggering hormones in *Drosophila melanogaster*. *FEBS Lett.* **463**, 133–138.

Park, Y., Filippov, V., Gill, S. S., and Adams, M. E. (2002a). Deletion of the ecdysis-triggering hormone gene leads to lethal ecdysis deficiency. *Development* **129**, 493–503.

Park, Y., Kim, Y. J., and Adams, M. E. (2002b). Identification of G protein-coupled receptors for *Drosophila* PRXamide peptides, CCAP, corazonin, and AKH supports a theory of ligand-receptor coevolution. *Proc. Natl. Acad. Sci. USA* **99**, 11423–11428.

Park, Y., Kim, Y. J., Dupriez, V., and Adams, M. E. (2003). Two subtypes of ecdysis-triggering hormone receptor in *Drosophila melanogaster*. *J. Biol. Chem.* **278**, 17710–17715.

Reynolds, S. E., Taghert, P. H., and Truman, J. W. (1979). Eclosion hormone and bursicon titres and the onset of hormonal responsiveness during the last day of adult development in *Manduca sexta* (L.). *J. Exp. Biol.* **78**, 77–86.

Reynolds, S. E. (1980). Integration of behaviour and physiology in ecdysis. *Adv. Insect Physiol.* **15**, 475–595.

Robinow, S., Talbot, W. S., Hogness, D. S., and Truman, J. W. (1993). Programmed cell death in the *Drosophila* CNS is ecdysone-regulated and coupled with a specific ecdysone receptor. *Development* **119**, 1251–1259.

Schwartz, L. M., and Truman, J. W. (1983). Hormonal control of rates of metamorphic development in the tobacco hornworm *Manduca sexta*. *Dev. Biol.* **99**, 103–114.

Shiga, S., Davis, N. T., and Hildebrand, J. G. (2003). Role of neurosecretory cells in the photoperiodic induction of pupal diapause of the tobacco hornworm *Manduca sexta*. *J. Comp. Neurol.* **462**, 275–285.

Sláma, K. (1980). Homeostatic function of ecdysteroids in ecdysis and oviposition. *Acta Entomol. Bohoemosl.* **73**, 65–75.

Stangier, J., Hilbich, C., Dircksen, H., and Keller, R. (1988). Distribution of a novel cardioactive neuropeptide (CCAP) in the nervous system of the shore crab *Carcinus maenas*. *Peptides* **9**, 795–800.

Taghert, P. H., and Truman, J. W. (1982). Identification of the bursicon-containing neurons in abdominal ganglia of the tobacco hornworm, *Manduca sexta*. *J. Exp. Biol.* **98**, 385–401.

Truman, J. W. (1971). Physiology of insect ecdysis. I. The eclosion behaviour of saturniid moths and its hormonal release. *J. Exp. Biol.* **54**, 805–814.

Truman, J. W. (1978). Hormonal release of stereotyped motor programmes from the isolated nervous system of the Cecropia silkmoth. *J. Exp. Biol.* **74**, 151–174.

Truman, J. W., and Copenhaver, P. F. (1989). The larval eclosion hormone neurones in *Manduca sexta*: Identification of the brain-proctodeal neurosecretory system. *J. Exp. Biol.* **147**, 457–470.

Truman, J. W., and Riddiford, L. M. (1970). Neuroendocrine control of ecdysis in silk moths. *Science* **167**, 1624–1626.

Truman, J. W., Rountree, D. B., Reiss, S. E., and Schwartz, L. M. (1983). Ecdysteroids regulate the release and action of eclosion hormone in the tobacco hornworm, *Manduca sexta* (L). *J. Insect Physiol.* **29**, 895–900.

Tublitz, N. J., Copenhaver, P. F., Taghert, P. H., and Truman, J. W. (1986). Peptidergic regulation of behavior: An identified neuron approach. *Trends Neuro. Sci.* **9**, 359–363.

Vafopoulou, X., and Steel, C. G. (1991). Circadian regulation of synthesis of ecdysteroids by prothoracic glands of the insect *Rhodnius prolixus*: Evidence of a dual oscillator system. *Gen. Comp. Endocrinol.* **83**, 27–34.

Veenstra, J. A. (1989). Isolation and structure of corazonin, a cardioactive peptide from the American cockroach. *FEBS Lett.* **250**, 231–234.

Weeks, J. C., and Truman, J. W. (1984). Neural organization of peptide-activated ecdysis behaviors during metamorphosis in *Manduca sexta*. I. Conservation of the peristalsis motor pattern at the larval-pupal transformation. *J. Comp. Physiol.* **155**, 407–422.

Zitnan, D., and Adams, M. E. (2000). Excitatory and inhibitory roles of central ganglia in initiation of the insect ecdysis behavioural sequence. *J. Exp. Biol.* **203**, 1329–1340.

Zitnan, D., Kingan, T. G., Hermesman, J. L., and Adams, M. E. (1996). Identification of ecdysis-triggering hormone from an epitracheal endocrine system. *Science* **271**, 88–91.

Zitnan, D., Ross, L. S., Zitnanova, I., Hermesman, J. L., Gill, S. S., and Adams, M. E. (1999). Steroid induction of a peptide hormone gene leads to orchestration of a defined behavioral sequence. *Neuron* **23**, 523–535.

Zitnan, D., Zitnanova, I., Spalovska, I., Takac, P., Park, Y., and Adams, M. E. (2003). Conservation of ecdysis-triggering hormone signaling in insects. *J. Exp. Biol.* **206**, 1275–1289.

Zitnanova, I., Adams, M. E., and Zitnan, D. (2001). Dual ecdysteroid action on the epitracheal glands and central nervous system preceding ecdysis of *Manduca sexta*. *J. Exp. Biol.* **204**, 3483–3495.

2

A Molecular Genetic Approach to the Biosynthesis of the Insect Steroid Molting Hormone

Lawrence I. Gilbert and James T. Warren

Department of Biology, University of North Carolina, Chapel Hill North Carolina 27599

I. Introduction
II. Ecdysteroids
    A. The Route to Cholesterol and 7-Dehydrocholesterol
    B. Low Ecdysone Mutants
III. Neuropeptide Control of Ecdysteroidogenesis
IV. Subcellular Translocation of Ecdysteroid Intermediates
V. The "Black Box"
VI. Terminal Hydroxylations (From the Diketol to 20-Hydroxyecdysone)
    A. Introduction
    B. The Halloween Genes
VII. Epilogue
References

Insect growth, development, and molting depend upon a critical titer of the principal molting hormone of arthropods, 20-hydroxyecdysone (20E). Although the structure of 20E as a polyhydroxylated steroid was determined more than five decades ago, the exact steps in its biosynthesis have eluded identification. Over the past several years, the use of the fly database and the techniques and paradigms of biochemistry, analytical chemistry, and molecular genetics have allowed the cloning and sequencing of four genes in the Halloween gene family of *Drosophila melanogaster*, all of them encoding cytochrome P450 (CYP) enzymes, each of which mediates one of the four terminal hydroxylation steps in 20E biosynthesis. Further, the sequence of these hydroxylations has been determined, and developmental alterations in the expression of each of these genes have been quantified during both embryonic and postembryonic life. © 2005 Elsevier Inc.

I. INTRODUCTION

When the phylogenetic lines between insects and vertebrates diverged from a common ancestor some 700 million years ago, most of the biochemical and genetic pathways and blueprints now seen in these two groups had been ordained. It is now known that about 70% of the approximately 20,000 human genes are present in the fruit fly, *D. melanogaster* and virtually every housekeeping pathway, as well as the multitude of signal transduction pathways, are identical in these groups, notwithstanding the fact that insects arose some 400–500 millions of years ago. These pathways must have endowed these groups with selective advantages which have withstood very frequent mutations to elements of the pathways over the millennia. This chapter discusses the results of an experimental paradigm that has allowed us to begin to elucidate a biosynthetic pathway leading to the insect steroid molting hormone, a pathway that is a requisite for the great success that insects have displayed over the past half billion years, but also to demonstrate the close biochemical similarity between mammals and insects (i.e., the utilization of steroid hormones) but for completely different purposes. Thus, it appears that it is not so much the chemistry of the molecule that has changed during evolution, but rather the way it is utilized by a particular animal group (Schneiderman and Gilbert, 1964). Since this chapter is concerned with the insect molting hormone, the term molting will be defined.

One reason that insects have been so successful on this planet is the protection afforded to them by the hard exoskeleton (cuticle) in which they are encased as they inhabit almost every available ecological niche. However, as is accepted by researchers that "the solution to one problem always seems to elicit new problems." In the case of the insect, the almost inflexible cuticle is a barrier to growth and for the insect to increase in mass, this cuticle must be shed periodically and a new one of greater size must be synthesized and then hardened (tanned). The process known as molting consists of the synthesis of the new cuticle interior to the old one, digestion of most of the old cuticle by specific enzymes and recycling of the hydrolytic products, and finally shedding of the remnants of the old cuticle (ecdysis). This very complex series of events allows the insect to grow and is initiated and regulated by the principal molting hormone, the polyhydroxylated sterol 20-hydroxyecdysone (20E) (Fig. 1), synthesized by a variety of tissues from the precursor ecdysone (E), which in many insects (e.g., the fruit fly *D. melanogaster*) is synthesized by cells comprising the prothoracic gland. In the specific case of higher flies, these cells are part of the composite gland, or the ring gland, so named because it circles the anterior portion of the digestive system and lies very close to the brain (Dai and Gilbert, 1991). In addition to the prothoracic gland cells that synthesize ecdysone, the ring gland houses cells of the corpus allatum (source of another growth hormone, the juvenile hormone (see Dubrovsky, this volume), and cells of the corpus cardiacum consisting in part of neurosecretory cell endings from the brain. This chapter is concerned with the biosynthesis of the molting hormone 20E, the molecule that regulates insect molting and is responsible for the ability of insects to grow and surely has contributed significantly to the evolutionary success of insects.

II. ECDYSTEROIDS

A. THE ROUTE TO CHOLESTEROL AND 7-DEHYDROCHOLESTEROL

The term ecdysteroid is used for a family of steroidal compounds closely related to ecdysone (Fig. 1), both natural and those developed by the synthetic chemist. To begin to understand the individual links in the biosynthesis of these molecules, one must realize that insects do not have the ability to synthesize cholesterol from simple precursors, but rather must ingest one of a variety of steroids with their food. In the case of plant-eating insects (phytophagous), those molecules are primarily sitosterol and campesterol and must be dealkylated by the insect to cholesterol. However, most plants also contain a small amount of cholesterol, which can be concentrated in some insects in specific tissues (e.g., the ring gland), and *Drosophila* appears to be one of those species. Of course, carnivorous insects obtain their cholesterol neat (Gilbert *et al.*, 2002). In many studies of the subject, it was shown that most insects can dealkylate and reduce the common 24(*R*)-alkyl plant sterols, such as sitosterol and campesterol, to cholesterol, although some, including *Drosophila*, apparently cannot (Gilbert *et al.*, 2002). The insect gut is the site of dealkylation, and in contrast to the synthesis of ecdysone (Fig. 1), the details of the dealkylation process have been known for some time. Sitosterol is first oxidized to fucosterol via an as yet

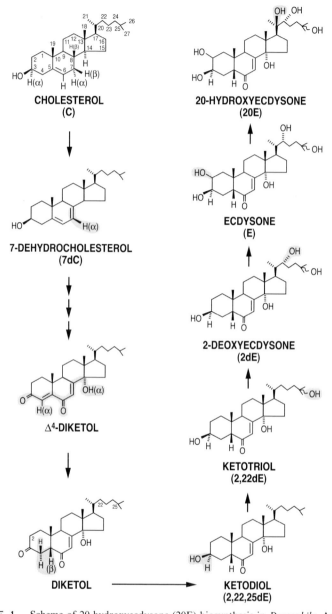

FIGURE 1. Scheme of 20-hydroxyecdysone (20E) biosynthesis in *Drosophila*. As many as six (or more) cytochrome P450 (CYP) enzymes may be involved, starting from cholesterol (C). Multiple arrows indicate an uncharacterized pathway, perhaps involving more than one biochemical transformation. The Δ^4-diketol is cholesta-4,7-diene-3,6-dione-14α-ol, the diketol is 5β[H]-cholesta-7-ene-3,6-dione-14α-ol, the ketodiol is 5β[H]-cholesta-7-ene-6-one-3β,14α-diol (2,22,25-trideoxyecdysone or 2,22,25dE) and herein the generic term ketotriol represents 5β[H]-cholesta-7-ene-6-one-3β, 14α, 25-triol (2,22-dideoxyecdysone or 2,22dE). The uncharacterized

uncharacterized P450 oxygenase or other enzymes, which is then epoxidized and the product subsequently subjected to lyase activity. Additional oxidation then proceeds resulting in the production of desmosterol after loss of the two carbon units. The latter is a common reaction in all sterol dealkylation pathways resulting in desmosterol, which is C24–25 unsaturated but is then reduced by a specific reductase to cholesterol. Although ecdysone was characterized from a ton or so of silkworm pupae nearly half a century ago and 20E characterized a short time later, the exact biosynthetic pathway from cholesterol to 20E has remained enigmatic despite efforts by a handful of laboratories since the 1950s. This failure was a result of both the lability of intermediates and their extremely low concentration. One might have thought that the advent of *in vitro* techniques for the endocrine glands that synthesize ecdysteroids would facilitate the availability of metabolic intermediates, but it has not although the endocrine glands (prothoracic glands and ring glands, etc.) appear to release their ecdysteroid product as soon as it is synthesized. Despite this, past years have seen the use of molecular genetics applied to the problem with dramatic success (see Section VI.B).

Figure 1 shows what was known with some certainty a few years ago regarding the biosynthesis of ecdysteroids. Basically, cholesterol is first converted to 7-dehydrocholesterol (7dC) through the mediation of a cholesterol 7,8 dehydrogenase, in contrast to the events in mammals where 7dC is a precursor of cholesterol and vitamin D and is derived from lanosterol. Not all insects can make this conversion and as is their "habit" during evolution, many species have devised unique ways to survive. For example, there is a desert species of *Drosophila* (*D. pachea*) that feeds only on a cactus, which provides a required Δ7-sterol lathosterol (Heed and Kircher, 1965), and the insect has lost its ability to 7,8 dehydrogenate cholesterol. The formation of 7dC has been studied for several decades, and it has been shown that 7dC can support the growth and development of *D. pachea* when they are fed on a sterol-free diet. The reaction yielding 7dC involves the stereospecific removal of both the 7β and 8β hydrogens from the sterically hindered β top face of the planar cholesterol molecule. Further, the reaction is basically irreversible and is catalyzed by a microsomal P450 enzyme restricted to the prothoracic glands and the substrate specificity of this enzyme appears to be quite broad (Gilbert *et al.*, 2002).

---

reactions between 7-dehydrocholesterol (7dC) and the Δ^4-diketol are commonly referred to as the "Black Box." Yellow shade delineates the areas of the sterol backbone that are involved in the indicated transformations. The Δ^4-diketol has neither been isolated from insects, nor shown to be a product of radio labeled cholesterol *in vitro* metabolism in insects. It has, however, been shown in crabs (Blais *et al.*, 1996) to be reduced to the diketol, which is then converted into ecdysone. (See Color Insert.)

B. LOW ECDYSONE MUTANTS

This chapter will focus on the use of *Drosophila* low ecdysone mutants in an attempt to elucidate the post-7dC steps in 20E biosynthesis, but will briefly diverge since the use of low ecdysone mutants has added important data to the knowledge of the formation of 7dC from cholesterol. In the *without children* (*woc*) mutant the animal dies at the end of the last larval instar (stage), and therefore does not molt to the pupa and then to the adult fly (i.e., does not undergo metamorphosis) (Wismar *et al.*, 2000). The *woc* gene was cloned and characterized and is believed to be a transcription factor (Wismar *et al.*, 2000). These mutant larvae could be rescued by including 7dC in the diet, indicating that the transcription factor may play a role in modulating the activity of the 7,8 dehydrogenase (Warren *et al.*, 2001). Thus, these mutants have a phenotype analogous to the normal *D. pachea* discussed in an earlier section. When the source of ecdysone, the ring glands, was studied *in vitro*, it was shown that the ability to convert cholesterol to 7dC was absent in homozygous mutant larval glands. Because of this defect, they could not synthesize ecdysone, but this inhibition was reversed by prior 7dC feeding. Although it was disappointing that the *woc* gene did not encode the cholesterol 7,8 dehydrogenase, the working hypothesis is that the putative transcription factor upregulates the dehydrogenase that mediates the first step in the ecdysteroidogenic pathway, similar perhaps to SF1 and other orphan transcription factors in mammals. The most important lesson was that the use of more appropriate low ecdysone mutants could possibly allow the identification and characterization of enzymes that mediate specific reactions in the ecdysteroidogenesis pathway. This possibility has borne fruit as will be seen subsequently in Section VI.B.

Many years ago it was realized that one way to approach the problem of ecdysteroid biosynthesis was to investigate how the prothoracic glands are turned on and then be able to manipulate the glands *in vitro* to attain maximal rates of ecdysteroidogenesis, which in turn could provide us with larger amounts of intermediates. This led to more than 20 years of research on the prothoracic glands of the tobacco hornworm, *Manduca sexta*.

III. NEUROPEPTIDE CONTROL OF ECDYSTEROIDOGENESIS

The work on this subject has been summarized in another volume of this series (Henrich *et al.*, 1999), and the entire field is critically scrutinized by Rybczynski (2005), but the major points will be discussed briefly as background for this chapter. It has been known for more than 80 years that the insect brain was the source of a growth hormone involved in the control of insect molting. That demonstration by Kopeć was the first to show that nervous tissue displayed characteristics of an endocrine gland (i.e., initiated the field of neuroendocrinology). Later observations showed that the implantation of "active" brains and the injection of brain extracts into recipients devoid of brains would trigger events leading to molting and that the target of this "brain hormone" was the prothoracic glands. Then it was demonstrated in the 1980s that two large neurosecretory cells on either side of the *Manduca* brain synthesized a peptide hormone, now termed prothoracicotropic hormone (PTTH), which was responsible for the stimulation of the prothoracic glands, and that the degree of stimulation of ecdysteroidogenesis could be quantified by use of an *in vitro* assay in which prothoracic glands were maintained in culture medium and exposed to brain extract and later recombinant PTTH and the amount of ecdysteroid then measured by a variety of radioimmunoassay (RIAs) (see Gilbert *et al.*, 2000).

The PTTH of *Manduca* is a homodimeric molecule of about 30 kDa with a single intermonomeric cysteine–cysteine bond and three intramonomeric cysteine–cysteine bonds (Rybczynski, 2005). The composite data indicate that there is a PTTH receptor on the surface of the cells of the prothoracic glands, although it has yet to be characterized, and the ligand–receptor interaction leads to the opening of a calcium channel and then the influx of extracellular calcium. This is followed by a rise in cAMP, presumably due to stimulation of a calcium-calmodulin-dependent adenylyl cyclase followed by PKA activation, and then a series of rapid protein phosphorylations dependent on the mediation of PKA, MAPKs, and probably several other kinases. It is likely that complex cross-talk takes place between these transductory pathways and the action of a 70 kDa S6 kinase, leading to the multiple phosphorylation of the ribosomal protein S6, which is believed to stimulate translation initiation in the cells comprising the prothoracic glands. This leads to the increased rate of ecdysteroid biosynthesis, although the steps between S6 phosphorylation and ecdysteroidogenesis have not yet been elucidated. These data are no doubt the tip of the iceberg, and with the history of many transductory systems it is most likely that further research will reveal increased complexity. For example, it is now known that PTTH action also requires a small population of phosphorylated ERK molecules suggesting transcriptional targets as well (Gilbert *et al.*, 2002).

The details of these transductory events as well as others in analogous systems and the recognition of cross talk between pathways will no doubt be solved in time. The ultimate question for students of insect endocrinology is where in the ecdysteroidogenic pathway does the final product of the transduction pathway intersect and exert its effect on a rate-limiting step in the synthesis of 20E. To date it is believed that the phosphorylated form of S6 is the key effector, but it is essential that the steps in ecdysteroid biosynthesis be elucidated so that the effect of S6 on individual reactions can be examined before the final conclusion is reached on how PTTH stimulates ecdysteroidogenesis in the cells of the prothoracic glands.

IV. SUBCELLULAR TRANSLOCATION OF ECDYSTEROID INTERMEDIATES

Since the cholesterol to 7dC step was known through the work on *woc* (discussed in earlier section), a series of experiments was conducted using both homogenized and perturbed prothoracic glands from *Manduca* larvae based on the hypothesis that trafficking would be analogous to what is accepted for ACTH control of cholesterol transport within the adrenal cortical mitochondria of mammals (Stocco and Clark, 1997; Strauss *et al.*, 2003; Thomson, 2003). In that regard, the working model was the possibility that the nonpolar 7dC moves from the ER where it is synthesized and apparently not further metabolized, and then to the mitochondria (or another subcellular organelle) where it undergoes conversion to an intermediate, which is a more immediate precursor of ecdysone (Warren and Gilbert, 1996). When labeled cholesterol was added to these systems the kinetics of conversion of 7dC to ecdysteroid was found to be slower than when pure 7dC was added directly to gland homogenates (i.e., the $\Delta^{5,7}$sterol synthesized in the ER transited more slowly to the interior of the mitochondria than did exogenous 7dC added to homogenates).

Radiochemical studies with ring glands of the low ecdysteroid mutant *ecdysoneless* (*ecd1*) suggested that the mutation may be affecting 7dC transit to the mitochondria (Warren *et al.*, 1996), but the protein described for the cell autonomous roles of *ecd1*, although evolutionarily conserved (Gaziova *et al.*, 2004), does not have the characteristics of a transit protein. Such carrier or transit proteins have not been studied in depth in insects, although the *Drosophila* gene *start1*, analogous to the original steroidogenic regulatory protein StAR (Strauss *et al.*, 2003), has been shown to have the characteristics of a cholesterol transporter involved in ecdysteroid biosynthesis (Roth *et al.*, 2004). It is widely distributed in the fly larva when analyzed by RT-PCR but expressed mainly in the prothoracic gland cells of the ring gland as viewed by *in situ* hybridization. Whether the start1 protein is involved with 7dC movement was not studied and is conjectural as yet.

The question of insect steroid trafficking proteins remains hypothetical, but it is of obvious importance if the insect steroid hormone biosynthesis is to be truly understood. As will be seen subsequently in Section VI , P450 enzymes associated with ecdysteroid biosynthesis are found in either the ER or mitochondria, indicating that substrate-transit between these and other membrane-bound subcellular organelles is probably involved. In the case of mammals, it is generally accepted that intracellular trafficking of cholesterol in cells responsible for steroid hormone biosynthesis is quite important and that there are a family of proteins related to StAR, which are involved in the intramitochondrial movement of cholesterol (Christenson and Strauss, 2000; Strauss *et al.*, 2003). Even the presence of a sterol carrier protein in insects has been elusive despite its obvious importance and the almost ubiquitous presence of such molecules (e.g., SCP-2 in higher animals). A degree of optimism occurred due to the very promising work of Lan *et al.* on an insect SCP-2 analogue and the gene that encodes it. (Dyer *et al.*, 2003; Krebs and Lan, 2003; Lan and Wessely, 2004). Although the exact function of this protein remains to be elucidated, there is a possibility that it may be involved in the intercellular or intracellular trafficking of ecdysone sterol precursors, perhaps cholesterol or 7dC. This field of study suffers from a paucity of researchers in the field, but we should not be disheartened by the fact that trafficking is not well understood in the insect model as compared to the mammalian model (adrenal cortex in most cases), which has now been studied for two decades by numerous research groups although "a precise mechanism of how cholesterol passes from the outer to the inner membrane, however, still remains elusive" (Thomson, 2003).

V. THE "BLACK BOX"

The definition of what is euphemistically called the "Black Box" is that series of hypothetical and unproven mitochondrial- (or perhaps ER or peroxisome) mediated reactions that result in the oxidation of 7dC, first to the Δ^4-diketol (cholesta-4,7-diene-3,6-dione-14a-ol), and then to the diketol (5β[H]-cholesta-7-ene-3,6-dione or 2,22,25-tridexoxyecdysone) (Fig. 1), which for many years have been considered to be true intermediates in ecdysone biosynthesis (Gilbert *et al.*, 2002; Lafont *et al.*, 2005). However, despite many attempts, several laboratories have been totally frustrated in their attempts to isolate intermediate compounds between 7dC and the first molecules having the characteristic ecdysteroid structure, namely the *cis*-5β(H)-3β-ol or 3-dehydro-6-one-7-ene-14α-ol functionalities. In contrast to what is known about mammalian systems, reactions in the insect system must proceed at a considerably more rapid pace or the intermediates must be much more labile than those resulting from the repetitive hydroxylations and oxidations, which result in either cholesterol side chain cleavage or lanosterol 14-demethylation (both mediated by P450 enzymes). In both cases the intermediates have been isolated and characterized decades ago. Little is known about analogous reactions in insects, and they are the last frontier in understanding the details of ecdysteroid biosynthesis, as will be seen subsequently.

The next challenge for the insect is to convert the Δ^4-diketol to the diketol having a *cis*-A, B-ring fusion (Fig. 1), presumably mediated by a 5β(H)-reductase (Blais *et al.*, 1996). In contrast, analogous transformations required for mammalian steroidogenesis are mediated by 5α-reductases yielding hormones having a *trans*-A, B-ring junction. In general, the 5β- configuration in mammals typically yields inactive steroids. After the mammalian steroids have been saturated, the A-ring 3-ketone group can be reversibly reduced to the 3β-alcohol using 3β-hydroxysteroid dehydrogenases (Brown, 1998).

Similarly, insects use analogous enzyme systems for both the reversible and irreversible interchange of 3-keto and either 3β-ol (active) or 3α-ol (inactive) functionalities in downstream ecdysteroid intermediates, but exactly where in the biosynthetic scheme these reactions are most important appears to differ among insects (see later section).

A discussion of the chemistry of 7dC was presented previously (Gilbert et al., 2002) with the hope that this might suggest further experiments to elucidate the nature of the "Black Box." It did not, although it remains likely that a 3-dehydro-Δ^4 sterol intermediate plays a role in ecdysone biosynthesis. This would be consistent with findings that the prothoracic glands of a large number of insects, but not all, secrete 3-dehydroecdysone (3DE) (*Manduca* does but *Drosophila* does not). In insects that do, the 3DE is rapidly reduced to ecdysone in the hemolymph (Warren et al., 1988; Kiriishi et al., 1990). In *Drosophila* the ring gland mainly secretes ecdysone and the details of the 3β-ecdysteroid reduction are not known. Also unclear is how early in ecdysone biosynthesis the putative oxidation of the 3β-hydroxyl group of the sterol A-ring occurs (i.e., whether it is cholesterol or 7dC that is so dehydrogenated). The site of this reaction is also unknown (i.e., in the ER, mitochondria, or another organelle, such as the peroxisome). Nevertheless, it is likely that any 3-dehydro-7dC is formed within, as well as remains within, a non-aqueous environment because of its extreme lability. Thus, its expected facile isomerization product was not found, the 3-dehydro-$\Delta^{4,7}$-diene after *Manduca* prothoracic glands were incubated in the presence of either labeled cholesterol, which itself is converted under these conditions in this system to 7dC (Warren and Gilbert, 1996) or labeled 7dC itself.

Nevertheless, our working hypothesis remains that 3-dehdyro-7dC is an integral and immediate precedent intermediate leading to the Δ^4-diketol (Fig. 1) (see Gilbert et al., 2002). Proton abstraction from C14 of this intermediate could allow for the subsequent activation (by extensive electron delocalization) of the entire leading surface of the A–D rings of the steroid molecule and so prime them for the multiple oxidations that would follow. The formation of such a high energy transient-state molecule could facilitate further oxidation, either mediated by a P450 enzyme or the result of autooxidation by atmospheric oxygen or endogenous hydrogen peroxide. In either scenario, the initial formation of a classic, stable 6-membered high energy epidioxide would lead to fragmentation and stabilization as the Δ^4-diketol. Another scenario might involve P450-catalyzed rapid, sequential hydroxylations of such a transition state molecule, first at C14 and then at C6, followed by further oxidations and rearrangements resulting in the Δ^4-diketol, perhaps via high energy epoxide intermediates.

Blais et al. (1996) have described a 5β[H]-reductase in the crustacean Y-organ (analogous to the insect prothoracic gland), which specifically converts the Δ^4-diketol to the 5β(H)-diketol, which subsequently is converted to ecdysone. Since the reaction requires NADPH, it appears to be similar to the 3-oxo-Δ^4-steroid 5β(H)-reductases of mammals, which are involved in the degradation of steroids and the synthesis of bile acids (Gilbert et al., 2002). This crustacean Δ^4-diketol-5β(H)-reductase has not yet been characterized for insects, but it is an important enzyme and should be investigated in order to truly understand ecdysone biosynthesis. As can be seen from the discussion in the earlier section, the Black Box remains a mystery, but may at last be prone to dissection using the biochemical and molecular genetic techniques employed to elucidate the enzymes involved in the last four terminal hydroxylation steps that yield 20E.

VI. TERMINAL HYDROXYLATIONS (FROM THE DIKETOL TO 20-HYDROXYECDYSONE)

A. INTRODUCTION

As seen in Fig. 1 the diketol must undergo reduction at C3 to form the ketodiol. In *Drosophila* it is converted to the ketotriol (2,22-dideoxyecdysone) by hydroxylation at C25. To ultimately synthesize 20E, successive hydroxylations must also occur at C22, C2, and C20. It is important to note that Fig. 1 applies to *Drosophila* and several other higher flies, but there are a few subtle modifications for the prothoracic glands of *Manduca*, many other species of Lepidoptera, and other insects that synthesize 3-dehydroecdysone (3DE). For *Manduca* it is the 3DE that is secreted into the hemolymph where it is converted to ecdysone, whereas for *Drosophila* ecdysone itself is synthesized in the prothoracic gland cells of the ring gland and then secreted into the hemolymph. Exactly where in the *Drosophila* biosynthetic pathway the 3-keto function undergoes 3β-reduction is not known with certainty. Nevertheless, in both cases, ecdysone is then transported in the hemolymph to peripheral tissues where it is converted to the principal molting hormone, 20E.

Over the years there have been numerous studies of hydroxylases in a wide variety of insect tissues, but for those enzymes involved in ecdysteroid biosynthesis, only tissue homogenates or crude microsomal and mitochondrial fractions were utilized. A great deal of useful information was obtained, but there were always questions of contamination of these subcellular fractions and whether the enzyme was specific for the substrate assayed. In mammalian systems, the P450s involved in steroid metabolism are generally but not always quite specific, but those involved in drug metabolism are not specific and can mediate the hydroxylation of a diversity of compounds (Poulos, 2003). What was learned from past studies using insect tissue homogenates or subcellular fractions was that most of the enzymes involved in ecdysteroid biosynthesis were cytochrome P450 enzymes (CYPs), which required molecular oxygen and NADPH, and were inhibited by the common CYP inhibitors, such as carbon monoxide, metyrapone, and piperonyl butoxide (see Feyereisen [2005] for a critical review of these and other studies on insect CYPs). The nomenclature can be confusing because the terms hydroxylase and monooxygenase are used interchangeably for the same reaction and both are accepted. The term hydroxylase is used in this chapter.

It was not until the seminal paper by Chávez *et al.* (2000) appeared that the real progress regarding terminal hydroxylases and ecdysone biosynthesis in general was achieved. The paper was based on a series of elegant studies in the 1980s that investigated a family of genes in *Drosophila*, several of which when mutated, resulted in embryonic lethality (i.e., the embryos developed somewhat but died before embryogenesis was complete). These mutant embryos had a particular phenotype including the inability to lay down normal cuticle (Jürgens *et al.*, 1984; Nüsslein-Volhard *et al.*, 1984; Wieschaus *et al.*, 1984). This family of genes was named the Halloween gene family, and Chavez *et al.* concentrated on the gene *disembodied* (*dib*) and confirmed that when mutated, the embryos died, apparently as the result of a lower than normal ecdysteroid titer. They suggested on the basis of their data that *dib* encoded an enzyme involved in ecdysone biosynthesis. Further, they localized this gene to a single transcription unit defined by a 1.7 kb cDNA isolated from an embryonic *Drosophila* library and it appeared to code for a cytochrome P450 enzyme. When this clone was inserted into the germ line of mutant flies, the embryos were rescued (i.e., they completed embryonic development in a normal manner). The deduced gene product Dib was found to be a new member of the insect CYP superfamily, designated as CYP302a1, and *in situ* analysis revealed that *dib* expression was restricted to the prothoracic gland cells of the *Drosophila* ring gland. Additionally, other members of the Halloween gene family were identified, namely *shadow* (*sad*), *shade* (*shd*), *phantom* (*phm*), *shroud* (*shr*), and *spook* (*spo*) (see Gilbert, 2004).

B. THE HALLOWEEN GENES

1. Methodology

The main objective of our laboratory and that of our collaborators at the University of Minnesota and Université Pierre et Marie Curie was to conduct the functional genomics of the Halloween Gene Family in the hope that each member encodes a separate enzyme in the ecdysteroid biosynthetic scheme. In this endeavor we were most fortunate in three ways. First, the previous studies cited in earlier section in the 1980s elucidated the cytogenetics of these genes (i.e., they mapped the position of each gene on the readily observable and remarkable giant polytene chromosomes of *Drosophila*). Second, as noted previously, there was reasonable, although not definitive, evidence that at least one of these genes (*dib*) coded for a P450 enzyme. Third, we had the great advantage of access to the sequenced *Drosophila* genome (i.e., the fly database). For example, for *shade* (*shd*) we knew that the gene mapped to position 70D2-E8, which according to the fly database was in the close vicinity of a P450 gene, *CYP314a1*, located at position 70E4. A full length cDNA coding for CYP314A1 was then amplified from an embryonic cDNA library (Petryk *et al.*, 2003). To confirm that this gene was in fact altered in the various *shd* mutant alleles, corresponding genomic DNA was amplified by PCR from the appropriate heterozygous mutant stock. After sequencing, the mutant lesion was identified by a specific base change at a particular location. The resulting composite data showed that one *shd* allele had a stop codon at position 136, indicating a change in the first base from C to T, that a second allele showed a point mutation due to an amino acid change, glutamic acid to lysine at position 225, and the third allele showed a mutated acceptor site of intron 1, demonstrating that *CYP314a1* was a gene product of *shd*. Similar protocols were used to identify the remaining genes discussed in a later section.

To characterize any P450 hydroxylase, specific enzymatic activity using known substrates must be demonstrated unequivocally. For the functional analyses of the Halloween genes, the cDNA or the coding region of each gene was ligated into an expression vector and then transfected into a *Drosophila* cell culture line (S2 cells). As controls, similar promoter constructs that constitutively express GFP were also transfected and then exposed to the same conditions as the cells receiving the Halloween gene cDNAs. Three days after transfection, the S2 cells were incubated with known tritiated or nonradiolabeled substrates, the products were analyzed, first by RP-HPLC and TLC (employing UV, RIA, and/or radiotracer analysis), and when necessary by LC coupled to electrospray ionization (ESI) mass spectroscopy and for one of the four genes, by high field NMR. Additionally, gene expression was examined by *in situ* hybridization, RT-PCR, and Northern analysis and both gene- and tissue-specific rescue using wild type sequences linked to either ubiquitous or tissue-specific promoters.

2. *Disembodied (dib)*

For the mutant homozygous embryos, development progresses normally until about midway through embryogenesis when the ecdysteroid titer normally peaks, whereupon abnormal characteristics begin to be noticed. These include a failure of head involution and dorsal closure, abnormal looping of the hindgut, undifferentiated cuticle (cuticle does form during normal embryonic development), and the embryo appears much more compact than the normal wild type (Warren *et al.*, 2002), followed by death. The ecdysteroid titer is well below that of normal embryos, corroborating the results of Chavez *et al.* (2000). If the *dib* gene product does play a role in ecdysone biosynthesis, one would expect it to be expressed in the cells that are responsible for ecdysteroidogenesis. In this case *in situ* analysis revealed that

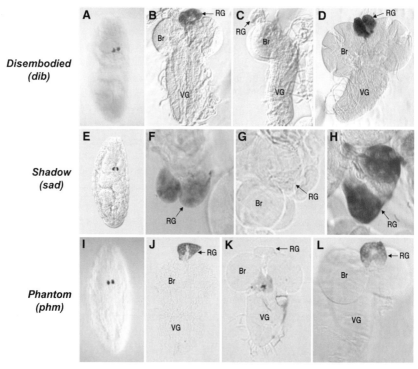

FIGURE 2. *In situ* expression of the Halloween genes *disembodied* (*dib*), *shadow* (*sad*) and *phantom* (*phm*) within the brain–ring gland complex during late embryonic and larval stages. Shown are stage 17 embryos (A, E, and I), late second instar (B, F, and J), and both early (C, G, and K) and late (D, H, and L) third instar brain–ring gland complexes. Note the down regulation of the expression of all three genes between the late second and early third instars and their subsequent up regulation between the early and late third instars. RG, ring gland; Br, brain; VG, ventral ganglion. (Data on embryonic *dib* expression is from Chavez *et al.*, 2000 and larval *dib* and *sad* expression is from Warren *et al.*, 2002. *Phm* expression data is from Warren *et al.*, 2004.) (See Color Insert.)

dib is clearly and unequivocally expressed in the prothoracic gland cells of the embryonic and larval ring glands (Fig. 2). (For the general reader, when *dib* is written in the italicized form it denotes the gene; when Dib is written in the roman form, this denotes the protein gene product.) During early embryogenesis when the ecdysteroid titer peaks, the result of *de novo* biosynthesis before the ring gland has formed, this gene is expressed in the cells comprising the epidermal stripes of the embryo, suggesting that the epidermis is possibly capable of ecdysone synthesis during embryonic development as others have found the epidermis to be during the postembryonic development of several insect species. Since the ecdysteroid titer varies during insect postembryonic development and these changes are required for normal growth, development, and metamorphosis (Henrich *et al.*, 1999), we examined gene activity during larval development semiquantitatively by *in situ* hybridization and found that *dib* was expressed strongly in the prothoracic glands cells of the ring gland during the second and third (last) larval instars (stages) (Fig. 2A–D). Of interest was the observation that this expression is downregulated dramatically just after ecdysis to the third instar when the ecdysteroid titer also falls. This observation suggested that this system would be a model for the examination of those factors regulating the downregulation of *dib*. These studies indicated a vital role for this Halloween gene in the synthesis of ecdysone and then 20E, and it became even more important to determine if the gene product coded for an enzyme.

S2 cells were transfected with *dib* as described earlier and the cells incubated in the presence of high specific activity [^{3}H]ketotriol. The medium was then extracted and subjected to the analytical procedures noted previously. Figure 3B reveals that the cells converted the substrate only to [^{3}H] 2-deoxyecdysone (2dE) with a rate of conversion of 82% over 4–6 h. The GFP control cells showed no conversion. The data thus demonstrated that the *dib* gene product is the **C22-hydroxylase** and that *dib* encodes this P450 enzyme (CYP302A1), the first gene characterized that has a specific function during ecdysone biosynthesis (Warren *et al.*, 2002). It was also ascertained that Dib is a mitochondrial P450 enzyme and this observation will be discussed subsequently.

3. *Shadow (sad)*

The phenotype of homozygous *sad* mutants was essentially the same as that of the *dib* mutants, and they were also shown to be low ecdysteroid mutants. The *in situ* hybridization results were also very similar and indicated that the *sad* gene product, another mitochondrial P450, was also involved in ecdysteroidogenesis and underwent dramatic changes in expression during late larval development (Fig. 2E–H). When the S2 cells were transfected with the *sad* gene and incubated with the labeled ketotriol and the resulting products analyzed, it was clear from the resulting data (Fig. 3A) that Sad had mediated the conversion of the ketotriol to 22-deoxyecdysone (22dE). Again, the GFP control did not metabolize the ketotriol significantly. These data indicated strongly that *sad* encoded the **C2-hydroxylase**. To validate this identification, the S2 cells transfected with *sad* were incubated in the presence of 2dE, the product of Dib-mediated hydroxylation of the ketotriol at the C22 position, and the sole product was ecdysone as determined by mass spectroscopy. Thus, the ketotriol, which is 2,22-dideoxyecdysone, must be hydroxylated at the C2 and C22 positions to be converted to ecdysone. Therefore, if one can cotransfect S2 cells with both *dib* and *sad*, and then expose them to the ketotriol, the result should be the synthesis of ecdysone. When we conducted that experiment, the [3H]ketotriol was converted to [^{3}H]ecdysone with a 57% yield, in addition to traces of 2dE and 22dE (Fig. 3C; Warren *et al.*, 2002). It is of interest that although Dib can

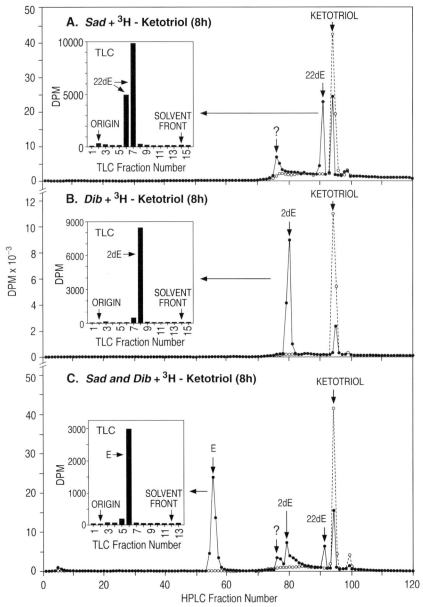

FIGURE 3. RP-HPLC/TLC analysis of ecdysteroids (30–100% methanol gradient) after *sad* (A), *dib* (B), *sad* and *dib* (C), or GFP-transfected S2 cell incubations with the [^{3}H]ketotriol (2,22dE) substrate. (A) *sad* (filled circles) or GFP (open circles) + [^{3}H]ketotriol. ?, Unidentified metabolite conjugate. (Inset) TLC analysis of [^{3}H]22-deoxyecdysone (22dE) product. (B) *dib* (filled circles) or GFP (open circles) + [^{3}H]ketotriol. (Inset) TLC of [^{3}H]2-deoxyecdysone (2dE) product. (C) *sad* and *dib* (filled circles) or GFP (open circles) + [^{3}H]ketotriol. (Inset) TLC of [^{3}H]ecdysone (E) product. (From Warren et al., 2002.)

mediate the hydroxylation of the ketotriol at C22, it cannot do the same when presented with labeled 22dE, indicating substrate specificity for the 2,22-dideoxyecdysone and suggesting strongly that *in vivo*, Dib acts before Sad (i.e., hydroxylation at C22 precedes hydroxylation at C2 during the biosynthesis of ecdysone). Ecdysone is not the arthropod molting hormone but rather its precursor and therefore must be hydroxylated at the C20 position to give the final product, 20E, the molecule that interacts with the ecdysteroid receptor to elicit and regulate manifold developmental processes during the molting cycle and metamorphosis (Henrich, 2005; Henrich *et al.*, 1999; Cherbas, this volume). This brings us to *shade*.

4. Shade (shd)

The cytogenetic studies in the 1980s noted previously, revealed that *shd* mapped to chromosomal location 70D2-E8 and the fly database revealed the presence of a P450 (*CYP314a1*) at position 70E34. Again, the genomic DNA from heterozygote mutant stocks was sequenced after amplification by PCR and three alleles were analyzed with the results demonstrating that CYP314A1 was the product of the *shd* gene (Petryk *et al.*, 2003). The phenotype of the *shd* mutant is very similar to that of the *dib* and *sad* mutants described in earlier sections, and with the Halloween genes in general, these mutant embryos do not exhibit normal expression of the 20E-responsive gene IMP-1, showing that all these genes are involved in the synthesis of ecdysteroids.

When the various tissues of the *shd* mutant embryos and last stage larvae were analyzed by *in situ* hybridization, it was found that *shd* was expressed later in embryonic life than *dib* and *sad*, was not expressed in the ring gland but rather in other tissues including the insect fat body (analogous to vertebrate liver) but neither in the nerve cord nor in the prothoracic gland cells of the ring gland. These observations were consistent with many studies of crude preparations of the enzyme, which mediates the conversion of ecdysone to 20E, the ecdysone 20-monooxygenase (herein called the 20-hydroxylase for consistency with the P450s encoded by the other Halloween genes). This is an extremely critical enzyme for insect and crustacean growth and metamorphosis since 20E affects virtually every tissue of the insect and in that sense acts as a general morphogen. The 20-hydroxylase was shown many years ago to be a classic P450 enzyme (e.g., Smith *et al.*, 1979). Further, since the prothoracic gland cells of the ring gland synthesize ecdysone, which is transported to peripheral tissues to be converted to 20E, and the 20E then interacts with the ecdysone receptor (EcR), our results showing the extra-ring gland distribution of the Shd gene product is in accord with its putative identification as the 20-hydroxylase. The absence of this enzyme from the nerve cord and ring gland has been known for almost three decades (Bollenbacher *et al.*, 1977), suggesting that for the metamorphosis of the nervous system, 20E is synthesized elsewhere and transported in the hemolymph to its site of action in the nervous system (or synthesized in these tissues in amounts below the resolving power of available assays).

The earlier noted physiological correlations are just that, correlations. We then set out to prove that *shd* encodes the 20-hydroxylase using the paradigm described in an earlier section utilizing the S2 cell transfection technique; but also demonstrated transgenic rescue of *shd* mutants using the Gal4-UAS expression system with two different promoters and the absence of 20E in the mutated embryos, whereas the wild type had significant quantities. When the S2 cells transfected with *shd* were incubated in the presence of high specific activity [^{3}H]ecdysone and nonradiolabeled ecdysone, there was a greater than 20% conversion of ecdysone to 20E. The 20E was identified by a variety of analytical techniques, including ESI-mass spectrometry, the latter showing unequivocally the molecular ion at 481 and the sequential loss of four molecules of water characteristic of 20E (Fig. 4). Thus, *shd* encodes the **20-hydroxylase.**

The data on *shd* are of particular interest because the analysis of the Halloween gene products in embryos showed that Dib and Sad are present

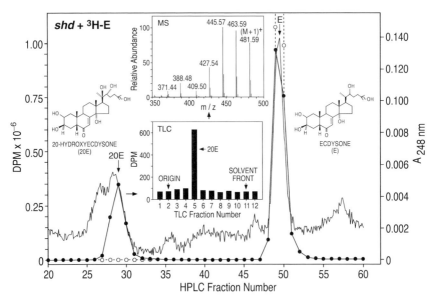

FIGURE 4. RP-HPLC/TLC/MS analysis (30–100% methanol gradient) of ecdysteroids after *shd*- or GFP-transfected S2 cell homogenate incubation (8 h) with [^{3}H]ecdysone (E) containing nonradiolabeled E (1 μg) and NADPH (0.5 mM). Radioactivity was measured after incubations with *shd* (solid circles) or GFP (open circles) with substrate. UV absorption was measured at 248 nm (solid line). (Lower inset) TLC (chloroform/ethanol) of RP-HPLC-purified 20-hydroxyecdysone (20E) product (1/1000 of total sample). (Upper inset) RP-HPLC (aqueous ACN)/ESI-MS on an TSQ Quantum (Thermo Finnigan, San Jose, California) of the TLC-purified 20E product. Structures of ecdysone (E) and 20-hydroxyecdysone (20E) are shown. (From Petryk *et al.*, 2003.)

prior to Shd, indicating that *Drosophila* embryos express *dib* and *sad* at the blastoderm stage and several hours later express *shd*. These data are consistent with past studies on ecdysteroid titers during embryogenesis and fit the now classic scheme in which ecdysone acts as the substrate for the true molting hormone of insects. Finally, in regard to these enzymes, we noted previously that Dib and Sad were mitochondrial P450s based on confocal microscopic analysis of the C-terminal HA-tagged proteins and on the charged segment at the N-terminus of the sequence (Warren et al., 2002). In transfected S2 cells the tagged Shd protein also exhibited a mitochondrial location, but surprisingly its predicted N-terminus contained not only a hydrophobic signal-type sequence typical of microsomal CYPs, but also an interior charged segment consistent with a mitochondrial targeting sequence (Petryk et al., 2003). Perhaps this explains the large number of literature references to the 20-hydroxylase being mitochondrial, microsomal, or both, depending on the tissue or species of insect being investigated (Bollenbacher et al., 1977; Feyereisen and Durst, 1978; Lafont et al., 2005; Weirich, 1997). We suggest that the Shd enzyme can reside in either cellular location as a result of differential posttranslational modifications (Petryk et al., 2003), precisely as has been described for the two mammalian P450 enzymes CYP1A1 and CYP2E1 (Addya et al., 1997; Feyereisen, 2005). However, there is one problem with these explanations. The cofactors associated with mitochondrial and microsomal P450s differ. Adrenodoxin and adrenodoxin reductase act together to transfer electrons to mitochondrial enzymes, while NADPH cytochrome P450 reductase plays this role in microsomal oxidations. At present, the amino acid motifs that are thought to interact with these different cofactors and across which the electrons must flow are not completely understood. It therefore remains possible that some P450 enzymes can utilize both electron donor systems. Alternatively, perhaps 20-hydroxylase activity has evolved in a different fashion in different insect taxa and the primary structure of the protein differs in those insects having a microsomal 20-hydroxylase. This can only be settled by the unequivocal characterization of the various tissue-specific 20-hydroxylases from several diverse species of insects, perhaps by the use of monoclonal antibodies.

The previous discussion has demonstrated the existence, nature, and function of three Halloween genes that encode the three terminal hydroxylases in the biosynthetic pathway to the molting hormone, 20E. What the astute reader will notice is that in all the functional genomic studies, the substrate utilized was the [^{3}H]ketotriol. An obvious question when considering the terminal hydroxylases concerns the origin, identification, and characterization of the 25-hydroxylase, which would mediate the conversion of the ketodiol to the ketotriol (Fig. 1). The hydroxyl group at C25 is an important constituent of the ecdysone and 20E molecules and is the subject of the functional genomic study of the Halloween genes.

5. Phantom (phm)

For the elucidation of the *phm* gene and gene product we were fortunate that Dr. Tetsuro Shinoda and his colleagues were studying several genes in the prothoracic glands of the commercial silkworm, *Bombyx mori*. One gene turned out to be an ortholog of *Drosophila phm* (Dm*phm*), and both species will be considered in this section. It should be mentioned that for the *Bombyx* work, RNA templates from early and late last instar larval prothoracic glands were used with a fluorescent mRNA differential display technique (FDD) to obtain an initial cDNA fragment of the *Bombyx phantom* (Bm*phm*) gene (Warren *et al.*, 2004).

As in the cases of the previous analyses, we utilized the cytogenetic data from 20 years ago (Wieschaus *et al.*, 1984), which showed that the Dm*phm* was located at the cytological interval of 17C5-D2 of the X chromosome, and *CYP306a1* at position 17D1 in the fly database was estimated to be a candidate for the 25-hydroxylase. The genomic DNA from an allele of the Dm*phm* mutant was sequenced, and a base alteration was identified at amino acid 286 that led to a stop codon. The Bm*phm* encoded a protein that was 37% identical to the *Drosophila* ortholog and it had a variety of structural characteristics, indicating that it was a CYP, including heme binding domain, microsomal import region, etc. (see Feyereisen [2005] for a discussion of these structural characteristics). The Dm*phm* mutant embryonic phenotype was identical to that of the other Halloween mutants as assessed by microscopic examination, the sum of the observations indicating that Phm was involved in ecdysteroidogenesis.

After transfection of the S2 cells and incubation with the ketodiol (Fig. 5), we demonstrated by our usual analytical chemical methodology, with the addition of NMR, that the major conversion product mediated by the Dm*phm* enzyme was the ketotriol (more than 25% conversion) and the same for the Bm*phm* (50% conversion), while the GFP controls showed no conversion. The NMR data were unequivocal for the identification of the product as the ketotriol and Phm as the *25-hydroxylase* (Fig. 5). It should be noted that at the same time the Kataoka laboratory used a different experimental paradigm (subtractive hybridization of early and late fifth instar *Bombyx* prothoracic glands) to identify Bm*phm* and basically the same analytical chemical protocol to show that the *Bombyx* gene, and by close analogy the fruit fly gene, encoded the 25-hydroxylase (Niwa *et al.*, 2004). The final proof for the veracity of all this research on the Halloween genes and the fact that we had achieved the identification of all four of the genes encoding the P450s responsible for mediating the four terminal steps in 20E biosynthesis (Fig. 6) was our ability to simultaneously transfect S2 cells with the *Drosophila phm, dib, sad,* and *shd* genes in the presence of ketodiol and determine that significant quantities of 20E were synthesized, or to substitute the Bm*phm* gene for the Dm*phm* gene and achieve the same result

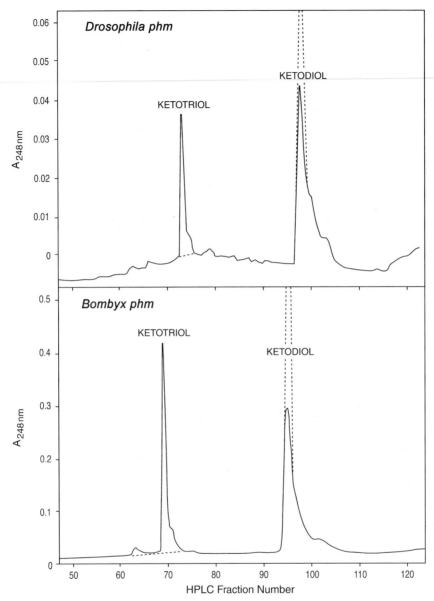

FIGURE 5. RP-HPLC analysis (30–100% methanol gradient; UV 248 nm) of the ketotriol (2,22-dideoxyecdysone) product formation following the incubation of the ketodiol (2,22,25-trideoxyecdysone) substrate with S2 cells transfected with either the *Drosophila phm* gene (upper solid line), the *Bombyx phm* gene (lower solid line) or the GFP control protein (dotted lines). (From Warren et al., 2004.)

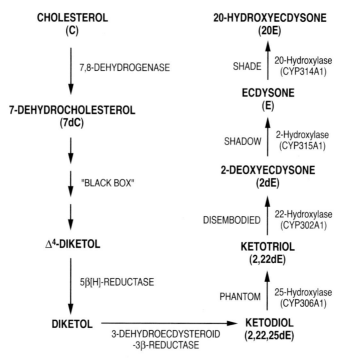

FIGURE 6. Scheme of 20-hydroxyecdysone (20E) biosynthesis in *Drosophila*. The identity of the terminal four sequential hydroxylation reactions of the ecdysteroid backbone at carbons 25, 22, 2, and 20 are shown along with three partially characterized enzymatic transformations (i.e., the cholesterol 7,8-dehydrogenase, the Δ^4-diketol 5β[H]-reductase and the 3β-dehydroecdysteroid-3β-reductase), in addition to the uncharacterized reactions of the "Black Box."

(Warren et al., 2004). Data on the substrate specificity of the *Drosophila* Phm enzyme also suggested that the sequential order of terminal hydroxylations is at C25, C22, C2, and then C20 (Fig. 6). Having completed this aspect of the biosynthesis conundrum, two major problems remain: (1) to dissect in an analogous manner the cholesterol to 7dC reaction and the oxidations of the "Black Box" in which 7dC is converted to the Δ^4-diketol and/or diketol, and (2) to understand the regulation of the Halloween genes, which is a requisite for comprehending the developmental events that are controlled by both increases and decreases in the ecdysteroid titer. We are actively working on the first (see Section II on *woc*) and have a few insights on the second.

For the latter, *phm* will be discussed a bit further since some of the data already obtained have provided a base for investigation. Developmental studies of the fruit fly embryo showed that embryonic expression of *phm* was noted as early as the early blastoderm stage as with *dib* and *sad* and that later during the peak ecdysteroid titer, expression was localized to the epidermal stripes. This expression pattern dissipated several hours later during germ band retraction, but a few hours afterward (by the time the primordium of the ring gland had developed), expression was restricted to this structure and then during larval life only in the prothoracic gland cells of the ring gland (Fig. 2I–L), as was also observed with *dib* and *sad*. These changes in expression suggest regulation of these genes as part of a developmental hierarchy, but the staging of the embryos and obtaining the correct minute structures preclude facile analysis of the regulatory mechanisms for these genes during embryogenesis at this time. On the other hand, analogous studies with larvae revealed that *phm* expression is high at the end of the second instar (stage), declines after ecdysis to the third instar, but then increases dramatically at the end of the third instar when the larvae begin to leave the food (wander) and are searching for a site to pupariate and prepare for the first metamorphic molt to the pupa (Fig. 2J–L). These data, as well as confirmatory analyses by RT-PCR, Western, and Northern studies, could be correlated with the major peak of 20E elucidated in a long and arduous RP-HPLC analysis of extracts of extremely well-timed third instar larvae (Warren *et al.*, 2005).

In the case of the RT-PCR studies, faint *phm* expression, along with that for *dib* and *sad*, was found in several other tissues, including brain and fat body (Lafont *et al.*, 2005). These findings are confirmatory of several biochemical studies and especially of 25-hydroxylase activity in tissue extracts two decades ago (Meister *et al.*, 1985, 1987), and suggest that one or more of these CYPs may have other actions in tissues which do not synthesize ecdysone, or perhaps that these tissues such as the brain, synthesize localized quantities of ecdysteroid as has been shown for the lepidopteran testis. If the former is true, it would be a major difference between the CYPs involved in insect and mammalian steroid hormone metabolism, since as mentioned previously, in the latter case there is usually extreme substrate specificity. It is important to note that faint expression of *phm* is also seen in tissues other than the prothoracic glands in *Bombyx* although in this case, the capacity for 25-hydroxylation changes during development (Warren *et al.*, 2004). In general the alterations in Bm*phm* expression appear analogous to those observed for *Drosophila*, although the silkworm has five larval instars and the fruit fly has three. In *Bombyx* the last stage larva spins a silken cocoon within which pupation and then adult development occurs, whereas for *Drosophila* the last stage larva contracts and the cuticle hardens and tans to produce the puparium, within which pupation and adult development ensue. In each case, there are critical changes in *phm* expression during the last larval instar that are correlated with the peaks in ecdysteroid concentration, and in both instances, the preponderance of gene activity is in the cells of the prothoracic gland. The composite data leave no doubt that *phm* encodes the **25-hydroxylase** and that this gene product is intimately involved in the biosynthesis of ecdysone and therefore the molting hormone 20E. The developmental studies reveal that because there are drastic alterations in 25-hydroxylase activity (i.e., *phm* expression), this gene, as well as *dib, sad,* and *shd*, could be an excellent target for studying both gene regulation and the regulation of ecdysteroidogenesis, using the last and penultimate instars of either the fruit fly or silkworm. Finally, these studies on *phm* and the other Halloween genes corroborate the view that the use of *Drosophila* mutants and the fly database are of value for studies, such as those elaborated herein, as well as for answering many other biological questions. These studies also underscore the fact that similar results can be obtained by using the new and more sensitive techniques of FDD and quantitative RT-PCR as was done for the silkworm gene.

VII. EPILOGUE

Our ability to clone and characterize the Halloween genes that play a crucial role in the synthesis of the arthropod molting hormone portends in our view that the marriage of biochemistry and molecular genetics will shortly result in the long-awaited elucidation of the complete pathway between cholesterol and 20E. In a more general sense, this work demonstrates that analogous experiments could be conducted for illuminating pathways involved with the synthesis of other complex hormones, particularly for an organism whose genome has been sequenced (e.g., human, mouse, mosquito, honeybee, the commercial silkworm, *C. elegans*, and *Arabodopsis*).

ACKNOWLEDGMENTS

We are grateful to our colleagues and collaborators over the years, but for this article especially to our collaborators on the Halloween gene project, Dr. Michael O'Connor and his colleagues at the University of Minnesota and Dr. Chantal Dauphin-Villemant and her colleagues at Université P. and M. Curie. The research from our laboratory on the Halloween genes has been generously supported by National Science Foundation grant IBN 0130825.

REFERENCES

Addya, S., Anandatheerthavarada, H., Biswas, G., Bhagwat, S., Mullick, J., and Avadhani, N. (1997). Targeting of NH2-terminal-processed microsomal protein to mitochondria: A novel pathway for the biogenesis of hepatic mitochondrial P450MT2. *J. Cell Biol.* **139**, 589–599.

Blais, C., Dauphin-Villemant, C., Kovganko, N., Girault, J.-P., Descoins, C., Jr., and Lafont, R. (1996). Evidence for the involvement of 3-oxo-Δ^4 intermediates in ecdysteroid biosynthesis. *Biochem. J.* **320,** 413–419.

Bollenbacher, W., Smith, S., Wielgus, J., and Gilbert, L. I. (1977). Evidence for an α-ecdysone cytochrome P-450 mixed function oxidase in insect fat body mitochondria. *Nature* **268,** 660–663.

Brown, G. D. (1998). The biosynthesis of steroids and triterpenoids. *Nat. Prod. Rep.* 653–696.

Chávez, V. M., Marqués, G., Delbecque, J. P., Kobayashi, K., Hollingsworth, M., Burr, J., Nätzle, J. E., and O'Connor, M. B. (2000). The *Drosophila disembodied* gene controls late embryonic morphogenesis and codes for a cytochrome P450 enzyme that regulates embryonic ecdysone levels. *Development* **127,** 4115–4126.

Christenson, L. K., and Strauss, J. F. (2000). Steroidogenic acute regulatory protein (StAR) is a sterol transfer protein. *J. Biol. Chem.* **273,** 26285–26288.

Dai, J.-D., and Gilbert, L. I. (1991). Metamorphosis of the corpus allatum and degeneration of the prothoracic glands during the larval-pupal-adult transformation of *Drosophila melanogaster*: A cytophysiological analysis of the ring gland. *Dev. Biol.* **144,** 309–326.

Dyer, D. H., Lovell, S., Thoden, J. B., Holden, H. M., Rayment, I., and Lan, Q. (2003). The structural determination of an insect sterol carrier protein-2 with a ligand bound C16 fatty acid at 1.35 Å resolution. *J. Biol. Chem.* **40,** 39085–39091.

Feyereisen, R. (2005). Insect Cytochrome P450. *In* "Comprehensive Molecular Insect Science" (L. I. Gilbert, K. Iatrou, and S. Gill, Eds.), Vol. 4, pp. 1–77. Elsevier, Oxford.

Feyereisen, R., and Durst, F. (1978). Ecdysterone biosynthesis: A microsomal cytochrome P-450 linked ecdysone 20-monooxygenase from tissues of the African migratory locust. *Eur. J. Biochem.* **88,** 37–47.

Gaziova, I., Bonnette, P., Henrich, V., and Jindra, M. (2004). Cell-autonomous roles of the *ecdysoneless* gene in *Drosophila* development and oogenesis. *Development* **13,** 1–11.

Gilbert, L. I. (2004). Halloween genes encode P450 enzymes that mediate steroid hormone biosynthesis in *Drosophila melanogaster*. *Mol. Cell. Endocrinol.* **215,** 1–10.

Gilbert, L. I., Rybczynski, R., Song, Q., Mizoguchi, A., Morreale, R., Smith, W., Matubayashi, H., Shinoya, M., Nagata, R., and Kataoka, H. (2000). Dynamic regulation of prothoracic gland ecdysteroidogenesis: *Manduca sexta* recombinant prothoracicotropic hormone and brain extracts have identical effects. *Insect Biochem. Mol. Biol.* **30,** 1079–1089.

Gilbert, L. I., Rybczynski, R., and Warren, J. T. (2002). Control and biochemical nature of the ecdysteroidogenic pathway. *Ann. Rev. Entomol.* **47,** 883–916.

Heed, W. B., and Kircher, H. W. (1965). Unique sterol in the ecology and nutrition of *Drosophila pachea*. *Science* **149,** 758–761.

Henrich, V. (2005). The ecdysteroid receptor. *In* "Comprehensive Molecular Insect Science" (L. I. Gilbert, K. Iatrou, and S. Gill, Eds.), Vol. 3, pp. 243–285. Elsevier, Oxford.

Henrich, V. C., Rybczynski, R., and Gilbert, L. I. (1999). Peptide hormones, steroid hormones and puffs: Mechanisms and models in insect development. *In* "Vitamins and Hormones" (G. Litwack, Ed.), Vol. 5, pp. 73–125. Academic Press, San Diego.

Jürgens, G., Wieschaus, E., Nüsslein-Volhard, C., and Kluding, H. (1984). Mutations affecting the pattern of the larval cuticle in *Drosophila melanogaster*. II. Zygotic loci on the third chromosome. *Roux's Arch. Dev. Biol.* **193,** 283–295.

Kiriishi, S., Rountree, D. B., Sakurai, S., and Gilbert, L. I. (1990). Prothoracic gland synthesis of 3-dehydroecdysone and its hemolymph 3β-reductase mediated conversion to ecdysone in representative insects. *Experientia* **46,** 716–721.

Krebs, K., and Lan, Q. (2003). Isolation and expression of a sterol carrier protein-2 gene from the yellow fever mosquito, *Aedes aegypti*. *Insect Mol. Biol.* **12,** 51–60.

Lafont, R., Dauphin-Villemant, C., Warren, J. T., and Rees, H. (2005). Ecdysteroid chemistry and biochemistry. *In* "Comprehensive Molecular Insect Science" (L. I. Gilbert, K. Iatrou, and S. Gill, Eds.), Vol. 3, pp. 125–195. Elsevier, Oxford.

Lan, Q., and Wessely, V. (2004). Expression of a sterol carrier protein-x gene in the yellow fever mosquito, *Aedes aegypti*. *Insect Mol. Biol.* **13**, 519–529.

Meister, M. F., Dimarcq, J.-L., Kappler, C., Hétru, C., Lagueux, M., Lanot, R., Luu, B., and Hoffmann, J. A. (1985). Conversion of a labelled ecdysone precursor, 2,22,25-trideoxyecdysone, by embryonic and larval tissues of *Locusta migratoria*. *Mol. Cell. Endocrinol.* **41**, 27–44.

Meister, M. F., Brandtner, H., Koolman, J., and Hoffmann, J. (1987). Conversion of a radiolabeled putative ecdysone precursor, 2,22,25-trideoxyecdysone (5β-ketodiol) in larvae and pupae of *Calliphora vicina*. *Int. J. Invert. Reprod Dev.* **12**, 13–28.

Niwa, R., Matsuda, T., Yoshiyama, T., Namiki, T., Mita, K., Fujimoto, Y., and Kataoka, H. (2004). CYP306A1, a cytochrome P450 enzyme, is essential for ecdysteroid biosynthesis in the prothoracic glands of *Bombyx* and *Drosophila*. *J. Biol. Chem.* **279**, 35942–35949.

Nüsslein-Volhard, C., Wieschaus, E., and Kluding, H. (1984). Mutations affecting the pattern of the larval cuticle in *Drosophila melanogaster*. I. Zygotic loci on the second chromosome. *Roux's Arch. Dev. Biol.* **183**, 267–282.

Petryk, A., Warren, J. T., Marqués, G., Jarcho, M. P., Gilbert, L. I., Parvy, J.-P., Dauphin-Villemant, C., and O'Connor, M. B. (2003). Shade: The *Drosophila* P450 enzyme that mediates the hydroxylation of ecdysone to the steroid insect molting hormone 20-hydroxyecdysone. *Proc. Natl. Acad. Sci. USA* **100**, 13773–13778.

Poulos, T. (2003). Cytochrome P450 flexibility. *Proc. Natl. Acad. Sci. USA* **100**, 13121–13122.

Roth, G., Gierl, M., Vollborn, L., Meise, M., Lintermann, R., and Korge, G. (2004). The *Drosophila* gene *Start1*: A putative cholesterol transporter and key regulator of ecdysteroid synthesis. *Proc. Natl. Acad. Sci. USA* **101**, 1601–1606.

Rybczynski, R. (2005). The prothoracicotropic hormone. In "Comprehensive Molecular Insect Science" (L. I. Gilbert, K. Iatrou, and S. Gill, Eds.), Vol. 3, pp. 61–123. Elsevier, Oxford.

Schneiderman, H. A., and Gilbert, L. I. (1964). Control of growth and development in insects. *Science* **143**, 325–333.

Smith, S. L., Bollenbacher, W. E., Cooper, D., Schleyer, H., Wielgus, J., and Gilbert, L. I. (1979). Ecdysone 20-monooxygenase: Characterization of an insect cytochrome P-450 dependent steroid hydroxylase. *Mol. Cell. Endocrinol.* **15**, 111–133.

Stocco, D. M., and Clark, B. J. (1997). Regulation of the acute production of steroids in steroidogenic cells. *Endocrin. Rev.* **17**, 221–244.

Strauss, J. F., Kishida, T., Christenson, L. K., Fujimoto, T., and Hiroi, H. (2003). START domain proteins and the intracellular trafficking of cholesterol in steroidogenic cells. *Mol. Cell. Endocrinol.* **202**, 59–65.

Thomson, M. (2003). Does cholesterol use the mitochondrial contact site as a conduit to the steroidogenic pathway? *BioEssays* **25**, 252–258.

Warren, J. T., and Gilbert, L. I. (1996). Metabolism *in vitro* of cholesterol and 25-hydroxycholesterol by the larval prothoracic glands of *Manduca sexta*. *Insect Biochem. Mol. Biol.* **26**, 917–929.

Warren, J. T., Sakurai, S., Rountree, D. R., Gilbert, L. I., Lee, S.-S., and Nakanishi, K. (1988). Regulation of the ecdysteroid titer of *Manduca sexta*: Reappraisal of the role of the prothoracic glands. *Proc. Natl. Acad. Sci. USA* **85**, 958–962.

Warren, J. T., Bachman, J. S., Dai, J. D., and Gilbert, L. I. (1996). Differential incorporation of cholesterol and cholesterol derivatives into ecdysteroids by the larval ring glands and adult ovaries of *Drosophila melanogaster*: A putative explanation for the *l(3)ecd1* mutation. *Insect Biochem. Mol. Biol.* **26**, 931–943.

Warren, J. T., Wismar, J., Subrahmanyam, B., and Gilbert, L. I. (2001). *Woc (without children)* gene control of ecdysone biosynthesis in *Drosophila melanogaster*. *Mol. Cell. Endocrinol.* **181**, 1–14.

Warren, J. T., Petryk, A., Marqués, G., Jarcho, M., Parvy, J. P., Dauphin-Villemant, C., O'Connor, M. B., and Gilbert, L. I. (2002). Molecular and biochemical characterization of two P450 enzymes in the ecdysteroidogenic pathway of *Drosophila melanogaster*. *Proc. Natl. Acad. Sci. USA* **99**, 11043–11048.

Warren, J. T., Petryk, A., Marqués, G., Parvy, J.-P., Shinoda, T., Itoyama, K., Kobayashi, J., Jarcho, M., Li, Y., O' Connor, M., Dauphin-Villemant, C., and Gilbert, L. I. (2004). *Phantom* encodes the 25-hydroxylase of *Drosophila melanogaster* and *Bombyx mori*: A P450 enzyme critical in ecdysone biosynthesis. *Insect Biochem. Mol. Biol.* **34**, 991–1010.

Warren, J. T., Yerushalmi, Y., Shimell, M. J., O'Connor, M. B., Restifo, L., and Gilbert, L. I. (2005). Discrete pulses of molting hormone, 20-hydroxyecdysone, during late larval development of *Drosophila melanogaster*: Correlations with changes in gene activity. *Dev. Dynamics* (in press).

Weirich, G. (1997). Ecdysone 20-hydroxylation in *Manduca sexta*. (Lepidoptera: Sphingidae) midgut: Development-related changes of mitochondrial and microsomal ecdysone 20-monooxygenase activities in the fifth larval instar. *Eur. J. Entomol.* **94**, 57–65.

Wieschaus, E., Nüsslein-Volhard, C., and Jürgens, G. (1984). Mutations affecting the pattern of the larval cuticle in *Drosophila melanogaster*. III. Zygotic loci on the X-chromosome and fourth chromosome. *Roux's Arch. Dev. Biol.* **193**, 296–307.

Wismar, J., Habtemichael, N., Warren, J. T., Dai, J. D., Gilbert, L. I., and Gateff, E. (2000). The mutation *without children* (*rgl*) causes ecdysteroid deficiency in third-instar larvae of *Drosophila melanogaster*. *Dev. Biol.* **226**, 1–17.

3

ECDYSTEROID RECEPTORS AND THEIR APPLICATIONS IN AGRICULTURE AND MEDICINE

SUBBA R. PALLI,* ROBERT E. HORMANN,[†]
UWE SCHLATTNER,[‡] AND MARKUS LEZZI[‡]

*Department of Entomology, College of Agriculture, University of Kentucky
Lexington, Kentucky 40546
[†]RheoGene Inc., Norristown, Pennsylvania 19403
[‡]Institute of Cell Biology, Swiss Federal Institute of Technology (ETH), CH-8093
Zurich, Switzerland

I. Introduction
II. Ecdysteroid Receptors
   A. Cellular Locale
   B. Construction of Ligand-Controlled Transcription Factors
   C. Ligand-Binding Domain
   D. DNA Binding Domain and Hinge Region
   E. N- and C-Terminal Modulator Domains
   F. Sequence of Events
III. Applications
   A. Medicine
   B. Agriculture
IV. Conclusions and Perspectives
   References

I. INTRODUCTION

Ecdysteroids (Ec) are signaling molecules widespread in the animal as well as in the plant kingdom (Lafont and Wilson, 1992). However, they do not occur naturally in vertebrates, a feature that makes them suitable as ligands in medical gene switch applications due to the reduced likelihood of pleiotropic effects. Ecdysteroids fulfill diverse tasks because they serve as hormones, pheromones, or insect deterrents (Nijhout, 1994). Their most frequent and prominent role, however, is their function as "molting hormones," thereby controlling not only insect and arthropod development but also reproduction and other physiological processes (Spindler, 1997). The action of Ec on target cells, as is the case with steroids in general, may be divided into fast and ephemeral (Tomaschko, 1999), as well as slow and systemic, effects. As will be discussed in the Section II.F., the two modes of Ec action may eventually converge, giving rise to the integral cellular response. Fast Ec effects are generally traced back to a not yet well-defined target on the cell membrane. On the other hand, the systemic effects involve an intracellular receptor, namely the ecdysteroid receptor (EcR) (Koelle et al., 1991).

The EcR is a member of the nuclear receptor superfamily (Mangelsdorf and Evans, 1995; Mangelsdorf et al., 1995) and exhibits the typical modular structure composed of the N-terminal A/B domain, the DNA-binding C domain, the hinge (D) region, the ligand-binding E domain, and the C-terminal F domain. The ligand-binding domain is multifunctional and includes ligand-dependent dimerization and transactivation functions, while ligand-independent transactivation and dimerization functions are found in the terminal domains and in the region spanning the DNA binding domain and the N-terminal region of the hinge, respectively. The EcR heterodimerizes with other members of the nuclear receptor superfamily (Henrich, 2004), noticeably with the ultraspiracle protein (USP), which is an orthologue of the vertebrate retinoic acid X receptor (RXR) (Thomas et al., 1993; Yao et al., 1992, 1993). The EcR/USP heterodimers bind to the Ec response elements (EcRE) present in the promoter regions of Ec response genes and regulate their transcription. Most of the nuclear hormone receptors, including EcR, function as ligand-controlled transcription factors, a characteristic that renders these receptors or their key regions (i.e., the ligand- and DNA-binding domains) especially suitable as constituents of gene switches. Several nuclear receptors, including glucocorticoid receptor (GR), progesterone receptor (PR), estrogen receptor (ER), and EcR, are being used to develop gene switches for applications in medicine and agriculture. Since the EcR and its ligands are not found in vertebrates, they are attractive targets for the development of gene switches to be used in humans.

This chapter is divided into Section I, which summarizes the present knowledge of the EcR's structure and function with special emphasis on those aspects that are relevant for gene switch applications, and Section II, which describes gene switch technology with a focus on EcR-based gene switches. For a comprehensive and comparative overview on EcR's role within the gene control network and during development and evolution, the reader is referred to Henrich (2004).

II. ECDYSTEROID RECEPTORS

A. CELLULAR LOCALE

By analogy with other heterodimer-forming nuclear receptors, it is hypothesized that the EcR/USP heterodimer is bound to its target genes in its apo-conformation (Mangelsdorf and Evans, 1995). Transactivation and immunolocalization studies with EcR-deficient insect cells overexpressing foreign EcR (Hu et al., 2003) and vertebrate cells co-transfected with EcR and USP (Lezzi et al., 1993), respectively, support this hypothesis. On the other hand, immunofluorescence studies on nontransfected insect cells yield a picture for EcR localization that confirms with that of other homodimerizing, steroid hormone receptors. In naive tissue culture cells of *Chironomus*, such studies show a more or less equal EcR distribution in cytoplasm and nucleus. However, the balance is shifted toward a nuclear localization by an Ec treatment of these cells (Lammerding-Köppel et al., 1998). An even more clear-cut cytoplasmic versus nuclear (re)distribution occurs within the daily (circadian) rhythm governing epidermis and brain cell function of the insect, *Rhodnius prolix* (Vafopoulou and Steel, 2005), which parallels extreme fluctuations of hormonal titres (Vafopoulou and Steel, 2001). In oligopausing (diapausing) larvae of *Chironomus tentans* or in stage L1 larvae of *Trichosia pubescens*, both stages of which exhibit an extremely low Ec titer, almost no EcR can be detected at their polytene salivary gland chromosomes. Exposure of the glands to Ec-containing medium causes EcR to appear at the specific Ec-inducible puff-sites (Fig. 1; Stocker et al., 1997; Wegmann et al., 1995), suggesting that the hormone drives EcR to these loci. Co-localization studies demonstrated that under normal physiological conditions EcR is always accompanied by USP and is associated with a high transcription activity of the respective loci (Wegmann et al., 1995). These general observations are not agreeable with gene repression by the apo- or holo-EcR (Ashburner et al., 1974; Hu et al., 2003) and dimerization of EcR and USP with alternative partners. It is only upon heat shock treatment of salivary glands without extra hormone administration that unbalanced EcR/USP ratios are found and that EcR accumulates in certain loci (Lezzi, 1996). In gene switch experiments, it is not known where the various EcR-based fusion proteins reside prior to induction by ligand, and the possibility that they interact with endogenous dimerization partners cannot be excluded. However, overexpression and stress conditions most likely will influence localization and performance of these fusion proteins.

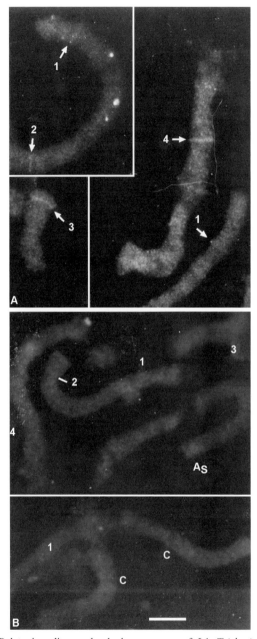

FIGURE 1. Polytenic salivary gland chromosomes of L1 *Trichosia pubescens* larvae. Glands were incubated for 1 h in hemolymph of L7 larvae with highest (A) or of L1 larvae with lowest ecdysteroid titres (B), respectively. Immunofluorescence reaction revealing ecdysteroid receptor bound to primary puff sites (1, 2, 3, and 4) after incubation with ecdysteroid (A versus B). A_s and C, designation of chromosomes. Bar: 20 μm. From Stocker *et al.* (1997).

B. CONSTRUCTION OF LIGAND-CONTROLLED TRANSCRIPTION FACTORS

The EcR may be regarded a ligand-controlled transcription factor (lcTF). Theoretically, one can envision two modes on how ligand puts an lcTF into action. (1) The lcTF is constitutively bound at the target gene. Interaction with ligand allosterically alters a specific surface (S1) of the factor, which thereupon recruits co-activators to activate the target gene. (2) The lcTF is constitutively capable of transactivation but it is not bound to DNA. Ligand binding alters another surface (S2), which enables the transcription factor to bind to and to activate the target gene. Ecdysteroid receptor-based gene switches of the two-hybrid type work according to this second mode of action, where the endogenous constitutive transactivation functions are replaced by very strong exogenous ones (e.g., GAL4-AD or VP16). Depending on the host cell type employed, these fused-on activation functions can be over 100 times more potent than any activation or repression functions remaining in the EcR fragment (e.g., Hu *et al.*, 2003; No *et al.*, 1996). With such truncated EcR fragments, these residual functions are not even measurable in yeast cells (Tran *et al.*, 2001b).

A minimal model for ligand action on lcTF that combines both activation modes consists of four components: (1) the ligand-binding pocket; (2) the co-activator interacting-surface (S1) or -groove; (3) the surface, which enables lcTF to become tethered to the target gene (S2, known as the dimerization interface for nuclear receptors); and (4) the internal "machinery," which transduces signals between those functional regions. The corresponding parameters relevant for the processes involved are the ligand **b**inding ability of the binding pocket (B), the **c**o-factor binding ability of the groove (C), the **d**imerization ability of the interface (D), and the ability of the internal "machinery" to bring about **a**llosteric changes (A). Thus, the ligand-controlled action of such a transcription factor is a function of A, B, C, and D. It is reasonable to assume that these parameters act synergistically rather than independently of each other. In fact, there is correlative evidence for cooperative interaction between B and D (Fig. 2). In an ideal ligand-controlled gene switch, the values for B and A should be high in order to achieve a maximal change of D upon ligand induction, whereas the values for D and C in the apo-conformation should be low in order to avoid background. In heterodimeric switches, the silent partner should exhibit a high ability to dimerize (D), particularly with the holo-conformation of the counterpart.

With natural lcTFs, such as the EcR, the situation is more complex. First, EcR makes use of both activation modes outlined above (Hu *et al.*, 2003). Constitutive and ligand-induced activation functions (called AF-1 and AF-2, respectively) act in a more or less additive manner. However, they may exhibit a pseudosynergism since both depend on ligand-induced EcR/USP

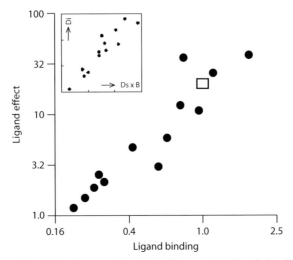

FIGURE 2. Correlation between ligand binding to the ligand binding domain of *Drosophila melanogaster* ecdysteroid receptor (B_{EcR}) and ligand effect on heterodimerization (Ef_{lig}) between the ligand binding domains of the ecdysteroid receptor and ultraspiracle peptide fused with the GAL4 activation or DNA binding domain, respectively, and expressed in yeast cells. *Dots*: values obtained with mutations in various regions of the EcR LBD. Open square: wild-type. Inset: ligand-induced heterodimerization (D_i) correlates with the product of spontaneous dimerization (D_s) and ligand binding to EcR LBD ($B = B_{EcR}$) suggesting cooperativity. Reproduced with permission from © The Biochemical Society.

dimerization. If both AF types in EcR share common transcription factors (TIF2, in the case of the ER [Benecke et al., 2000]) then a true (but limited) synergism might be expected as well. Second, EcR must heterodimerize with USP for stabilization, firm binding to the ligand, and target gene recognition (Lezzi et al., 2002). The surface that becomes allosterically altered by ligand binding (S2) thus corresponds to the heterodimerization interface in the ligand binding domain of EcR. Third, ligand-induced recruitment of co-activators most likely is accompanied by release of co-repressors as is the case with other nuclear receptors.

C. LIGAND-BINDING DOMAIN

1. General Characteristics

The ligand-binding domain (LBD) is the central regulatory unit of the EcR and of any Ec-controlled gene switch. Its molecular architecture follows the canonical structure of nuclear hormone receptors (Wurtz et al., 1996) in that it is composed of 10–12 α-helices and a β-sheet and is folded into three layers (Billas et al., 2003). There are two characteristics by which the EcR and particularly its LBD appears to deviate considerably from other receptors: its enormous instability and adaptability (Billas et al., 2003; Kumar *et al.*, 2004). It is only by binding to ligand and/or the dimerization partner (USP) that the LBD of EcR adopts a more or less stable conformation.

It is known from studies with isolated and purified LBDs of other nuclear receptors that ligand binding brings about stabilization, compaction due to desolvation of the ligand-binding pocket, and overall reduction in surface hydrophobicity (Apriletti *et al.*, 1995; Billas *et al.*, 2003; Egea *et al.*, 2000; Pissois *et al.*, 2000). Indirect evidence suggests that in the case of the EcR LBD, the effect of ligand on these general characteristics is particularly pronounced and possibly constitutes the dominant effect of ligand on the EcR altogether. For gene switch purposes, it might be desirable to decrease surface hydrophobicity by site-directed mutagenesis because such a modification would increase intracellular solubility of the EcR protein and facilitate intermolecular interactions (Bergman *et al.*, 2004; Billas *et al.*, 2003).

Ultraspiracle protein is replaced most often by RXR in gene switch applications because EcR and USP heterodimers are constitutively active in mammalian cells (Palli *et al.*, 2003). For this reason and because USP LBD does not bind Ec, USP will be discussed briefly in this chapter. The USP LBD exhibits the canonical structure closely resembling that of RXR, with the exception of the *Drosophila* USP LBD, which contains a rather long and variable insertion (Clayton *et al.*, 2001). Purified and crystallized USP LBDs from *H. virescens* and *D. melanogaster* exhibit a lipid in their ligand-binding pocket, a fortuitous circumstance, which may not be physiologically meaningful. Most likely, the lipid is sequestered from the host cell (*Escherichia coli*) in which the USP LBD was expressed (Billas *et al.*, 2001; Clayton *et al.*, 2001). This finding together with reports on juvenile hormone binding and action (Jones *et al.*, 2001) raises serious questions concerning the genuine structure, function, and ligand of USP LBD (Sasorith *et al.*, 2002).

2. Ligand-Binding Pocket

The ligand-binding pocket of the EcR is lined essentially by residues of α-helices 3, 5, 7, 11 and 12 and the β-sheet (Billas *et al.*, 2003). The precise three-dimensional structure of the ligand-binding pocket while devoid of ligand is not known, perhaps as a consequence of the extraordinary lability of the EcR LBD overall. From its remarkable capacity to bind different ligand types (Fig. 3; Billas *et al.*, 2003) it is inferred that the apo-conformation of this pocket is variable and unstable. The ligand-binding pocket of apo-EcR is perhaps likened to a formless sack, which acquires its shape from its contents (i.e., the ligand). Nevertheless, the interaction of Ec with the pocket in EcR is specific even in the absence of a stabilizing dimerization partner. As a case in point, *Drosophila* EcR LBD, expressed in yeast cells as fusion protein with the GAL4-activation domain, binds ^{3}H-ponasterone A with high specificity but low affinity (1% of that of heterodimerized EcR LBD) (Grebe *et al.*, 2003, 2004; Lezzi *et al.*, 2002). Since yeast cells are not considered to contain nuclear hormone receptor-like proteins, it is assumed in these studies that the fusion

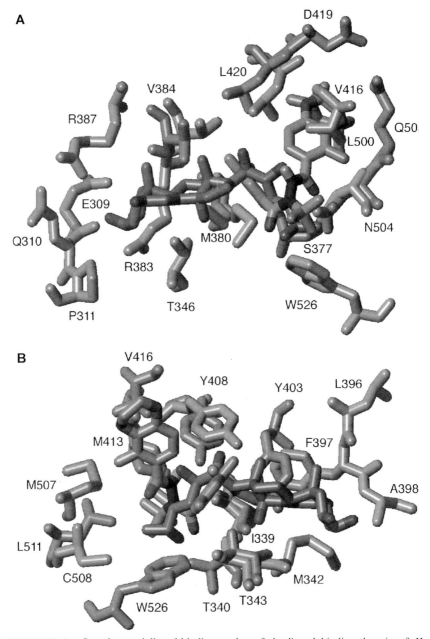

FIGURE 3. Superimposed ligand-binding pocket of the ligand binding domain of *H. virescens* crystallized with ponasterone A (PDB structure 1R1K, residues = yellow, ponasterone A = orange) and crystallized with BY-106830 (PDB structure 1R20, residues, green; BY-106830, cyan). Depicted residues lie within 4 Å of the ligand in each case. (A) Orientation as in Fig. 5 (EcR). (B) 180° rotation of the image (A) along an approximately vertical axis (Billas et al., 2003). (See Color Insert.)

protein exists as a monomer (see Section II.E) participating in a large complex ("molecular chaperone-containing complex," see Section II.F) and also comprising factors interacting with the GAL4 fragment. Mutation of ligand-contacting residues (homologous to I339, M380, and R387 in *H. virescens*; Fig. 3) negatively affects the binding ^{3}H-ponasterone A (Grebe *et al.*, 2003, 2004) as well as the normal course of ligand-controlled functions of the *Drosophila* LBD (e.g., heterodimerization; Bergman *et al.* [2004]). Mutations of residues making no ligand contact may also affect ligand binding and ligand-dependent functions (Table I), sometimes even positively (Bergman *et al.*, 2004; Grebe *et al.*, 2003). The effects of these mutations are considered indirect and support the notion of substantial flexibility of the pocket considered alone and the entire LBD domain considered as a whole.

The presence of USP LBD increases the extent of ligand binding to EcR LBD (Grebe *et al.*, 2003, 2004; Lezzi *et al.*, 2002), most likely by heterodimerization and a stabilization of the conformation of the EcR pocket. The rate of ^{3}H-ponasterone A association increases and that of dissociation decreases (Grebe *et al.*, 2004). Mutations situated both inside or outside the pocket affect ligand binding to the heterodimer as well as to the EcR LBD alone but with different reaction patterns regarding level and dynamics of ligand binding (Grebe *et al.*, 2004). Binding values higher than wild-type could not be seen in presence of USP LBD because the ratios of USP to EcR LBD used in the respective studies were chosen to ensure maximal ^{3}H-ponasterone A binding to the wild-type heterodimer (Grebe *et al.*, 2003). The effect of USP LBD on ^{3}H-ponasterone A binding (Ef_{usp}) to a mutated EcR LBD can be estimated by dividing the value for ligand binding to the heterodimer ($B_{EcR+USP}$) by that to EcR LBD alone (B_{EcR}): $Ef_{USP} = B_{EcR+USP}/B_{EcR}$. The parameter Ef_{USP} should relate to the intrinsic ability of an EcR LBD to heterodimerize. In fact, a tendency for such a correlation can be observed (Table I). Obvious deviations may be explained not only by the intentional constraint of saturation in the experiment but possibly also by variations in the capability of mutated EcR LBD to transmit signals from its dimerization interface to the ligand-binding pocket. The structural characteristics and selective interaction of the ligand-binding pocket of various EcR types with nonsteroidal ligands will be discussed in Section III.

3. Dimerization Interface

a. Apo-EcR

The basic architecture of the dimerization interface is rather well conserved, both among various nuclear receptor types as well as in different conformational states of nuclear receptors. It is composed mainly of helix 10 with contributions of helices 7, 8, 9, 11 and loops 8–9, 9–10, and 10–11 (Billas *et al.*, 2003). Due to the high lability, the exact conformation of the interface in the apo-EcR is not known. In the apo-conformation, only about 1% of all EcR LBDs heterodimerize *in vivo* with USP LBD (Lezzi *et al.*,

TABLE I. Impact of Mutations in the Ligand Binding Domain of the Ecdysteroid Receptor of *Drosophila melanogaster* on Dimerization, Ligand Binding, and their Mutual Effects—Comparison between *In Vivo* (Yeast Two-Hybrid) and *In Vitro* (Ligand Binding) Assays (Bergman et al., 2004; Grebe et al., 2003)

	Main impact[a]			
	Two hybrid[b]		Ligand binding[c]	
EcR mutations	Dimerization[d]	Ligand effect[e]	USP effect[f]	Ligand binding[g]
DIF[h] mutations				
S553A	•[l]		•[l,m]	•[l]
L615A	•		•	[l]
T619A	•[l]	[l]	•[l,m]	•[l]
LBP[i] mutations				
I463T		•	[l,m]	•
M504R		•	•	•
R511Q		•	•[l,m]	•
MT[j] mutations				
K497A	•[l]	•	•[l]	•
E647R		•	•[l,m]	•
E648K		•	[l]	•
Other[k] mutations				
E476A		•		•
A612V	•		•	
I617A	[l]	•	•[l,m]	

[a]Fold increase or decrease in comparison to wild-type as defined by Bergman et al. (2004).
[b]Ligand binding domains of USP and EcR were fused to GAL4 DNA binding and activation region, respectively, and assayed for β-galactosidase induction after co-expression in yeast cells.
[c]^{3}H-ponasterone A binding to yeast-expressed GAL4-fusion proteins.
[d]Spontaneous dimerization (in the absence of ligand).
[e]Ef_{lig}: ligand-induced dimerization/spontaneous dimerization (D_s/D_i).
[f]Ef_{USP}: ligand binding to EcR LBD + USP LBD/ligand binding to EcR LBD ($B_{EcR+USP}/B_{EcR}$).
[g]Binding to EcR LBD alone (B_{EcR}).
[h]Dimerization interface (see Fig. 4).
[i]Ligand-binding pocket (Fig. 3).
[j]Mechanism transducing signals within EcR LBD.
[k]At sites of unknown or uncertain functions.
[l]Value higher than wild-type.
[m]Value for Ef_{USP} underestimated because $B_{EcR+USP}$ was determined under saturating conditions.

2002). However, mutational screening followed by dimerization assays *in vivo* (yeast two-hybrid method) supports the assumption that residues of the structural elements listed make contacts with USP also in the case of apo-EcR (Bergman *et al.*, 2004), with the possible exception of sites lying C-terminal of helix 10 (DrmeEcR N626, Section II.C.4, and Bergman *et al.*, 2004). However, mutations outside the dimerization interface may affect ligand-independent dimerization as well, albeit with a lower impact compared to that of interface mutations (for definition of "impact," see Table I). As was observed with ligand binding, a number of mutations have positive effects, whether they are situated inside or outside the interface. They increase the value for spontaneous heterodimerization beyond that of wild-type (Bergman *et al.*, 2004), a phenomenon termed "hyperdimerization," (Bergman *et al.*, 2004), which has been observed as early as 10 years ago with interface mutations of RXRα by use of an electrophoretic mobility shift assay (Zhang *et al.*, 1994). Estimation of Ef_{USP}, a measure of heterodimerization, corroborates the notion of "hyperdimerization" occurring with certain EcR LBD mutations (see Table I).

b. Holo-EcR

Since ligand binding stabilizes the EcR LBD and increases its interaction with USP LBD, the dimerization interface of heterodimerized holo-EcR could be analyzed by x-ray crystallography (Billas *et al.*, 2003) and is depicted in Fig. 4. Its organization follows the canonical rule (Bourguet *et al.*, 2000) in that a central lipophilic patch is shielded by a ring of polar residues. The interface of *Drosophila* EcR LBD differs from that of *Heliothis* mostly by a polar exchange of a lipophilic methionine (Hevi M502) by an asparagine (Drme N626). Mutating the latter to lipophilic alanine (DrmeEcR N626A) increases ligand-dependent heterodimerization *in vivo* (Bergman *et al.*, 2004), most probably due to stronger interaction with the conserved lipophilic leucine (DrmeUSP L480) in the USP LBD. Mutants probing EcR residues contacting the USP LBD as well as of residues lying elsewhere in the EcR LBD exhibit a different pattern of enhancement or suppression of dimerization as compared to ligand-independent dimerization (Bergman *et al.*, 2004). Since ligand-induced dimerization possibly relates to not only the ability of the EcR to interact with USP but also with ligand, it is important to evaluate the contributions from each of the responsible structures (i.e., interface and pocket) separately. The effect of ligand on dimerization (Ef_{lig}) was calculated by dividing the value for ligand-induced dimerization (D_i) with that for spontaneous dimerization (D_s): $Ef_{lig}=D_i/D_s$. It was found that Ef_{lig} positively correlates with ligand binding to EcR LBD alone ($Ef_{lig} \sim B_{EcR}$, see Fig. 2). This unexpected and important finding indicates that (1) the ligand affects heterodimerization by specific binding of ligand to its pocket in EcR rather than by unspecific effects in the yeast system, and (2) the intrinsic ability of EcR LBD to bind ligand in the absence

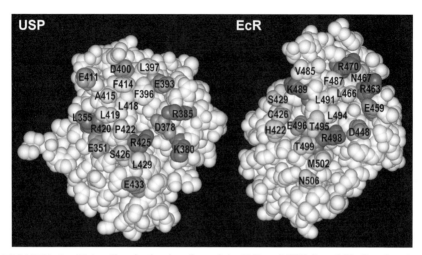

FIGURE 4. Heterodimerization interface of the EcR and USP ligand binding domains. CPK surface representation of the interacting faces of *H. virescens* of the USP LBD (left) and of the muristerone A-bound LBD of EcR (right) (PDB 1R1K). Residues involved in intermolecular contacts at the interfaces are colored (hydrophobic in yellow, polar in green, acidic in red; basic in blue). The model was prepared with WebLab Viewer Pro; contacts were derived with CSU software (Sobolev *et al.*, 1999). (See Color Insert.)

rather than presence of USP LBD constitutes a critical parameter for ligand-induced heterodimerization. Above correlations showing $D_i/D_s = Ef_{lig} \sim B_{EcR}$ may be transformed mathematically into $D_I \sim D_s x B_{EcR}$ indicating that ligand binding and heterodimerization act cooperatively (Fig. 2 inset and Bergman *et al.* [2004]). Note that the intrinsic ability of an EcR LBD to undergo an allosteric change (A) is not explicitly described by this formula.

Biochemical studies on ligand-induced EcR/USP LBD heterodimerization essentially confirm the notion that sites lying not only along the interface but also in the pocket or elsewhere in the LBD are important for ligand-induced dimerization (Grebe *et al.*, 2003; Hu *et al.*, 2003). Since with almost all published studies, spontaneous heterodimerization was not measured directly and independently, its specific contribution to ligand-induced dimerization is not known. Furthermore, high concentrations of EcR and/or USP LBDs often employed very likely prevent detection of positive mutational effects due to signal truncation resulting from saturation (Grebe *et al.*, 2003; Hu *et al.*, 2003).

c. USP

The physicochemical properties of the EcR dimerization partner USP should influence the kinetics and the extent of heterodimerization as well. A series of mutations at various sites in the LBD of *Drosophila* USP were constructed and assayed as fusion proteins with GAL4-DNA binding domain *in vivo* (two-hybrid gene expression) as well as *in vitro* (electrophoretic mobility shift) for ligand-independent and ligand-dependent dimerization (Bergman *et al.*, 2004; Przibilla *et al.*, 2004). Three mutations (L281Y, I323A, and C329A + S330N) in the putative ligand-binding pocket of USP LBD are especially remarkable in that they specifically either reduce or increase ligand-dependent dimerization with EcR LBD while assayed *in vivo* (Bergman *et al.*, 2004). Mutation DrmeUSP I323A in helix 5, in particular, yields an induction value (Ef_{lig}) for the Ec muristerone A, which is five to seven times greater than for wild-type USP and is therefore called "superinducer." This value more than doubles while paired with another "superinducer" [DrmeEcR interface mutation T619A in helix 10 (Bergman *et al.*, 2004)]. Since muristerone A does not bind to the ligand binding domain of USP, direct effect of ligand on USP LBD and on its capability to heterodimerize can be excluded. An "interface compatibility model" has been proposed (Lezzi, 2002) for interpreting the extraordinary effect of the USP I323A mutant by postulating an increased affinity of its interface for the corresponding interface of wild-type and particularly T619A-mutated EcR LBDs in their holo-conformation. Super-inducing mutants of the silent dimerization partner, like DrmeUSP I323A, are especially valuable in gene switches; this particular mutant has been shown to convey superinducibility to a vertebrate gene induction system as well [M.O. Imhof as cited by (Bergman *et al.*, 2004)].

4. Co-factor Interacting Groove

The surface of nuclear receptors that interacts with co-factors, is characterized by a hydrophobic groove formed mainly by helices 3, 4, and 5. In the apo-conformation or in the antagonist-binding conformation, co-repressors occupy this groove. However, in the active holo-conformation induced by binding to an agonist, helix 12 covers part of that groove thus forming a new surface (AF-2), which is capable of recruiting co-activators (Privalsky, 2004).

It is assumed that the EcR basically conforms to other nuclear receptors in terms of the groove's structure, co-factor binding property, and by its ability to undergo a conformational change upon ligand binding. The effects of three mutations at putative co-repressor or co-activator binding sites (A483T and K497A in helix 4 and F645A in helix 12, respectively) studied with *Drosophila* EcR LBD yield the expected results while probed in co-factor containing host cells for effects on transactivation [(Hu *et al.*, 2003; Tsai *et al.*, 1999); V. C. Henrich in the second Note added in proof in (Bergman *et al.*, 2004)]. While only A483T was experimentally tested for co-repressor (SMRTER) interaction, the complexity of cellular systems precludes a distinction between the effects of co-factor interaction on the one hand and ligand binding or heterodimerization on the other. One may conceive plausible explanations which do not involve co-repressor effects at all. For instance, residue K497 is also part of the locking salt-bridge, which holds helix 12 at a precise position (see Section II.C.4); perturbation of this residue may have multiple effects on individual components of the LBD.

The natural conformation of the co-factor interacting groove of USP is not known. Ligand binding domains of *Drosophila* and *Heliothis* USP expressed in *E. coli* contain a small lipid molecule in their putative ligand-binding pocket which forces helix 12 to attain an antagonist position (Billas *et al.*, 2001; Clayton *et al.*, 2001). Investigations suggest that juvenile hormone may exert a similar or the opposite effect depending on juvenoid structure and other parameters (Kethidi *et al.*, 2004; Maki *et al.*, 2004; Sasorith *et al.*, 2002; Wozniak *et al.*, 2004).

5. Mechanisms Transducing Signals between Functional Regions of the Ligand Binding Domain

Since the exact conformations of neither unliganded nor nonheterodimerized ligand binding domains of the EcR have been analyzed structurally as yet, the allosteric changes of EcR LBD elicited by interaction with ligand or USP are a matter of speculation. The mechanisms proposed are, in part, the outcome of sequence comparisons or of mutational analyses aimed at the elucidation of other EcR LBD functions and therefore should be (re)investigated with specifically designed methods and mutations. This would be profitable also for the conception of powerful gene switch constructs.

With other nuclear receptors, the principal ligand-induced allosteric changes involve the "lid," which closes the pocket upon ligand binding (Wurtz *et al.*, 1996). This lid mainly consists of helix 12 and other movable or flexible parts (i.e., helix 11 and the C'-end of helix 10). In the apo-conformation, helices 11 and 12 are joined and flexibly connected to helix 10 by a short loop. In the crystal structure of the apo-RXR, this joined helix 11–12 protrudes from the rest of the LBD but it may attain different positions in solution or in the living cell (Pogenberg *et al.*, 2004). The ligand, bound to the pocket, orients and coordinates these structures by "attracting" helices 10–12 so that helix 12 flaps over the pocket and nestles into the co-factor interacting groove. In this configuration, helix 12 is locked by a salt-bridge, which in DrmeEcR is formed between residue E648 and residue K497 in helix 4. In the holo-conformation, helix 11 joins helix 10 rather than 12, forming a continuous but bent α-helix 10–11 (Fig. 5). There are various parameters that influence the positioning of helix 10–11 [i.e., altered flexibility and curvature of its helical axis, type of ligand, and interaction with associated proteins (dimerization partners, co-factors)]. It is plausible that those parameters that enable or force α-helix 10–11 to bend further towards the ligand support positioning of helix 10–11 and thereby strengthen, in turn, ligand binding, which is beneficial to further ligand-coordinated restructurations (e.g., of helix 12) and subsequent ligand-dependent functions. The effects of mutations at two sites in the joined helix 10–11 are consistent

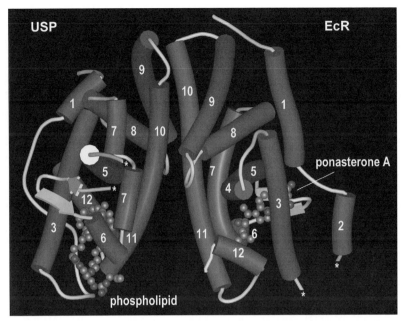

FIGURE 5. Molecular structure of the ligand binding domains (LBDs) of the ecdysteroid receptor (EcR) and ultraspiracle protein (USP). The heterodimer between the EcR LBDs and USP LBDs of *H. virescens* (PDB 1R1K) is shown in a schematic backbone representation with α-helices as red tubes and β-sheets as blue arrows. The α-helices are numbered from 1 to 12. Bound phospholipid and ponasterone A are shown in ball-and-stick representation colored according to atom type (C in grey, O in red, P in orange, N in blue). (*) Loops connecting EcR helices 9 and 10, as well as USP helix 5 and β-sheet ("extra loop") are not depicted (Billas *et al.*, 2003). The model was prepared with WebLab Viewer Pro. (See Color Insert.)

with this view. In the *Drosophila* EcR, an exchange of threonine by alanine at position 619 produces higher than wild-type values for ligand binding and ligand-induced dimerization (Table I). In RXR and USP, the homologous position is occupied by a proline that introduces a relatively sharp bent into helix 10–11 (Fig. 5 and Bourguet *et al.*, 1995). Replacement of this proline by another residue depresses ligand-dependent functions (Przibilla *et al.*, 2004; Zhang *et al.*, 1994). Mutations of asparagine 626 of the DrmeEcR enhance ligand-dependent dimerization and/or ligand binding. Studies demonstrate that, in particular, the rate of ligand dissociation is reduced by mutating N626 (Grebe *et al.*, 2004) as if to indicate that facilitated closure of the lid would impede the escape of the ligand from its pocket. Conversely, mutations that demolish the locking mechanism of the lid (DrmeEcR K497A and E648K) not only prevent stable ligand binding (Grebe *et al.*, 2003) but also promote EcR/USP LBD interaction (Table I, columns 2 and 4). A deletion comprising helices 11–12 has a similar effect (Lezzi *et al.*, 2002; M. Lezzi and

T. Bergman, unpublished observation). The uncertain position of helices 11 and 12 of the apo-EcR in solution and in the living cell precludes a mechanistic explanation of the phenomenon of enhanced EcR/USP LBD interaction. Signal exchange between ligand-binding pocket and dimerization interface in the helix 10–11 region is likely to occur. In particular, note that in the 17-residue stretch between position numbers 494 and 510 of the HeviEcR, two to four ligand contact sites and three USP-contact sites are intermingled (Fig. 4; Billas *et al.*, 2003, supplementary information). Likewise, helix 12 combines sites that interact with the ligand, with presumable co-factors, and by interhelical contacts, with the co-factor interacting groove (Billas *et al.*, 2003; Hu *et al.*, 2003). This renders helix 12 a candidate for communication between these three functional areas in the EcR LBD.

The C-terminal half of helix 7 in HeviEcR LBD is composed of three USP-contact sites (Fig. 4), whereas the N-terminal half is "attracted" by contacts with the ligand. This "attraction" and the resulting bent in the helix 7 axis is obviously more pronounced with the nonsteroidal agonist BY106830 than with ponasterone A (Billas *et al.*, 2003). Comparative crystal structure and sequence analyses of a large number of nuclear receptor LBDs reveal a salt-bridge between a glutamic acid residue in helix 5 (in DrmeEcR at position 502) and an arginine residue in the loop 8–9 (DrmeEcR R573), which is typical for heterodimer forming nuclear receptor types (Brelivet *et al.*, 2004). Similarly, contacts are made between helices 5 and 10 as well as between the β-sheet (S1) and helices 7–8. Since the involved residues are neighbors of ligand binding and USP binding sites, respectively, a communication between pocket and interface is feasible and, in the case of the salt-bridge, has been proposed by Brelivet *et al.* (2004). In fact, mutation of residue D572, which is adjacent to the R573 of the salt-bridge, not only affects heterodimerization but also diminishes, with a comparably high impact, the ligand effect on heterodimerization [Ef_{lig} (Bergman *et al.*, 2004)]. Therefore, this site was previously considered to fulfill an architectural task but it might also play an allosteric role.

D. DNA BINDING DOMAIN AND HINGE REGION

The DNA binding domain (DBD) of the EcR exhibits the canonical structure typical for nuclear hormone receptors which belong to the group of DNA binding proteins that contain two zinc fingers. Two reports describe the structure of EcR and USP DNA binding domains (Devarakonda *et al.*, 2003; Orlowski *et al.*, 2004). The outstanding features of EcR DBD are twofold: (1) *extreme instability*—it is only by interaction with a cognate response element in DNA and/or dimerization partner that this "intrinsically unstructured" protein domain adopts a defined and stable structure and (2) *extraordinary flexibility and adaptability*—full-length EcR/USP heterodimer can bind and act upon DNA sequences of a variety of configurations. Usually, the motif that is best recognized as an EcRE is a bipartite motif consisting of two half-sites organized as inverted repeat (palindrome) with one base spacing (IR-1) (Vögtli et al., 1998). However, longer spacing and arrangements of the half-sites as direct repeats are accepted as well (Devarakonda et al., 2003). Nucleotides adjacent to the strictly six base-pair long core half-site (a degenerated AGCTCA) are important for target gene recognition as well and have been shown to make contacts to amino acid residues of EcR residing C-terminally of the core DBD (e.g., in the so-called hinge region). Isolated DBDs of *Drosophila* EcR bind to degenerate IR-1 motifs either as monomers or, with higher affinities, in heterodimeric complexes with DBDs of USP, or as homodimers. Full-length EcR molecules hardly form homodimers (Grebe et al., 2004; Perera et al., 2005). However, when the DBD of EcR is exchanged by the strong DNA binding and dimerization regions of GAL4, LexA, or of the GR, homodimers the resulting EcR chimeras settle on the cognate bipartite DNA motif as homodimers (Martinez et al., 1999b; Padidam et al., 2003). This happens only to an appreciable extent when the EcR LBD is in the holo-conformation. The possibility of homodimerization, rather than heterodimerization, is interesting for gene switch use (Martinez et al., 1999b; Padidam et al., 2003) as it dispenses with the co-expression of a second hybrid construct. In contrast to the GR that binds as a homodimer to its inverted repeat-type response element, the original full-length EcR prefers the heterodimeric configuration regardless whether the half-sites of its EcRE are arranged as inverted or alternatively as direct repeats (Vögtli et al., 1998). Thus, the DNA binding domain of EcR must be able to cope with the different dimerization modes (head-to-head versus head-to-tail) imposed by the configuration of the repeats in the EcRE. Moreover, in the full-length EcR, the hinge region must possess a high rotational freedom.

E. N- AND C-TERMINAL MODULATOR DOMAINS

The amino acid sequences lying N-terminally to the DNA binding domain and C-terminally to the ligand binding domain not only exhibit the largest variability among EcRs of different species but also within the same organism. The latter situation gives rise to EcR isoforms. In *D. melanogaster*, three isoforms varying in their N-terminal domain have been identified (Talbot et al., 1993). The B1-isoform is most abundant among tissues. Its N-terminal modulatory domain carries transcription stimulating AF-1 and repressing functions as characterized by the yeast one-hybrid assay (Mouillet et al., 2001). The N-terminus of the B2-isoform is rather short and exhibits an AF-1, whereas that of the A-isoform is repressive (Mouillet et al., 2001). This differential behavior of the three isoforms may vary depending on the type of host cell. The C-terminal modulatory domain of all *D. melanogaster* isoforms is the same and deviates from that of other species by its enormous length. It exhibits a modest AF-1 (Hu *et al.*, 2003). All these activating or repressing functions in the N-terminal and C-terminal domains are constitutive, which means that ligand binding to the EcR pocket is not required to put them into action. However, in the context of the full-length receptor, they still may depend indirectly on the action of ligand, namely on ligand-induced EcR/USP dimerization.

F. SEQUENCE OF EVENTS

Ecdysteroid receptor is synthesized at the endoplasmatic reticulum in the cytoplasm and eventually resides in the neighborhood of a target gene controlling this gene's transcriptional activity. What happens in between these two events? Three plausible scenarios are presented.

1. Scenario 1

Like any other steroid hormone receptor, EcR is part of a larger complex ("molecular chaperone-containing heterocomplex" [MCH]) (Arbeitman and Hogness, 2000; Gehring, 1998; Pratt and Toft, 1997) before it encounters the ligand. This complex is located in the cytoplasm and is required for correct folding of the receptor peptide, for stabilization of the monomeric apo-conformation, and possibly for keeping the unliganded EcR out of action. Once ligand binds to the EcR monomer within the complex, an allosteric change leads to release or exchange of the complex's constituents. Subsequent steps, such as heterodimerization, nuclear translocation, DNA binding, and finally target gene activation, ensue deterministically from ligand binding. Although none of the sites at the EcR's surface with which the complex's components may interact have yet been determined and mapped, a decisive role of chaperones and their release by the action of ligand must be considered a possibility for EcR activation (Arbeitman and Hogness, 2000). If a gene switch were to operate according to Scenario 1, all relevant ligand-dependent allosteric changes of the EcR fragment would be confined to regions responsible for interaction with components of the MCH.

2. Scenario 2

Ultraspiracle protein rather than ligand liberates EcR from the cytoplasmic complex. The heterodimer between apo-EcR and USP translocates to the nucleus and binds to DNA independently of ligand. It is only the very last step (i.e., the activation of AF-2) that requires ligand binding (Hu *et al.*, 2003). Rather than being the singular general signaling pathway, this second route may be invoked primarily by stress conditions (see Section II.A.).

If an EcR-based gene switch were to function according to Scenario 2, the strong fused-on activation domain (e.g., GAL4 or VP16) would have to be functionally coupled with the (C-terminal) AF-2 in helix 12 of the EcR LBD.

In the case of fusions to the N'-end of the EcR fragment, such a coupling must be indirect and would depend on co-activators interacting with both AF-2 and the extraneous activation region.

3. Scenario 3

Ligand-induced dimerization with USP causes EcR to leave the cytoplasmic (MCH) complex. Ligand binding thus precedes USP interaction leading to an intermediary EcR conformation. In this conformation, EcR exhibits an increased affinity for USP. (A scenario in which ligand and USP would simultaneously interact with EcR is not considered for probalistic reasons.) The holo-EcR/USP heterodimer translocates to the nucleus, binds to DNA and co-activators, and eventually activates target genes. Immunostainings of polytenic chromosomes localize SMRTER to chromosome regions, which are puffed (Tsai et al., 1999), and therefore probably still active in transcription. Co-repressors could tune down EcR activity in the aftermath of a ligand-dependent gene activation cycle. They could conceivably stabilize apo-EcR/USP (Pissios et al., 2000) at the EcRE in the temporary absence of ligand for a fast response to the next ligand signal. For gene switches that might function according to Scenario 3, ligand-dependent allosteric changes of the dimerization interface in the EcR LBD play the key role in the switching process.

Scenarios that bring ligand into play at an early rather than a late step conform better with the immunohistochemical findings reported in Section II.A. and also with the observation of refractory states in Ec-induced activation of early puff-sites (Lezzi, 1996). There is high correlative evidence that this refractoriness of specific loci in certain developmental stages and diurnal phases is caused by a condensed chromatin conformation that somehow inhibits the interaction of the EcR/USP heterodimer with DNA (Lezzi, 1996). One view is that co-activators and co-repressors control chromatin conformation of target genes by co-activator-/co-repressor-coupled histone modifying enzyme activities [for EcR, see (Cakouros et al., 2004; Sawatsubashi et al., 2004; Sedkov et al., 2003)]. However, the immunohistochemically observed change in the condensational state of the puff-sites in question precedes noticeable EcR localization (Lezzi, 1996). Thus, it can hardly be attributed to EcR/co-activator-coupled enzyme activities. Instead, specific sensitivities to certain ionic conditions of these chromosome regions together with the parallel observation of modulation in nuclear ion activities (Lezzi, 1996), suggest an influence of the intracellular ionic milieu on the state of chromatin and refractoriness toward Ec. It is not known what causes the change in the intracellular ionic milieu; a role of Ec acting on the cell membrane may be considered (Tomaschko, 1999). Likewise, Ec-controlled pathways leading to altered concentrations of the second messengers Ca^{2+} or cAMP could merge with the conventional Ec action (e.g., at the level of EcR and USP phosphorylation) (Rauch et al., 1998; Song and Gilbert, 1998).

III. APPLICATIONS

A. MEDICINE

1. Gene Switch Technology for Use in Medicine

Regulated expression of genes is gaining importance in a variety of applications in functional genomics, gene therapy, therapeutic protein production, and tissue engineering (Albanese et al., 2002; Fussenegger, 2001; Weber and Fussenegger, 2002). Earlier versions of gene regulation systems used endogenous promoters regulated by heat, metal ions, hypoxia, and hormones. The *D. melanogaster* heat-shock protein 70 promoter-based gene regulation system was the first successful example of a regulated expression that used a heterologous gene (*c-myc*) in mammalian cells (Wurm et al., 1986). Later, the hypoxia-responsive promoter of the phosphoglycerate kinase gene was used to regulate human erythropoietin (hEpo) expression (Rinsch et al., 1997). The interferon inducible promoter of the Mx1 gene was employed to control gene targeting by Cre recombinase in mice (Kuhn et al., 1995). All of these systems support rapid induction of transgenes but potential pleiotropic effects resulting from induction of endogenous genes by heat-shock, hypoxia, and growth regulators limited the use of these gene regulation systems especially for *in vivo* applications. It quickly became apparent that interference with host regulatory networks would be a general deficiency of gene regulation systems derived from endogenous sources, and over time, *specificity* emerged as an essential design feature. To circumvent pleiotropic effects, the gene regulatory systems need to be either artificial and/or heterologous to the host organism. The following effector–responder pairs were employed to develop gene switches for use in medicine: tetracycline–antibiotic resistance operon of *Streptomyces* (Fussenegger et al., 2000; Gossen and Bujard, 1992); erythromycin/erythromycin-resistance regulon (Weber et al., 2002); lactose and glucose or analogs/*E. coli* lactose (lac) operon (Miller and Reznikoff, 1980); FK506, rapamycin, or cyclosporine/immunophilins (FKBP 12, cyclophilin) (Belshaw et al., 1996; Spencer et al., 1993); progesterone antagonist (RU486)/mutated PR (Wang et al., 1994); estrogen antagonist (tamoxifen)/mutant ER (Feil et al., 1996); and muristerone A/EcR (No et al., 1996). Various gene switches developed for applications in medicine have been reviewed (Fussenegger, 2001). The scope of this review is EcR-based gene switches.

2. EcR-Based Gene Switches

Ecdysteroids and EcR are found only in insects and related invertebrates (Riddiford et al., 2000). After initial reports (Christopherson et al., 1992; Yang et al., 1995) on the function of EcR as an Ec dependent transcription factor in cultured mammalian cells, No et al. (1996) used DrmeEcR and human RXR to develop an Ec inducible gene expression system that can

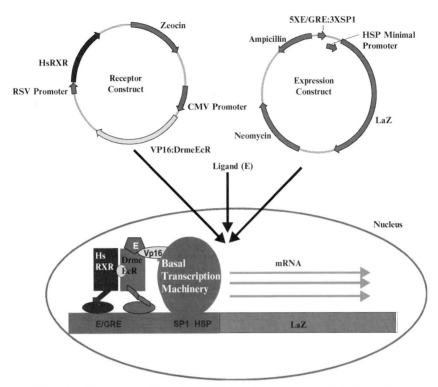

FIGURE 6. First commercial version of EcR-based gene switch marketed by Invitrogen. A fusion proteins of VP16 activation domain and *D. melanogaster* EcR CDEF domains (VP16: DrmeEcR) and complete human RXR are expressed under the control of CMV and RBS promoters respectively. The reporter gene or gene of interest is cloned under the control of heat shock protein (HSP) minimal promoter and 5XE/GRE and 3X SP1 response elements. These two plasmids are co-transfected into mammalian cells and the transfected cells are exposed to ecdysteroids (E), such as ponasterone A or muristerone A. (See Color Insert.)

function in mammalian cells and mice (Fig. 6). Later, Suhr *et al.* (1998) showed that the nonsteroidal Ec agonist, tebufenozide, induced a high level of transactivation of reporter genes in mammalian cells through *Bombyx mori* EcR (BomoEcR [Swevers *et al.*, 1995]) and endogenous RXR. Hoppe *et al.* (2000) combined DrmeEcR and BomoEcR systems and created a chimeric *Drosophila/Bombyx* EcR (DrBoEcR) that had combined positive aspects of both the systems (i.e., the chimeric receptor was capable of binding to modified EcRE and also functioned without exogenous RXR). Improvements to the EcR-based gene switch include expression of both EcR and RXR in a bicistronic vector (Wyborski *et al.*, 2001) and the discovery that the RXR ligands enhance the ligand-dependent activity of EcR-based gene switches (Saez *et al.*, 2000).

An optimal gene regulation system should exhibit the following characteristics: no expression of target gene in the absence of inducer, expression of target gene at a physiologically desirable (usually substantial) magnitude upon addition of inducer, induction of variable expression of target gene in a dose-dependent manner, sensitivity to low inducer concentrations, rapid on/off switching in response to addition and withdrawal of ligand, support of repeated on/off cycles, specific response to inducer with no pleiotropic effects, optimized transport across the placenta and/or blood–brain barrier, absence of host immune response toward inducer and receptor, and high modularity of switch components to allow efficient customization and small size of DNA constructs for convenient packaging into viruses for *in vivo* delivery.

Ecdysteroid receptor-based gene switches do not satisfy all of these criteria desirable for a generally useful gene regulation system. However, the EcR system possesses several fundamental features that confer great potential for enhancement. To improve the EcR-based switch, several combinations of GAL4 DNA-binding domain (GAL4 DBD), VP16 activation domain (VP16 AD), EcR, and RXR were tested. A two-hybrid format switch, in which GAL4 DBD was fused to ChfuEcR (DEF) and VP16 AD was fused to MmRXR (EF) was found to be the best combination (Fig. 7) in terms of low background levels of reporter gene activity in the absence of a ligand and high levels of reporter gene activity in the presence of a ligand (Palli *et al.*, 2003).

3. Gene Switches for Simultaneous and Independent Regulation of Multiple Genes

It is certainly possible to build a multiplex gene regulatory system by combining two or more of the distinctly different switches that were already developed. However, from a practical standpoint, difficulties would arise from licensing, logistics, and most importantly, lack of parallelism of engineering design. Nonetheless, regulated dual and multigene expression technologies have been described for tetracycline-based and streptogramin-based switches (Baron *et al.*, 1995; Moser *et al.*, 2000). Most desirable would be a multiplex system built on a single receptor platform. The great diversity and ligand specificity of EcRs from different species of insects as well as the high flexibility of their ligand-binding pocket provide such a platform for the development of gene switches that can be used for simultaneous and independent regulation of multiple genes. Development of a yeast system for creation and discovery of ligand–receptor pairs for transcriptional control with small molecules will also help in development of multiplex gene regulation system (Schwimmer *et al.*, 2004).

For simultaneous regulation of two genes in the same cell, two orthogonal EcR:Ec pairs are required ("orthogonal" means that there is no cross-talk between the two ligand/EcR pairs). Studies have shown that lepidopteran

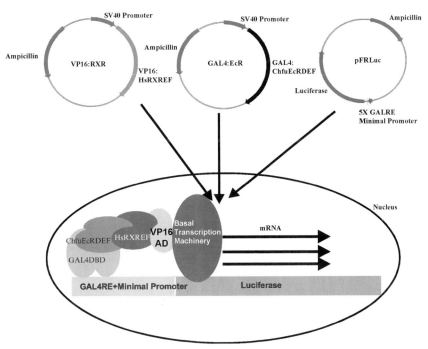

FIGURE 7. Two-hybrid version of EcR-based gene switch. The fusion proteins of GAL4 DNA binding domain and EcR DEF, and VP16 activation domain and RXR EF are expressed under the control of SV40 promoters in two different plasmids. The reporter gene or gene of interest is expressed under the control of synthetic TATAA minimal promoter and 5X GAL4 response elements. These three plasmids are co-transfected into cells, and the transfected cells are exposed to ecdysteroid analog, RG-102240. (See Color Insert.)

EcRs, such as ChfuEcR, MaseEcR, and BomoEcR, respond poorly to Ec, such as 20-hydroxyecdysone, ponasterone A, and muristerone A in mammalian cells, in contradistinction to EcRs cloned from nonlepidopteran insects (e.g., DrmeEcR and AeaeEcR), which respond well to such Ec. On the other hand, diacylhydrazine ligands function well through lepidopteran rather than other insects' EcR. Furthermore, some of the tetrahydroquinoline ligands work exclusively through AeaeEcR (Palli, unpublished). By homology modeling and mutagenesis, we found that exchanging alanine by proline at ChfuEcR position 110 or homologous positions in other EcRs abolishes responsiveness towards Ec but not diacylhydrazines (Kumar et al., 2002). Thus, it is possible to render a given EcR insensitive to Ec by introducing such mutation.

Moreover, site-directed mutation offers the opportunity to use one and the same EcR to construct gene switches for simultaneous regulation of multiple genes. Wild-type ChfuEcR does not bind well to tetrahydroquinolines, such as RG-120768. Changing a single amino acid residue (V128 to F or Y) in the LBD of ChfuEcR results in a mutant EcR that works quite well with RG-120768 rather than diacylhydrazine compounds, such as RG-102240 (Kumar *et al.*, 2004). This V128F mutant and wild-type ChfuEcR were used to demonstrate simultaneous regulation of two genes (Kumar *et al.*, 2004).

4. Applications of Gene Switches in Medicine

Expression of a transgene is an essential component of many gene-based applications in biopharmaceutical production, gene therapy, functional genomics, drug discovery, and tissue engineering. It is becoming evident that these applications require reliable systems for regulated expression of transgenes.

a. Expression of Therapeutic Proteins

Inducible gene regulation systems are indispensable for efficient expression of therapeutic proteins that are toxic to cells or that affect cell metabolism. Nonetheless economically important are applications in which the metabolism of cells is engineered to control cell proliferation and/or to improve the efficiency of protein production. Such applications are documented by the following three examples. Sekiguchi and Hunter (1998) regulated the production of the interferon-responsive factor in BHK21 cells by an ER switch. Mazur *et al.* (1998) used a tetracycline-based gene regulation system to control the expression of protein p27 in CHO cells and achieved a 15-fold increase in the production of this protein. Umana *et al.* (1999) could increase the specific activity of the antihuman neuroblastoma monoclonal antibody, chCE7, by production of this antibody under the regulation of a tetracycline-based system.

b. Gene Therapy

Inducible gene regulations systems allow precise regulation of genes that are used in replacement therapies to correct gene defects in patients. Progesterone receptor-based dimerization type switch was used to regulate the expression of erythropoietin and the human growth hormone genes (Magari *et al.*, 1997; Rivera *et al.*, 1996, 1999, 2000; Wang *et al.*, 1997). An EcR-based switch was used to regulate the expression of inwardly rectifying potassium channels in rat superior cervical ganglion neurons. Overexpression of these inwardly rectifying potassium channels following the application of Ec inhibited evoked and spontaneous activity in neurons (Johns *et al.*, 1999).

c. Functional Genomics

Inducible gene regulation systems are useful for investigating gene function as well as in developing animal models of human diseases. Temporal-specific and spatial-specific regulation of multiple target genes is essential for these studies. Some genes may play important roles both in early stages of development, such as embryogenesis, and also during later stages of development. Probing the function of such genes through mutation may lead to lethality at the early stages of development, making it impossible to study the function at later stages. To circumvent this dilemma, it is possible to place the wild-type gene or a mutant variant under the control of a gene switch. The presence of an inducer at the early stages of development will allow animals to develop; the withdrawal of the inducer at later stages of development will cause regulatable shut-down of expression and allow studies on functions of these genes at these stages. Gene regulation systems that use inducers, such as tetracycline (Gossen and Bujard, 1992), IPTG (Cronin et al., 2001), FK1012 (Spencer et al., 1996), and ponasterone A (No et al., 1996) have been used for inducible regulation of genes in transgenic mice.

Gene switches are unique tools for the identification of downstream target genes. They allow a timed and targeted knock-out of such genes by homologous recombination using, for instance the Cre-loxP system. As a case in point, tamoxifen-inducible ER-Cre fusion protein was employed to study the role of RXR during development (Feil et al., 1996). Gene switches are also useful for antisense strand expression in co-suppression studies or for the control of double-stranded RNA production in experiments investigating gene suppression by RNA interference (RNAi).

d. Drug Discovery

Inducible gene regulations systems have a great potential in high throughput screening assays to express target proteins that are toxic to host cells and would preclude cell viability before administration of the screening compound.

e. Tissue Engineering

Ex vivo production of organs from cells require precise programming of key regulatory pathways, such as differentiation and proliferation. By use of inducible gene switches, the expression of genes that play critical roles in these pathways can be regulated.

B. AGRICULTURE

1. Ecdysteroid Agonists: Species-Specific Insecticides

Over the years several attempts have been made to use Ec analogs for insect control. The most dramatic discovery occurred in the early 1980s while chemists at Rohm and Haas Company synthesized 1,2-diacyl-1-substituted hydrazines, which had potent insecticidal activity (Fig. 8). This class of compounds induces a precocious larval molt in the susceptible species acting through the EcR complex (Dhadialla et al., 1998; Wing, 1988). The

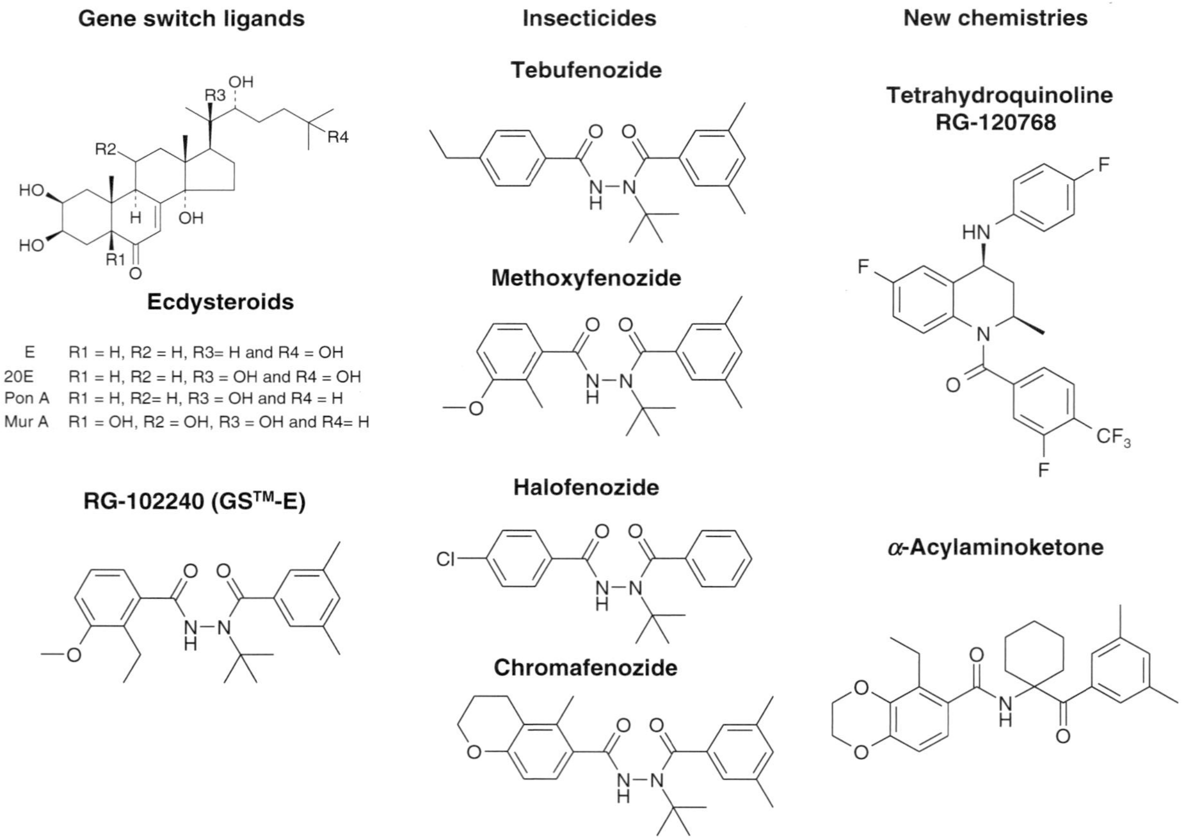

FIGURE 8. Ecdysteroids and stable nonsteroidal Ec agonists used in gene switch and pest management applications.

narrow spectrum of activity makes these compounds an excellent fit for integrated pest management programs. For example, tebufenozide is effective against many lepidopteran pests but has little or no effect on their hymenopteran parasites. Because of their safety profiles, Ec agonists are attractive for insect control in urban settings. Four Ec agonists belonging to the diacylhydrazine chemical class are commercially available for controlling lepidopteran and coleopteran pests (Fig. 8).

After the discovery and successful development of these Ec agonist insecticides, both industrial and academic laboratories continued in search of Ec agonists that might control insects, such as aphids, leaf hoppers, and white flies. Ecdysteroid receptor cDNAs cloned from *D. melanogaster* and other insects showed variability in the amino acid sequence of their LBD (Fig. 9). These cDNA clones were used to design high throughput screening assays in cell lines and yeast (Dixson *et al.*, 2000; Tran *et al.*, 2001a,b). High throughput screening of a large number of compounds together with conventional lead optimization resulted in the identification of new Ec agonists belonging to diacylhydrazine, tetrahydroquinoline, and alpha-acylamidoketone chemical classes (Smith *et al.*, 2003; Tice *et al.*, 2003a, b). In bioassays, some compounds belonging to the tetrahydroquinoline group showed higher potency in the mosquito *A. aegypti* while compared to their activity in a moth (*H. virescens*) (Palli *et al.*, 2005).

Crystal structures of the EcR protein from *H. virescens* show different and only partially overlapping binding cavities for Ec and diacylhydrazine ligands, thereby providing a three-dimensional spatial explanation for the differential activity of these ligands in insects (Fig. 3, Billas *et al.*, 2003). The modeling and mutational studies on Ec, diacylhydrazine, and tetrahydroquinoline ligands also showed somewhat different and only partially overlapping binding cavities for these three groups of ligands (Kumar *et al.*, 2002). In transactivation assays, tetrahydroquinoline compounds activate wild-type ChfuEcR only poorly but a change in a single amino acid in the LBD (valine 128 to either phenylalanine or tyrosine) shifted the ligand specificity of the aforementioned EcR (Kumar *et al.*, 2002) very much in favor of tetrahydroquinolines over diacylhydrazines. Similarly, the previously mentioned mutation of alanine 110 to proline in the ChfuEcR LBD completely eliminated Ec ligand binding, while the binding of diacylhydrazine ligands to this mutant receptor remained unaffected (Kumar *et al.*, 2002). Mutating a single amino acid residue in the LBD of EcR from *A. aegypti*, namely phenylalanine 529, resulted in altered Ec ligand specificity (Wang *et al.*, 2000). These findings point to a highly flexible ligand-binding pocket of EcR and suggest opportunity in designing target-specific insecticides. As more EcRs are identified from pest insects and as efficient high throughput screening assays are developed, it should become increasingly feasible to identify Ec agonists that safely and selectively control various pest insect groups.

		H1	H2	H3	H4	H5	s1 s2 s3 H6

```
            H1                          H2                              H3                                    H4                         H5                  s1    s2   s3   H6
Bian 107 LTANQQFFIARLVWYQDGYDQPSEEDLKRVTQTWQADTEEIGEASDLPFRQITEMTIITVQLIVEFAKGLPGFAKISQPDQITLLKACSSEVMELRVSRYIDMSTDSVMFANRQAYTRENYKAG
Juco 102 LTANQQSILARLVWYQDGEQPSEEDLKRITQTWQEADED.DEESDLPFRQITEMTIITVQLIVEFAKGLPGFAKISQPDQITLLKACSSEVMELRVARRYIDATTDSVLFANRRAYTRENYKAG
Chfu 284 LTANQQFFIARLIWYQDGEQPSEEDLKRITQTWQQADDE.NEESDTPFRQITEMTIITVQLIVEFAKGLPGFAKISQPDQITLLKACSSEVMELRVARRYIDAASDSVLFANNQAYTRDNYKAG
Chsu 290 LTANQQFFIARLVWYQDGEQPSEEDLKRVTQTWQSNEDE.EEETDLPFRQITEMTIITVQLIVEFAKGLPGFAKISQPDQITLLKACSSEVMELRVARRYIDAATDSVLFANNQAYTRDNYKAG
Plin 286 LTANQQFFIAGLVWYQDGEQPSEEDLKRVTQTWQEADED.DD..DMPFRQITEMTIITVQLIVEFAKGLPGFSKISQPDQITLLKACSSEVMELRVARRYIDAATDSVLFANNQAYTRDNYKAG
Bomo 349 LSANQKSILARLVWYQEGEQPSEEDLKRVTQS....DEE.DEESDLPFRQITEMTIITVQLIVEFAKGLPGFSKISQSDQITLLKASSSEVMELRVARRYIDAATDSVLFANNKAYTRDNYKAG
Mase 290 LTANQKSILARLVWYQDGEQPSEEDLKRVTQTWQLLEEEE.EEETDMPFRQITEMTIITVQLIVEFAKGLPGFSKISQSDQITLLKASSSEVMELRVARRYIDAATDSVLFANNQAYTRDNYKAG
Hevi 290 LTANQKSLIARLVWYQDGEQPSEEDLKRVTQTWQLEEEE.EEETDMPFRQITEMTIITVQLIVEFAKGLPGFAKISQSDQITLLKACSSEVMELRVARRYIDAATDSVLFANNQAYTRDNYKAG
Lucu 457 SSTRNQLAVIYKLIWYQDGYEQPSEEDLKRIM.S...SPDENESQHDVSFRHITETIITVQLIVEFAKGLPAFTKIPQEDQITLLKACSSEVMELRMAARRYTDHNSDSIFFANNRSYTRDSYKMAG
Drme 416 LSTNQLAVIYKLIWYQDGEQPSEEDLKRIM.S...SPDENESQHDASFRHITETIITVQLIVEFAKGLPAFTKPQEDQITLLKACSSEVMELRMAARRYTDHNSDSIFFANNRSYTRDSYKMAG
Drme 391 LTRNQLAVIYKLIWYQDGEQPSEEDLKRIM.S...QPDENESQTDVSFRHITETIITVQLIVEFAKGLPAFTKPQEDQITLLKACSSEVMELRMAARRYDHSSDSIFFANNRSYTRDSYKMAG
Aeae 434 LTANQMAVIYKLIWYQDGEQPSEEDLKRIMIG....SPNEEEDQHDVHFRHITETIITVQLIVEFAKGLPAFTKPQEDQITLLKACSSEVMELRMAARRYDHNSDSIFFANNRSYTRDSRMAG
Aeal 333 LTANQMAVIYKLIWYQDGEQPSEEDLKRIMIG....SPNEEEDQHDVHFRHITETIITVQLIVEFAKGLPAFTKPQEDQITLLKACSSEVMELRMAARRYDAETDSIFFANNRSYTRDSYMAG
Anga 280 LTANQMAVIYKLIWYQDGEQPSEEDLKRIMIN....SPNEEEDPHEIHFRHITETIITVQLIVEFAKGLPAFTKPQEDQITLLKACSSEVMELRMAARRYDHDSDSILFANNTAYTKQTQLAG
Chte 275 LTANQAVIYKLIWYQDGEQPSEEDLKRITT.....LEEEEEDQEHEANFRYITEVTIITVQLIVEFAKGLPAFIKPQEDQITLLKACSSEVMELRMAARRYDVQTDSIFANNQPYTKQSITVAG
Apmo 578 ISPEQEEIIHRLVYFQENEYESPSEEDLKRITNQ....PS.EGEDISDYKFRHITETIITVQLIVEFSKRLPGFDRMREDQIALLKACSSEVMELRMARKRYDVQDISIFANNQPYTKDSITVAG
Lomi 305 VSPEQEELIHRLVYFQENEVESPSEEDLRRVTSQ....PT.EGEDQSDVRFRHITETIITVQLIVEFAKRLPGFDKLLREDQKLLLQRDQIALLKACSSEVMELRMARKRYTVDNSDSLFVNNQPPRDSNLAG
Drme 260 PEQEELIIHRLVYFQNEYEHPSEEDVKRIINQ.....PI.DGEDQCEIRFRHITETIITVQLIVEFRHITETIITVQLIVEFAKGLPGFDKLLQRDQIALLKACSSEVMELRMARKRYDVQDISLFVNNQPPRDSNLAG
Drme 103 LTREQEELINTLVYFQEFQEPTEADIKKIRFT....F.DGGDTSDMRFHITEMTIITVQLIVEFSKQLPGLGTLQREDQITLLKACSSEVMELRVARRYFDAKTDCIVFGNTLPYTQSSYEFAG
Cama 282 LTREQEELIITLYVYQEEFEQPTEADVKKIREN....F.DGGDTSDMRFHITEMTIITVQLIVEFSKQLPEATLQREDQITLLKACSSEVMELRMAARRYDAKTDSIVFGNNTPYTQASVALAG
Cepu 324 LSSSQEDILNKLVYYQQEFESPSEEDMKKTTPF...PLGDSEEDNQRRFQHITETIITVQLIVEFKRVPGFDTLAREDQITLAGRKYDVKTDSIVFANNQPYTRDNYRSAS

            H7                          H8                              H9                                    H10                         H11                         H12
Bain 346 FGYVIENLLHFCRCHYAMSMDNVHYALLTAVVIFSDRPGLENPQLVEEIQRYIVNTRVYIINQQSSTRCPVVYGKILSVLSELRSLGMONSNMCISLKLNRKLPPFLEIHVD
Juco 339 MSYVIENLLHFCRCMYTMSMDNVHYALLITAVVIFSDRPGLEQPHIVEEIQRYIVNTLRVYIMNQLASSRCPVFGKILSILSELRTLGMONSNMCISLKLNRKLPPFLEIHVD 523
Chfu 523 MAYVIEDLLHFCRCMYSMALDNIHYALLITAVVIFSDRPGLEQPOIVEEIQRYIVNTLRVYIINQLSGSARSVIYGKILSILSELRTLGMONSNMCISLKLKNRKLPPFLEIHVD 529
Chsu 529 MAYVIEDLLHFCRCMYSLSMDNVHYALLITAVVIFSDRPGLEQPOIVEEIQRYIVNTLRVYIMNQHSASPRCAVLYAKILSVLTELRTLGMONSEMCFSLKLKNRKLPPFLEIHVD 523
Plin 523 MAYVIEDLLHFCRCHYSMTMDNVHYALLITAVVIFSDRPGLEQPQIVEEIQRYIVNTLRVYIMNQHSDSPRCAVVFGKILSVLTELRTLGMONSNMCISLKLKNRKLPPFLEIHVD 587
Bomo 587 MAYVIEDLLHFCRCHFAMGMDNVHYALLITAVVIFSDRPGLEQPSIVEEIQRYIVNTLRVYIINQNASSRCAVIYGRILSVLTELRTLGTONSNMCISLKLKNRKLPPFLEIHVD 529
Mase 529 MSYVIEDLLHFCRCHYSMSMDNVHYALLITAVVIFSDRPGLEQPLIVEEIQRYLKTLRVYIINQSASPRCAVLFGKILLGVLTELRTLGTONSNMCISLKLKNRKLPPFLEIHVD 528
Hevi 528 MAYVIEDLLHFCRCHYSMAMDNVHYALLITAVVIFSDRPGLEQPLIVEEIQRYIVNTLRVYIINQLAASPRGAVIFGEILGIITELRTLGMONSNMCISLKLKNRKLPPFLEIHVD 690
Lucu 690 MADNIEDLLHFCRCHYSMKVDNVEYALLITAVVIFSDRPGLEFEAEIVEAIQSYYIDTLRYIINRCGDPMSIVFFAKILSIITELRTLGNQNAEMCFSLKLKNRKLPKFLEIHVD 675
Drme 675 MADNIEDLLHFCRCHYSMKVDNVEYALLITAVVIFSDRPGLEEKADLVEAIQSYYIDTLRIYIINRCGDSMSIVFTAKILSIITELRTLGNAEMCFSLKLKNRKLPKFLEIHVD 652
Ceca 652 VADNIEDLLHFCRCHYSMKVDNVEYALLITAVVIFSDRPGLEKAQLVEEIQSYYIDTLRVYIINRCGDSMSIVFFAKILSIITELRTLGNQNAEMCFSLKLKNRKLPRFLEIHVD 627
Aeae 627 MADYIEDLLHFCRCMFSLTVDNVEYALLITAVVIFSDRPGLEQPEIVEHIQSYYIDTLRIYIINRHAGDPKCSVIFAKILSIVLTELRTLGMONSNMCISLKLKNRKLPPFLEIHVD 570
Aeal 570 MADTIEDLLHFCRCHMFSLTVDNVEYALLITAVVIFSDRPGLEQPEIVEHIQSYYIDTLRIYIINRHAGDPKCSVIFAKILSIILSELRTLGMONSEMCISLKLKNRKLPPFLEIHVD 570
Anga 570 MADTIEDLLHFCRCMFSLTVDNVEYALLITAVVIFSDRPGLEQEAELVETIQSYYYIDTLRIYIINRHGDPKCSVTFAKILSVLTELRTLGNQNSEMCFSLKLRNRKLPPFLEIHVD 517
Chte 517 MEETIDDLLHFCRCHYALSIDNVEYALLITAVVIFSDRPGLEQKAEVETIQSYYETLKVIYVNRHGESRCSVQFAKLLGILITELRTMGNRNSEMCFSLKLRNRGLPRFLEIHVD 512
Apmo 512 MGETIEDLLHFCRQHYAMKVNNAEYALLITAVVIFSERPNIEGHKVEKIQEIYIEALRAYVNRR..RPNPGTVFARLLSVLTELRTLGNQNSEMCFSLKFNKGLPVFLAEIHVD 812
Lomi 812 MGETIEDMLHFCRQHYAMKVDNAEYALLITAVVIFSERPSIVEGHKVEKIQEIYIEALRAYVDNRRP.RPKSGTIFAKLLSVLTELRTLGNQNSEMCFSLKLKNRKLPPFLAEIHVD 539
Temo 539 MGETIEDLLHFCRTMYSMKVDNAEYALLITAVVIFSERPSIIEGHKVEKIQEIYLEALRAYVDNRR.SPSRGTIFAKLLSVILTELRTLGNQNSEMCISLKLKNRKLPPFTDEIHVD 487
Cama 487 LGESSQIIFPRCRNLCRMKVDNAEYALLSAIIFSERPNLEKLQKVEKQEIYLDALRAYVCNQR..FPKPGMVFAKLNITLELRTLGRNISEVCFSLKLRNRKLPPFLAEIHVD 336
Cepu 336 LGESAEIILFRFCRSLCRMKVDNAEYALLAAIIFSERPNLEKLKVEKQEIYLEALKSYVENRR..LPRSNMVFAKLNIITLELRTLSLKNRLPPFLAEIHVD 515
Amam 515 VGDSADALFREFCRKMCQLRVDNAEYALLITAVVIFSERPSLVDPHKVERIQEYYYIETLRMYSENHR..PPGKN.YFARLLSITLELRTLGNMNAEMCFSLKVQNKLGLPPFLAEIHVD 558
```

2. Gene Switch Technology for Use in Agriculture

Transgenes in plants are expressed under the control of constitutive promoters (e.g., the cauliflower mosaic virus) (CaMV) 35S promoter. Such an "always-on" arrangement can cause metabolic waste and pleiotropic effects due to expression of transgenes at all stages in all tissues. Unpropitiously, the potential for gene escape and expression in undesirable plants increases when the transgenes are expressed under the control of a constitutive promoter. Moreover, the constitutive promoters cannot be used to regulate genes, which if overexpressed, might have toxic effects on the plant. Expression of such genes in transformed cells may lead to their death resulting in a blockage of plant regeneration. Attempts at solving some of these problems led to the development of gene regulation systems based on plant promoters that increase transgene transcription upon application of herbicide safeners (De Veylder *et al.*, 1997; Jepson *et al.*, 1994), plant hormones (Suehara *et al.*, 1996), or heat shock (Severin and Schoffl, 1990). These systems supported a rapid induction of transgenes, but also induced the expression of endogenous plant genes that responded to these signals. Accordingly, these gene regulation systems have been used only to a limited extent.

Gene regulation systems based on receptors or transcription factors that inactivate or activate transgenes upon addition of ligands have been developed. As outlined in Fig. 10, these gene switches contain two expression cassettes: one (receptor gene construct) that uses a constitutive promoter to express the receptor and a second cassette (expression gene construct), which expresses the transgene under the control of multiple copies of transcription factor binding sites fused to a minimal promoter. Upon transformation of both constructs into transgenic plants, the receptor protein is produced and may even bind to the binding sites present in the expression gene construct.

FIGURE 9. Alignment of amino acids in the ligand binding domain of ecdysteroid receptors. The amino acid residues present in all EcRs and the amino acids present in only lepidopteran and dipteran EcRs are boxed. The EcR sequences are from *Bicyclus anynana* (Bian, unpublished, gi:6580162), *Junonia coenia* (Juco, unpublished, gi:6580625), *Choristoneura fumiferana* (Chfu [Kothapalli *et al.*, 1995]), *Chilo suppressalis* (Chsu [Minakuchi *et al.*, 2002]), *Plodia interpunctella* (Plin [Siaussat *et al.*, 2004]), *B. mori* (Bomo [Kamimura *et al.*, 1996; Swevers *et al.*, 1995]), *Manduca sexta* (Mase [Fujiwara *et al.*, 1995]), *H. virescens* (Hevi [Billas *et al.*, 2003; Martinez *et al.*, 1999a]), *Calliphora vicina* (Cavi, unpublished, gi:12034940), *Lucilia cuprina* (Lucu [Hannan and Hill, 1997]), *Drosophila melanogaster* (Drme [Koelle *et al.*, 1991]), *Ceratitis capitata* (Ceca, Verras *et al.*, 1999), *A. aegypti* (Aeae [Cho *et al.*, 1995]), *Aedes albopictus* (Aeal [Jayachandran and Fallon, 2000]), *Anopheles gambiae* (Anga, unpublished, gi:55234452), *Chironomus tentans* (Chte [Imhof *et al.*, 1993]), *Apis mellifera* (Apme, unpublished, *XP_394760*), *Locusta migratoria* (Lomi [Saleh *et al.*, 1998]), *Tenebrio molitor* (Temo [Mouillet *et al.*, 1997]), *Carcinus maenas* (Cama unpublished, gi:40748295), *Celuca pugilator* (Cepu [Chung *et al.*, 1998]), and *Amblyomma americanum* (Amam [Guo *et al.*, 1997]). (See Color Insert.)

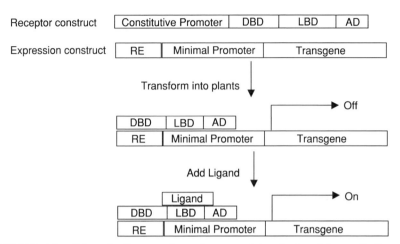

FIGURE 10. Schematic diagram of generic gene switches developed for applications in agriculture. A receptor fusion protein consisting of DNA-binding domain (DBD) and ligand-binding domain (LBD) and activation domain (AD) is expressed under the control of a constitutive promoter, such as cauliflower mosaic virus 35S [CaMV 35S] promoter. A reporter gene or a protein of interest is expressed under the control of a minimal promoter, such as CaMV 35S minimal promoter and multiple response elements (RE) that are recognized by DBD present in receptor fusion protein. The receptor proteins are expressed in plant cells and upon addition of chemical, the receptor fusion proteins bind to RE present in the promoter and turn on the expression of the reporter or gene of interest.

However, the transcription of the transgene is not activated until the appropriate ligand is applied to the plant (Fig. 10).

Gene switches based on *E. coli* tetracycline-resistant operon (Bohner et al., 1999; Gatz et al., 1992; Weinmann et al., 1994), an *Aspergillus* activator induced by ethanol (Caddick et al., 1998), GR (Aoyama and Chua, 1997; Bohner et al., 1999; Ouwerkerk et al., 2001; Tang and Newton, 2004a, b; Tang et al., 2004), ER (Bruce et al., 2000; Zuo and Chua, 2000; Zuo et al., 2000), EcR (Martinez et al., 1999b,c; Padidam et al., 2003; Unger et al., 2002), and the pristinamycin resistance operon of *Streptomyces coelicolor* (Frey et al., 2001) have all been developed and tested in *Arabidopsis*, tobacco and/or corn plants. While most of these gene switches are useful for regulation of a variety of genes for research applications in the laboratory, the ligands that are being used for development of these gene switches are inappropriate for field use in agriculture. Ligands that are suitable for field use are the commercially available nonsteroidal Ec agonists, tebufenozide (Dow AgroSciences), methoxyfenozide (Dow AgroSciences), halofenozide (Dow AgroSciences), and chromafenozide (Nippon Kayaku). Gene switches that are being developed for use in plants have been reviewed (Padidam, 2003; Tang et al., 2004). Only EcR-based gene switches will be discussed here.

3. EcR-Based Gene Switches

The first EcR-based gene switch for application in plants was developed by Martinez *et al.* (1999a,c), using hybrid transactivator containing sequences of the GR, the VP16 activation domain, the DNA binding domain of GR, and the ligand binding domain of EcR from *H. virescens*. In transgenic tobacco plants, tebufenozide induced the expression of a reporter gene regulated by this EcR-based gene switch to about 150% of the activity of the viral 35S promoter. Although the induction levels are remarkably high, the system is limited by the requirement of micromolar (μM) concentration of ligand for activation and by high background expression in the absence of ligand. The background activity of the transgene in the absence of ligand was later reduced by a replacement of the DNA binding domain from the GR receptor with that of GAL4 from yeast (Padidam *et al.*, 2003; Unger *et al.*, 2002). However, the continued requirement of this EcR gene switch for micromolar ligand concentrations is exacerbated by the generally poor systemicity of the diacylhydrazine ligands registered for agriculture, thereby limiting the use of this switch for large-scale field applications. Future successful use of gene switches in various plant biotechnology applications depends on the availability of EcR-based gene switches, which respond to nanomolar (nM) concentration of nonsteroidal Ec agonists that are systemic in plants and are registered for field use.

Ecdysteroid receptor-based gene switch technology provides the unique opportunity to develop multiplex gene switches. Ecdysteroid receptors from more than a dozen insect species have been cloned and characterized showing differential ligand specificities against the four commercially available nonsteroidal insecticides. Tebufenozide, methoxyfenozide, and chromofenozide bind with high affinity to lepidopteran rather than other EcRs; consequently these ligands function only in larvae of moths and butterflies (Palli and Retnakaran, 2000), whereas halofenozide is specific for a subgroup of coleopteran insects (e.g., Japanese beetle, *Papilio japonica* [Heller and Walker, 1996]). Once the EcR from *P. japonica* (PajaEcR) is cloned and its expected ligand specificity established, it would be possible to develop a multiplex switch by combining the two-switch pairs (e.g., methoxyfenozide/Chfu EcR and halofenozide/PajaEcR). Such combinations of gene switch pairs, especially if they are orthogonal (see Section III.A.3), will allow controlling two different traits in a single transgenic plant by the application of respective ligands with high specificity.

4. Applications of Gene Switches in Agriculture

The general application of gene switches is similar in agriculture and medicine (see Section III.A.4). This holds notably for the production of desired proteins, screening for drugs and ligands, functional genomics, and the rectification or improvement of a given phenotype by the control of transgene expression. In the following section, some specific examples of the use of gene switches in agriculture will be presented.

a. Functional Genomics

In *Arabidopsis*, the unusual floral organs (UFO) gene is required for flower development. Introduction of an ethanol-inducible expression system for the UFO gene into *ufo* loss-of-function plants allowed Laufs *et al.* (2003) to control the expression of this gene ubiquitously at different developmental stages and for various lengths of time, and thus to find floral phenotypes that could not be discovered otherwise. An ER-derived switch was used to regulate two transcription factors involved in the flavonoid pathway for identification of a number of downstream target genes (Bruce *et al.*, 2000). A GR-based switch was employed for dexamethasone-inducible expression of the bacterial avrRpt2 avirulence gene in transgenic plants. Treatment with dexamethasone led then to the cell-death response (McNellis *et al.*, 1998). An antisense strand of *Arabidopsis* CDC2b gene was expressed under the control of a GR-based switch. Induction of this antisense RNA by dexamethasone resulted in short hypocotyls and open cotyledons of the transgenic plants (Yoshizumi *et al.*, 1999).

b. Transgenic Plants

Because of public concerns about the use of antibiotic or herbicide resistance genes in transgenic plants and concerns about the decrease in transformation efficiency due to the effect of herbicides and antibiotics on multiplication and differentiation of transformed cells, the production of marker-free transgenic plants has become a major objective for plant biologists. Gene switches allow the selection of transgenic plants without the use of herbicides or antibiotics. For example, overexpression of the isopentenyltransferase (ipt) gene from the Ti-plasmid of *Agrobacterium tumefaciens* under the control of a dexamethasone/GR-based gene switch led to an increased generation of shoots, which developed into normal plants even after withdrawal of the inducer (Kunkel *et al.*, 1999). A chemical-inducible site-specific DNA excision system was developed using the Cre/*loxP* and an ER-based gene switch. When the plants were treated with inducer, the marker gene together with Cre and the ER gene switch were excised from the transgenic *Arabidopsis* plants due to their location between two *loxP* sites (Hare and Chua, 2002; Zuo *et al.*, 2001). Use of this system will eliminate some of the potential problems associated with expression antibiotic genes in genetically modified crops.

Gene switches permit an efficient spatial and temporal regulated expression of transgenes that can provide pest resistance, herbicide resistance, and trait improvement. One example is the application of an EcR-derived switch that controls male fertility in maize. The Ms45 maize gene that regulates microspore development was placed under the control of EcR-based switch and introduced into an otherwise sterile ms45 mutant. A nonsteroidal Ec agonist, methoxyfenozide, was able to restore male fertility in these transgenic plants (Unger et al., 2002). The EcR-based gene switch was used in transgenic *Arabidopsis* plants to regulate a coat protein gene from tobacco mosaic virus. Administration of ligand conferred virus resistance to these transgenic plants (Koo et al., 2004).

IV. CONCLUSIONS AND PERSPECTIVES

Progress in EcR research has shown that the ligand-binding pocket of EcR is highly flexible and adaptable. This property of EcR permits the development of receptor systems for practical application in medicine and agriculture. The application of EcR-based gene switch technology in cellular systems toward functional genomics, drug and ligand discovery, and therapeutic protein production has already commenced and will grow rapidly in near future. The application of EcR-based gene switches in plants has a great potential and benefits greatly from the fact that several potential ligands represent registered insecticides and have a track record of environmental safety. Ligand potency and switch background activity in the absence of ligand are the two major hurdles that need to be overcome. The hurdles for an application of EcR-based gene switches in humans are even higher. These are, for example, costs associated with registration of stable Ec ligands, potential pleiotropic effects of switch components and ligands, development of suitable application methods for switch components and ligands. Such hurdles constitute a real challenge for future gene switch technology based on the Ec/EcR system.

ACKNOWLEDGMENTS

We thank Xanthe Vafopoulou (York University, Toronto) for providing information prior to publication. The publisher's permission to reproduce the published pictures Figs. 1 and 2 is gratefully acknowledged. S. R. P. is supported by the Kentucky Agricultural Experimental Station USDA-CREES Hatch project, National Science Foundation (IBN-0421856), National Institute of Health (GM070559-02), National Research Initiative of the USDA-CSREES (2004-03070) and RheoGene Inc. This is contribution number 05–08–01 from the Kentucky Agricultural Experimental Station.

REFERENCES

Albanese, C., Hulit, J., Sakamaki, T., and Pestell, R. G. (2002). Recent advances in inducible expression in transgenic mice. *Semin. Cell. Dev. Biol.* **13,** 129–141.
Aoyama, T., and Chua, N.-H. (1997). A glucocorticoid-mediated transcriptional induction system in transgenic plants. *Plant J.* **11,** 605–612.

Apriletti, J. W., Baxter, J. D., Lau, K. H., and West, B. L. (1995). Expression of the rat α1 thyroid hormone receptor ligand binding domain in *Escherichia coli* and the use of a ligand-induced conformation change as a method for its purification to homogeneity. *Protein Expr. Purif.* **6,** 363–370.

Arbeitman, M. N., and Hogness, D. S. (2000). Molecular chaperones activate the *Drosophila* ecdysone receptor, an RXR heterodimer. *Cell* **101,** 67–77.

Ashburner, M., Chihara, C., Meltzer, P., and Richards, G. (1974). Temporal control of puffing activity in polytene chromosomes. *Cold Spring Harb. Symp. Quant. Biol.* **38,** 655–662.

Baron, U., Freundlieb, S., Gossen, M., and Bujard, H. (1995). Co-regulation of two gene activities by tetracycline via a bidirectional promoter. *Nucleic Acids Res.* **23,** 3605–3606.

Belshaw, P. J., Ho, S. N., Crabtree, G. R., and Schreiber, S. L. (1996). Controlling protein association and subcellular localization with a synthetic ligand that induces heterodimerization of proteins. *Proc. Natl. Acad. Sci. USA* **93,** 4604–4607.

Benecke, A., Chambon, P., and Gronemeyer, H. (2000). Synergy between estrogen receptor alpha activation functions AF1 and AF2 mediated by transcription intermediary factor TIF2. *EMBO Rep.* **1,** 151–157.

Bergman, T., Henrich, V. C., Schlattner, U., and Lezzi, M. (2004). Ligand control of interaction *in vivo* between ecdysteroid receptor and ultraspiracle ligand-binding domain. *Biochem. J.* **378,** 779–784.

Billas, I. M., Iwema, T., Garnier, J. M., Mitschler, A., Rochel, N., and Moras, D. (2003). Structural adaptability in the ligand-binding pocket of the ecdysone hormone receptor. *Nature* **426,** 91–96.

Billas, I. M., Moulinier, L., Rochel, N., and Moras, D. (2001). Crystal structure of the ligand-binding domain of the ultraspiracle protein USP, the ortholog of retinoid X receptors in insects. *J. Biol. Chem.* **276,** 7465–7474.

Bohner, S., Lenk, I. I., Rieping, M., Herold, M., and Gatz, C. (1999). Technical advance: Transcriptional activator TGV mediates dexamethasone-inducible and tetracycline-inactivatable gene expression. *Plant J.* **19,** 87–95.

Bourguet, W., Ruff, M., Chambon, P., Gronemeyer, H., and Moras, D. (1995). Crystal structure of the ligand-binding domain of the human nuclear receptor RXR-α. *Nature* **375,** 377–382.

Bourguet, W., Vivat, V., Wurtz, J.-M., Chambon, P., Gronemeyer, H., and Moras, D. (2000). Crystal structure of a heterodimeric complex of RAR and RXR ligand binding domains. *Mol. Cell* **5,** 289–298.

Brelivet, Y., Kammerer, S., Rochel, N., Poch, O., and Moras, D. (2004). Signature of the oligomeric behavior of nuclear receptors at the sequence and structural level. *EMBO Rep.* **5**(4), 1–7.

Bruce, W., Folkerts, O., Garnaat, C., Crasta, O., Roth, B., and Bowen, B. (2000). Expression profiling of the maize flavonoid pathway genes controlled by estradiol-inducible transcription factors CRC and P. *Plant Cell* **12,** 65–80.

Caddick, M. X., Greenland, A. J., Jepson, I., Krause, K. P., Qu, N., Riddell, K. V., Salter, M. G., Schuch, W., Sonnewald, U., and Tomsett, A. B. (1998). An ethanol inducible gene switch for plants used to manipulate carbon metabolism. *Nat. Biotechnol.* **16,** 177–180.

Cakouros, D., Daish, T. J., Mills, K., and Kumar, S. (2004). An arginine-histone methyltransferase, CARMER, coordinates ecdysone-mediated apoptosis in *Drosophila* cells. *J. Biol. Chem.* **279,** 18467–18471.

Cho, W. L., Kapitskaya, M. Z., and Raikhel, A. S. (1995). Mosquito ecdysteroid receptor: Analysis of the cDNA and expression during vitellogenesis. *Insect Biochem. Mol. Biol.* **25,** 19–27.

Christopherson, K. S., Mark, M. R., Bajaj, V., and Godowski, P. J. (1992). Ecdysteroid-dependent regulation of genes in mammalian cells by a *Drosophila* ecdysone receptor and chimeric transactivators. *Proc. Natl. Acad. Sci. USA* **89,** 6314–6318.

Chung, A. C., Durica, D. S., Clifton, S. W., Roe, B. A., and Hopkins, P. M. (1998). Cloning of crustacean ecdysteroid receptor and retinoid-X receptor gene homologs and elevation of retinoid-X receptor mRNA by retinoic acid. *Mol. Cell. Endocrinol.* **139**, 209–227.

Clayton, G. M., Peak-Chew, S. Y., Evans, R. M., and Schwabe, J. W. (2001). The structure of the ultraspiracle ligand-binding domain reveals a nuclear receptor locked in an inactive conformation. *Proc. Natl. Acad. Sci. USA* **98**, 1549–1554.

Cronin, C. A., Gluba, W., and Scrable, H. (2001). The lac operator-repressor system is functional in the mouse. *Genes Dev.* **15**, 1506–1517.

De Veylder, L., Van Montagu, M., and Inze, D. (1997). Herbicide safener-inducible gene expression in *Arabidopsis thaliana*. *Plant Cell Physiol.* **38**, 568–577.

Devarakonda, S., Harp, J. M., Kim, Y., Ozyhar, A., and Rastinejad, F. (2003). Structure of the heterodimeric ecdysone receptor DNA-binding complex. *EMBO J.* **22**, 5827–5840.

Dhadialla, T. S., Carlson, G. R., and Le, D. P. (1998). New insecticides with ecdysteroidal and juvenile hormone activity. *Annu. Rev. Entomol.* **43**, 545–569.

Dixson, J. A., Elshenawy, Z. M., Eldridge, J. R., Dungan, L. B., Chiu, G., Wowkun, G. S., and Wyle, M. J. (2000). A new class of potent ecdysone agonists: 4-Phenylamino-1,2,3,4-tetrahydroquinolin. *In* "ACS Mid Atlantic Regional Meeting." The University of Delaware, Newark, DE.

Egea, P. F., Mitschler, A., Rochel, N., Ruff, M., Chambon, P., and Moras, D. (2000). Crystal structure of the human RXRα ligand binding domain bound to its natural ligand: 9-cis Retinoic acid. *EMBO J.* **19**, 2592–2601.

Feil, R., Brocard, J., Mascrez, B., LeMeur, M., Metzger, D., and Chambon, P. (1996). Ligand-activated site-specific recombination in mice. *Proc. Natl. Acad. Sci. USA* **93**, 10887–10890.

Frey, A. D., Rimann, M., Bailey, J. E., Kallio, P. T., Thompson, C. J., and Fussenegger, M. (2001). Novel pristinamycin-responsive expression systems for plant cells. *Biotechnol. Bioeng.* **74**, 154–163.

Fujiwara, H., Jindra, M., Newitt, R., Palli, S. R., Hiruma, K., and Riddiford, L. M. (1995). Cloning of an ecdysone receptor homolog from *Manduca sexta* and the developmental profile of its mRNA in wings. *Insect. Biochem. Mol. Biol.* **25**, 845–856.

Fussenegger, M. (2001). The impact of mammalian gene regulation concepts on functional genomic research, metabolic engineering, and advanced gene therapies. *Biotechnol. Prog.* **17**, 1–51.

Fussenegger, M., Morris, R. P., Fux, C., Rimann, M., von Stockar, B., Thompson, C. J., and Bailey, J. E. (2000). Streptogramin-based gene regulation systems for mammalian cells. *Nat. Biotechnol.* **18**, 1203–1208.

Gatz, C., Frohberg, C., and Wendenburg, R. (1992). Stringent repression and homogeneous derepression by tetracycline of a modified CaMV 35S promoter in intact transgenic tobacco plants. *Plant J.* **2**, 397–404.

Gehring, U. (1998). Steroid hormone receptors and heat shock proteins. *Vitam. Horm.* **54**, 167–205.

Gossen, M., and Bujard, H. (1992). Tight control of gene expression in mammalian cells by tetracycline-responsive promoters. *Proc. Natl. Acad. Sci. USA* **89**, 5547–5551.

Grebe, M., Fauth, T., and Spindler-Barth, M. (2004). Dynamic of ligand binding to *Drosophila melanogaster* ecdysteroid receptor. *Insect. Biochem. Mol. Biol.* **34**, 981–989.

Grebe, M., Przibilla, S., Henrich, V. C., and Spindler-Barth, M. (2003). Characterization of the ligand-binding domain of the ecdysteroid receptor from *Drosophila melanogaster*. *Biol. Chem.* **384**, 105–116.

Guo, X., Harmon, M. A., Laudet, V., Mangelsdorf, D. J., and Palmer, M. J. (1997). Isolation of a functional ecdysteroid receptor homologue from the ixodid tick *Amblyomma americanum* (L.). *Insect Biochem. Mol. Biol.* **27**, 945–962.

Hannan, G. N., and Hill, R. J. (1997). Cloning and characterization of LcEcR: A functional ecdysone receptor from the sheep blowfly *Lucilia cuprina*. *Insect. Biochem. Mol. Biol.* **27**, 479–488.

Hare, P. D., and Chua, N. H. (2002). Excision of selectable marker genes from transgenic plants. *Nat. Biotechnol.* **20**, 575–580.

Heller, P., and Walker, R. (1996). Evaluation of RH-0345 and turcam 76WP for management of Japanese beetle grubs on a golf course fairway. *Arthropod Manage. Tests* **21**, 339.

Henrich, V. C. (2004). The ecdysteroid receptor. *In* "Comprehensive Molecular Insect Science" (L. I. Gilbert, K. Iatrou, and S. S. Gill, Eds.), vol. 3, pp. 243–285. Elsevier Ltd, Oxford, UK.

Hoppe, U. C., Marban, E., and Johns, D. C. (2000). Adenovirus-mediated inducible gene expression *in vivo* by a hybrid ecdysone receptor. *Mol. Ther.* **1**, 159–164.

Hu, X., Cherbas, L., and Cherbas, P. (2003). Transcription activation by the ecdysone receptor (EcR/USP): Identification of activation functions. *Mol. Endocrinol.* **17**, 716–731.

Imhof, M. O., Rusconi, S., and Lezzi, M. (1993). Cloning of a *Chironomus tentans* cDNA encoding a protein (cEcRH) homologous to *Drosophila melanogaster* ecdysteroid receptor (dEcR). *Insect Biochem. Mol. Biol.* **23**, 115–124.

Jayachandran, G., and Fallon, A. M. (2000). Evidence for expression of EcR and USP components of the 20-hydroxyecdysone receptor by a mosquito cell line. *Arch. Insect Biochem. Physiol.* **43**, 87–96.

Jepson, I., Lay, V. J., Holt, D. C., Bright, S. W., and Greenland, A. J. (1994). Cloning and characterization of maize herbicide safener-induced cDNAs encoding subunits of glutathione S-transferase isoforms I, II and IV. *Plant Mol. Biol.* **26**, 1855–1866.

Johns, D. C., Marx, R., Mains, R. E., O'Rourke, B., and Marban, E. (1999). Inducible genetic suppression of neuronal excitability. *J. Neurosci.* **19**, 1691–1697.

Jones, G., Wozniak, M., Chu, Y., Dhar, S., and Jones, D. (2001). Juvenile hormone III-dependent conformational changes of the nuclear receptor ultraspiracle. *Insect Biochem. Mol. Biol.* **32**, 33–49.

Kamimura, M., Tomita, S., and Fujiwara, H. (1996). Molecular cloning of an ecdysone receptor (B1 isoform) homologue from the silkworm, *Bombyx mori*, and its mRNA expression during wing disc development. *Comp. Biochem. Physiol. B Biochem. Mol. Biol.* **113**, 341–347.

Kethidi, D. R., Perera, S. C., Zheng, S., Feng, Q. L., Krell, P., Retnakaran, A., and Palli, S. R. (2004). Identification and characterization of a juvenile hormone (JH) response region in the JH esterase gene from the spruce budworm, *Choristoneura fumiferana*. *J. Biol. Chem.* **279**, 19634–19642.

Koelle, M. R., Talbot, W. S., Segraves, W. A., Bender, M. T., Cherbas, P., and Hogness, D. S. (1991). The *Drosophila* EcR gene encodes an ecdysone receptor, a new member of the steroid receptor superfamily. *Cell* **67**, 59–77.

Koo, J. C., Asurmendi, S., Bick, J., Woodford-Thomas, T., and Beachy, R. N. (2004). Ecdysone agonist-inducible expression of a coat protein gene from tobacco mosaic virus confers viral resistance in transgenic *Arabidopsis*. *Plant J.* **37**, 439–448.

Kothapalli, R., Palli, S. R., Ladd, T. R., Sohi, S. S., Cress, D., Dhadialla, T. S., Tzertzinis, G., and Retnakaran, A. (1995). Cloning and developmental expression of the ecdysone receptor gene from the spruce budworm, *Choristoneura fumiferana*. *Dev. Genet.* **17**, 319–330.

Kuhn, R., Schwenk, F., Aguet, M., and Rajewsky, K. (1995). Inducible gene targeting in mice. *Science* **269**, 1427–1429.

Kumar, M. B., Fujimoto, T., Potter, D. W., Deng, Q., and Palli, S. R. (2002). A single point mutation in ecdysone receptor leads to increased ligand specificity: Implications for gene switch applications. *Proc. Natl. Acad. Sci. USA* **99**, 14710–14715.

Kumar, M. B., Potter, D. W., Hormann, R. E., Edwards, A., Tice, C. M., Smith, H. C., Dipietro, M. A., Polley, M., Lawless, M., Wolohan, P. R., Kethidi, D. R., and Palli, S. R. (2004). Highly flexible ligand binding pocket of ecdysone receptor: A single amino acid

change leads to discrimination between two groups of nonsteroidal ecdysone agonists. *J. Biol. Chem.* **279**, 27211–27218.
Kunkel, T., Niu, Q. W., Chan, Y. S., and Chua, N. H. (1999). Inducible isopentenyl transferase as a high-efficiency marker for plant transformation. *Nat. Biotechnol.* **17**, 916–919.
Lafont, R. D., and Wilson, I. D. (1992). "The Ecdysone Handbook". Chromatographic Society, Nottingham, UK.
Lammerding-Köppel, M., Spindler-Barth, M., Lezzi, M., Drews, U., and Spindler, K.-D. (1998). Immunohistochemical localization of ecdysteroid receptor and ultraspiracle in the epithelial cell line from *Chironomus tentans* (Insecta, Diptera). *Tissue Cell* **30**, 187–194.
Laufs, P., Coen, E., Kronenberger, J., Traas, J., and Doonan, J. (2003). Separable roles of UFO during floral development revealed by conditional restoration of gene function. *Development* **130**, 785–796.
Lezzi, M. (1996). Chromosome puffing: Supramolecular aspects of ecdysone action. *In* "Metamorphosis, Postembryonic Reprogramming of Gene Expression in Amphibian and Insect Cells" (B. G. Atkinson, Ed.), pp. 145–173. Academic Press, San Diego.
Lezzi, M. (2002). Dimerization interface (DIF-) model four nuclear receptor interaction. (Abstracts from the XV International Ecdysone Workshop). *J. Insect Sci.* **2**, 16.
Lezzi, M., Bergman, T., Henrich, V. C., Vögtli, M., Frömel, C., Grebe, M., Przibilla, S., and Spindler-Barth, M. (2002). Ligand-induced heterodimerization between the ligand binding domains of the *Drosophila* ecdysteroid receptor and ultraspiracle. *Eur. J. Biochem.* **269**, 3237–3245.
Lezzi, M., Imhof, M.O., Rusconi, S., Vögtli, M. (1993). Localization of ecdysteroid receptor of *Chironomus tentans* expressed in vertebrate cells. (Abstracts from the 6th International Balbiani Ring Workshop. Jackson/Lake Tiak-O'Khata, Mississippi, U.S.A.).
Magari, S. R., Rivera, V. M., Iuliucci, J. D., Gilman, M., and Cerasoli, F., Jr. (1997). Pharmacologic control of a humanized gene therapy system implanted into nude mice. *J. Clin. Invest.* **100**, 2865–2872.
Maki, A., Sawatsubashi, S., Ito, S., Shirode, Y., Suzuki, E., Zhao, Y., Yamagata, K., Kouzmenko, A., Takeyama, K., and Kato, S. (2004). Juvenile hormones antagonize ecdysone actions through co-repressor recruitment to EcR/USP heterodimers. *Biochem. Biophys. Res. Commun.* **320**, 262–267.
Mangelsdorf, D. J., and Evans, R. M. (1995). The RXR heterodimers and orphan receptors. *Cell* **83**, 841–850.
Mangelsdorf, D. J., Thummel, C., Beato, M., Herrlich, P., Schutz, G., Umesono, K., Blumberg, B., Kastner, P., Mark, M., and Chambon, P. (1995). The nuclear receptor superfamily: The second decade. *Cell* **83**, 835–839.
Martinez, A., Scanlon, D., Gross, B., Perara, S. C., Palli, S. R., Greenland, A. J., Windass, J., Pongs, O., Broad, P., and Jepson, I. (1999a). Transcriptional activation of the cloned *H. virescens* (Lepidoptera) ecdysone receptor (HvEcR) by muristerone A. *Insect Biochem. Mol. Biol.* **29**, 915–930.
Martinez, A., Sparks, C., Drayton, P., Thompson, J., Greenland, A., and Jepson, I. (1999b). Creation of ecdysone receptor chimeras in plants for controlled regulation of gene expression. *Mol. Gen. Genet.* **261**, 546–552.
Martinez, A., Sparks, C., Hart, C. A., Thompson, J., and Jepson, I. (1999c). Ecdysone agonist inducible transcription in transgenic tobacco plants. *Plant J.* **19**, 97–106.
Mazur, X., Fussenegger, M., Renner, W. A., and Bailey, J. E. (1998). Higher productivity of growth-arrested Chinese hamster ovary cells expressing the cyclin-dependent kinase inhibitor p27. *Biotechnol. Prog.* **14**, 705–713.
McNellis, T. W., Mudgett, M. B., Li, K., Aoyama, T., Horvath, D., Chua, N. H., and Staskawicz, B. J. (1998). Glucocorticoid-inducible expression of a bacterial avirulence gene in transgenic *Arabidopsis* induces hypersensitive cell death. *Plant J.* **14**, 247–257.

Miller, J. H., and Reznikoff, W. S. (1980). "The Operon." Cold Spring Harbor Laboratory Press, New York.
Minakuchi, C., Nakagawa, Y., Kiuchi, M., Tomita, S., and Kamimura, M. (2002). Molecular cloning, expression analysis and functional confirmation of two ecdysone receptor isoforms from the rice stem borer *Chilo suppressalis*. *Insect Biochem. Mol. Biol.* **32**, 999–1008.
Moser, S., Schlatter, S., Fux, C., Rimann, M., Bailey, J. E., and Fussenegger, M. (2000). An update of pTRIDENT multicistronic expression vectors: pTRIDENTs containing novel streptogramin-responsive promoters. *Biotechnol. Prog.* **16**, 724–735.
Mouillet, J. F., Delbecque, J. P., Quennedey, B., and Delachambre, J. (1997). Cloning of two putative ecdysteroid receptor isoforms from *Tenebrio molitor* and their developmental expression in the epidermis during metamorphosis. *Eur. J. Biochem.* **248**, 856–863.
Mouillet, J. F., Henrich, V. C., Lezzi, M., and Vögtli, M. (2001). Differential control of gene activity by isoforms A, B1 and B2 of the *Drosophila* ecdysone receptor. *Eur. J. Biochem.* **268**, 1811–1819.
Nijhout, H. F. (1994). "Insect Hormones." Princeton University Press, Princeton.
No, D., Yao, T. P., and Evans, R. M. (1996). Ecdysone-inducible gene expression in mammalian cells and transgenic mice. *Proc. Natl. Acad. Sci. USA* **93**, 3346–3351.
Orlowski, M., Szyszka, M., Kowalska, A., Grad, I., Zoglowek, A., Rymarczyk, G., Dobryszycki, P., Krowarsch, D., Rastinejad, F., Kochman, M., and Ozyhar, A. (2004). Plasticity of the ecdysone receptor DNA binding domain. *Mol. Endocrinol.* **18**, 2166–2184.
Ouwerkerk, P. B. F., de Kam, R. J., Hodge, J. H. C., and Meijer, A. H. (2001). Glucocorticoid-inducible gene expression in rice. *Planta* **213**, 370–378.
Padidam, M. (2003). Chemically regulated gene expression in plants. *Curr. Opin. Plant Biol.* **6**, 169–177.
Padidam, M., Gore, M., Lu, D. L., and Smirnova, O. (2003). Chemical-inducible, ecdysone receptor-based gene expression system for plants. *Transgenic Res.* **12**, 101–109.
Palli, S. R., Kapitskaya, M. Z., Kumar, M. B., and Cress, D. E. (2003). Improved ecdysone receptor-based inducible gene regulation system. *Eur. J. Biochem.* **270**, 1308–1315.
Palli, S. R., Margam, V. M., and Clark, A. M. (2005). Biochemical mode of action and differential activity of new ecdysone agonists against mosquitoes and moths. *Arch. Insect Biochem. Physiol.*, **58**, 234–242.
Palli, S. R., and Retnakaran, A. (2000). Ecdysteroid and juvenile hormone receptors: Properties and importance in developing novel insecticides. *In* "Biochemical Sites of Insecticide Action and Resistance" (I. Ishaaya, Ed.), pp. 107–132. Springer, Berlin.
Perera, S. C., Zheng, S., Feng, Q. L., Krell, P. J., Retnakaran, A., and Palli, S. R. (2005). Heterodimerization of ecdysone receptor and ultraspiracle on symmetric and asymmetric response elements. *Arch. Insect Biochem. Physiol.* **60**, 55–70.
Pissios, P., Tzameli, I., Kushner, P. J., and Moore, D. D. (2000). Dynamic stabilization of nuclear receptor ligand binding domains by hormone or corepressor binding. *Mol. Cell* **6**, 245–253.
Pogenberg, V., Guichon, J.-F., Vivat-Hannah, V., Kammere, S., Pérez, E., Germain, P., de Lera, A. R., Gronemeyer, H., Royer, C. A., and Bourguet, W. (2004). Characterization of the interaction between RAR/RXR heterodimers and transcription coactivators through structural and fluorescence anisotropy studies. *J. Biol. Chem.* **280**, 1625–1633.
Pratt, W. B., and Toft, D. O. (1997). Steroid receptor interactions with heat shock protein and immunophilin chaperones. *Endocr. Rev.* **18**, 306–360.
Privalsky, M. L. (2004). The role of corepressors in transcriptional regulation by nuclear hormone receptors. *Ann. Rev. Physiol.* **66**, 315–360.
Przibilla, S., Hitchcock, W. W., Szecsi, M., Grebe, M., Beatty, J., Henrich, V. C., and Spindler-Barth, M. (2004). Functional studies on the ligand-binding domain of Ultraspiracle from *Drosophila melanogaster*. *Biol. Chem.* **385**, 21–30.

Rauch, P., Grebe, M., Elke, C., Spindler, K. D., and Spindler-Barth, M. (1998). Ecdysteroid receptor and ultraspiracle from *Chironomus tentans* (Insecta) are phosphoproteins and are regulated differently by molting hormone. *Insect. Biochem. Mol. Biol.* **28**, 265–275.

Riddiford, L. M., Cherbas, P., and Truman, J. W. (2000). Ecdysone receptors and their biological actions. *Vitam. Horm.* **60**, 1–73.

Rinsch, C., Regulier, E., Deglon, N., Dalle, B., Beuzard, Y., and Aebischer, P. (1997). A gene therapy approach to regulated delivery of erythropoietin as a function of oxygen tension. *Hum. Gene Ther.* **8**, 1881–1889.

Rivera, V. M., Clackson, T., Natesan, S., Pollock, R., Amara, J. F., Keenan, T., Magari, S. R., Phillips, T., Courage, N. L., Cerasoli, F., Jr., Holt, D. A., and Gilman, M. (1996). A humanized system for pharmacologic control of gene expression. *Nat. Med.* **2**, 1028–1032.

Rivera, V. M., Ye, X., Courage, N. L., Sachar, J., Cerasoli, F., Jr., Wilson, J. M., and Gilman, M. (1999). Long-term regulated expression of growth hormone in mice after intramuscular gene transfer. *Proc. Natl. Acad. Sci. USA* **96**, 8657–8662.

Rivera, V. M., Wang, X., Wardwell, S., Courage, N. L., Volchuk, A., Keenan, T., Holt, D. A., Gilman, M., Orci, L., Cerasoli, F., Jr., Rothman, J. E., and Clackson, T. (2000). Regulation of protein secretion through controlled aggregation in the endoplasmic reticulum. *Science* **287**, 826–830.

Saez, E., Nelson, M. C., Eshelman, B., Banayo, E., Koder, A., Cho, G. J., and Evans, R. M. (2000). Identification of ligands and coligands for the ecdysone-regulated gene switch. *Proc. Natl. Acad. Sci. USA* **97**, 14512–14517.

Saleh, D. S., Zhang, J., Wyatt, G. R., and Walker, V. K. (1998). Cloning and characterization of an ecdysone receptor cDNA from *Locusta migratoria*. *Mol. Cell Endocrinol.* **143**, 91–99.

Sasorith, S., Billas, I. M., Iwema, T., Moras, D., and Wurtz, J. M. (2002). Structure-based analysis of the ultraspiracle protein and docking studies of putative ligands. *J. Insect Sci.* **2**, 25.

Sawatsubashi, S., Maki, A., Ito, S., Shirode, Y., Suzuki, E., Zhao, Y., Yamagata, K., Kouzmenko, A., Takeyama, K., and Kato, S. (2004). Ecdysone receptor-dependent gene regulation mediates histone poly(ADP-ribosyl)ation. *Biochem. Biophys. Res. Commun.* **320**, 268–272.

Schwimmer, L. J., Rohatgi, P., Azizi, B., Seley, K. L., and Doyle, D. F. (2004). Creation and discovery of ligand-receptor pairs for transcriptional control with small molecules. *Proc. Natl. Acad. Sci. USA* **101**, 14707–14712.

Sedkov, Y., Cho, E., Petruk, S., Cherbas, L., Smith, S. T., Jones, R. S., Cherbas, P., Canaani, E., Jaynes, J. B., and Mazo, A. (2003). Methylation at lysine 4 of histone H3 in ecdysone-dependent development of *Drosophila*. *Nature* **426**, 78–83.

Sekiguchi, T., and Hunter, T. (1998). Induction of growth arrest and cell death by over-expression of the cyclin-Cdk inhibitor p21 in hamster BHK21 cells. *Oncogene* **16**, 369–380.

Severin, K., and Schöffl, F. (1990). Heat-inducible hygromycin resistance in transgenic tobacco. *Plant Mol. Biol.* **15**, 827–833.

Siaussat, D., Bozzolan, F., Queguiner, I., Porcheron, P., and Debernard, S. (2004). Effects of juvenile hormone on 20-hydroxyecdysone-inducible EcR, HR3, E75 gene expression in imaginal wing cells of *Plodia interpunctella* (Lepidoptera). *Eur. J. Biochem.* **271**, 3017–3027.

Smith, H. C., Cavanaugh, C. K., Friz, J. L., Thompson, C. S., Saggers, J. A., Michelotti, E. L., Garcia, J., and Tice, C. M. (2003). Synthesis and SAR of cis-1-benzoyl-1,2,3,4-tetrahydroquinoline ligands for control of gene expression in ecdysone responsive systems. *Bioorg. Med. Chem. Lett.* **13**, 1943–1946.

Sobolev, V., Sorokine, A., Prilusky, J., Abola, E. E., and Edelman, M. (1999). Automated analysis of interatomic contacts in proteins. *Bioinformatics* **15**, 327–332.

Song, Q., and Gilbert, L. I. (1998). Alterations in ultraspiracle (USP) content and phosphorylation state accompany feedback regulation of ecdysone synthesis in the insect prothoracic gland. *Insect Biochem. Mol. Biol.* **28**, 849–860.

Spencer, D. M., Belshaw, P. J., Chen, L., Ho, S. N., Randazzo, F., Crabtree, G. R., and Schreiber, S. L. (1996). Functional analysis of Fas signaling *in vivo* using synthetic inducers of dimerization. *Curr. Biol.* **6**, 839–847.

Spencer, D. M., Wandless, T. J., Schreiber, S. L., and Crabtree, G. R. (1993). Controlling signal transduction with synthetic ligands. *Science* **262**, 1019–1024.

Spindler, K.-D. (1997). "Vergleichende Endokrinologie." Thieme Verlag, Stuttgart.

Stocker, A. J., Amabis, J. M., Gorab, E., Elke, C., and Lezzi, M. (1997). Antibodies against the D-domain of a *Chironomus* ecdysone receptor protein react with DNA puff sites in *Trichosia pubescens*. *Chromosoma* **106**, 456–464.

Suehara, K.-I., Takao, S., Nakamura, K., Uozumi, N., and Kobayashi, T. (1996). Optimal expression of GUS gene from methyl jasmonate-inducible promoter in high density cultures of transformed tobaco cell line By-2. *J. Ferm Bioeng.* **82**, 51–55.

Suhr, S. T., Gil, E. B., Senut, M. C., and Gage, F. H. (1998). High level transactivation by a modified *Bombyx* ecdysone receptor in mammalian cells without exogenous retinoid X receptor. *Proc. Natl. Acad. Sci. USA* **95**, 7999–8004.

Swevers, L., Drevet, J. R., Lunke, M. D., and Iatrou, K. (1995). The silkmoth homolog of the *Drosophila* ecdysone receptor (B1 isoform): Cloning and analysis of expression during follicular cell differentiation. *Insect Biochem. Mol. Biol.* **25**, 857–866.

Talbot, W. S., Swyryd, E. A., and Hogness, D. S. (1993). *Drosophila* tissues with different metamorphic responses to ecdysone express different ecdysone receptor isoforms. *Cell* **73**, 1323–1337.

Tang, W., Luo, X., and Samuels, V. (2004). Regulated gene expression with promoters responding to inducers. *Plant Sci.* **166**, 827–834.

Tang, W., and Newton, R. J. (2004a). Glucocorticoid-inducible transgene expression in loblolly pine (*Pinus taeda* L.) cell suspension cultures. *Plant Sci.* **166**, 1351–1358.

Tang, W., and Newton, R. J. (2004b). Regulated gene expression by glucocorticoids in cultured Virginia pine (*Pinus virginiana* Mill.) cells. *J. Exp. Bot.* **55**, 1499–1508.

Thomas, H. E., Stunnenberg, H. G., and Stewart, A. F. (1993). Heterodimerization of the *Drosophila* ecdysone receptor with retinoid X receptor and ultraspiracle. *Nature* **362**, 471–475.

Tice, C. M., Hormann, R. E., Thompson, C. S., Friz, J. L., Cavanaugh, C. K., Michelotti, E. L., Garcia, J., Nicolas, E., and Albericio, F. (2003a). Synthesis and SAR of alpha-acylaminoketone ligands for control of gene expression. *Bioorg. Med. Chem. Lett.* **13**, 475–478.

Tice, C. M., Hormann, R. E., Thompson, C. S., Friz, J. L., Cavanaugh, C. K., and Saggers, J. A. (2003b). Optimization of alpha-acylaminoketone ecdysone agonists for control of gene expression. *Bioorg. Med. Chem. Lett.* **13**, 1883–1886.

Tomaschko, K.-H. (1999). Nongenomic effects of ecdysteroids. *Arch. Insect Biochem. Physiol.* **41**, 89–98.

Tran, H. T., Askari, H. B., Shaaban, S., Price, L., Palli, S. R., Dhadialla, T. S., Carlson, G. R., and Butt, T. R. (2001a). Reconstruction of ligand-dependent transactivation of *Choristoneura fumiferana* ecdysone receptor in yeast. *Mol. Endocrinol.* **15**, 1140–1153.

Tran, H. T., Shaaban, S., Askari, H. B., Walfish, P. G., Raikhel, A. S., and Butt, T. R. (2001b). Requirement of co-factors for the ligand-mediated activity of the insect ecdysteroid receptor in yeast. *J. Mol. Endocrinol.* **27**, 191–209.

Tsai, C. C., Kao, H. Y., Yao, T. P., McKeown, M., and Evans, R. M. (1999). SMRTER, a *Drosophila* nuclear receptor coregulator, reveals that EcR-mediated repression is critical for development. *Mol. Cell* **4**, 175–186.

Umana, P., Jean-Mairet, J., Moudry, R., Amstutz, H., and Bailey, J. E. (1999). Engineered glycoforms of an antineuroblastoma IgG1 with optimized antibody-dependent cellular cytotoxic activity. *Nat. Biotechnol.* **17**, 176–180.

Unger, E., Cigan, A. M., Trimnell, M., Xu, R. J., Kendall, T., Roth, B., and Albertsen, M. (2002). A chimeric ecdysone receptor facilitates methoxyfenozide-dependent restoration of male fertility in ms45 maize. *Transgenic Res.* **11,** 455–465.

Vafopoulou, X., and Steel, C. G. H. (2001). Induction of rhythmicity in prothoracicotropic hormone and ecdysteroids in *Rhodnius prolixus*: Roles of photic and neuroendocrine Zeitgebers. *J. Insect Physiol.* **47,** 935–941.

Vafopoulou, X., and Steel, C. G. H. (2005). Ecdysteroid hormone nuclear receptor (EcR) exhibits circadian cycling in certain tissues, but not others, during development in *Rhodnius prolixus* (Hemiptera). *Cell Tissue Res.,* (in press).

Verras, M., Mavroidis, M., Kokolakis, G., Gourzi, P., Zacharopoulou, A., and Mintzas, A. C. (1999). Cloning and characterization of CcEcR. An ecdysone receptor homolog from the mediterranean fruit fly *Ceratitis capitata*. *Eur. J. Biochem.* **265,** 798–808.

Vögtli, M., Elke, C., Imhof, M. O., and Lezzi, M. (1998). High level transactivation by the ecdysone receptor complex at the core recognition motif. *Nucleic Acids Res.* **26,** 2407–2414.

Wang, S. F., Ayer, S., Segraves, W. A., Williams, D. R., and Raikhel, A. S. (2000). Molecular determinants of differential ligand sensitivities of insect ecdysteroid receptors. *Mol. Cell. Biol.* **20,** 3870–3879.

Wang, Y., DeMayo, F. J., Tsai, S. Y., and O'Malley, B. W. (1997). Ligand-inducible and liver-specific target gene expression in transgenic mice. *Nat. Biotechnol.* **15,** 239–243.

Wang, Y., O'Malley, B. W., Jr., Tsai, S. Y., and O' Malley, B. W. (1994). A regulatory system for use in gene transfer. *Proc. Natl. Acad. Sci. USA* **91,** 8180–8184.

Weber, W., and Fussenegger, M. (2002). Artificial mammalian gene regulation networks: Novel approaches for gene therapy and bioengineering. *J. Biotechnol.* **98,** 161–187.

Weber, W., Fux, C., Daoud-el Baba, M., Keller, B., Weber, C. C., Kramer, B. P., Heinzen, C., Aubel, D., Bailey, J. E., and Fussenegger, M. (2002). Macrolide-based transgene control in mammalian cells and mice. *Nat. Biotechnol.* **20,** 901–907.

Wegmann, I. S., Quack, S., Spindler, K. D., Dorsch-Häsler, K., Vögtli, M., and Lezzi, M. (1995). Immunological studies on the developmental and chromosomal distribution of ecdysteroid receptor protein in *Chironomus tentans*. *Arch. Insect Biochem. Physiol.* **30,** 95–114.

Weinmann, P., Gossen, M., Hillen, W., Bujard, H., and Gatz, C. (1994). A chimeric transactivator allows tetracycline-responsive gene expression in whole plants. *Plant J.* **5,** 559–569.

Wing, K. D. (1988). RH 5849, a nonsteroidal ecdysone agonist: Effects on a *Drosophila* cell line. *Science* **241,** 467–469.

Wozniak, M., Chu, Y., Fang, F., Xu, Y., Riddiford, L., Jones, D., and Jones, G. (2004). Alternative farnesoid structures induce different conformational outcomes upon the *Drosophila* ortholog of the retinoid X receptor, ultraspiracle. *Insect Biochem. Mol. Biol.* **34,** 1147–1162.

Wurm, F. M., Gwinn, K. A., and Kingston, R. E. (1986). Inducible overproduction of the mouse c-myc protein in mammalian cells. *Proc. Natl. Acad. Sci. USA* **83,** 5414–5418.

Wurtz, J. M., Bourguet, W., Renaud, J. P., Vivat, V., Chambon, P., Moras, D., and Gronemeyer, H. (1996). A canonical structure for the ligand-binding domain of nuclear receptors. *Nat. Struct. Biol.* **3,** 206.

Wyborski, D. L., Bauer, J. C., and Vaillancourt, P. (2001). Bicistronic expression of ecdysone-inducible receptors in mammalian cells. *Biotechniques* **31,** 618–620, 622, 624.

Yang, G., Hannan, G., Lockett, T., and Hill, R. (1995). Functional transfer of an elementary ecdysone gene regulatory system to mammalian cells-transient transfections and stable cell-lines. *Eur. J. Entomol.* **92,** 379–389.

Yao, T. P., Forman, B. M., Jiang, Z., Cherbas, L., Chen, J. D., McKeown, M., Cherbas, P., and Evans, R. M. (1993). Functional ecdysone receptor is the product of EcR and Ultraspiracle genes. *Nature* **366,** 476–479.

Yao, T. P., Segraves, W. A., Oro, A. E., McKeown, M., and Evans, R. M. (1992). *Drosophila* ultraspiracle modulates ecdysone receptor function via heterodimer formation. *Cell* **71,** 63–72.

Yoshizumi, T., Nagata, N., Shimada, H., and Matsui, M. (1999). An *Arabidopsis* cell cycle-dependent kinase-related gene, CDC2b, plays a role in regulating seedling growth in darkness. *Plant Cell* **11,** 1883–1896.

Zhang, X. K., Salbert, G., Lee, M. O., and Pfahl, M. (1994). Mutations that alter ligand-induced switches and dimerization activities in the retinoid X receptor. *Mol. Cell Biol.* **14,** 4311–4323.

Zuo, J., and Chua, N. H. (2000). Chemical-inducible systems for regulated expression of plant genes. *Curr. Opin. Biotechnol.* **11,** 146–151.

Zuo, J., Niu, Q. W., and Chua, N. H. (2000). Technical advance: An estrogen receptor-based transactivator XVE mediates highly inducible gene expression in transgenic plants. *Plant J.* **24,** 265–273.

Zuo, J., Niu, Q. W., Moller, S. G., and Chua, N. H. (2001). Chemical-regulated, site-specific DNA excision in transgenic plants. *Nat. Biotechnol.* **19,** 157–161.

4

LIGAND-BINDING POCKET OF THE ECDYSONE RECEPTOR

ISABELLE M. L. BILLAS AND DINO MORAS

*IGBMC, Laboratoire de génomique et Biologie Structurales, CNRS/INSERM/
Université Louis Pasteur, Parc d'Innovation BP10142
67404 Illkirch cedex, France*

I. Introduction
II. Sequence Comparison of Invertebrate EcR
   Ligand-Binding Domains
III. Crystal Structures of the EcR
   Ligand-Binding Domain
   A. Introduction
   B. Dimeric Arrangement of the EcR-LBD/USP-LBD
      Complexes
   C. The EcR-LBD Structure
   D. Adaptability in the Ligand-Binding Pocket of EcR
   E. Comparison with other NR LBD Structures
   F. Validation of the Structural Data by In Vitro
      Assays
IV. Conclusions
   References

The ecdysone receptor (EcR) belongs to the superfamily of nuclear receptors (NRs) that are ligand-dependent transcription factors. Ecdysone receptor is present only in invertebrates and plays a central role in regulating the expression of a vast array of genes during development and reproduction. The functional entity is a heterodimer composed of EcR and the ultraspiracle protein (USP)—the orthologue of the vertebrate retinoid X receptor (RXR). Ecdysone receptor is the molecular target of ecdysteroids—the endogenous steroidal molting hormones found in arthropods and nonarthropod invertebrates. In addition, EcR is the target of the environmentally safe bisacylhydrazine insecticides used against pests, such as caterpillars, that cause severe damage to agriculture. The crystal structures of the ligand-binding domains (LBDs) of the EcR/USP heterodimer, complexed to the ecdysteroid ponasterone A (ponA) and to the lepidopteran specific bisacylhydrazine BYI06830 used in the agrochemical pest control, provide the first insight at atomic level for these important functional complexes. The EcR/USP heterodimer has a shape similar to that seen for the known vertebrate heterodimer complexes with a conserved main interface, but with features, that are specific to this invertebrate heterodimer. The two EcR-LBD structures in complex with steroidal and nonsteroidal ligands reveal substantial differences. The adaptability of EcR to its ligand results in two radically different and only partially overlapping ligand-binding pockets with different residues involved in ligand recognition. The concept brought by these structural studies of a ligand-dependent binding pocket has potential applications for other NRs. © 2005 Elsevier Inc.

I. INTRODUCTION

Nuclear receptors form a superfamily of allosteric transcription factors that regulate transcription in response to extracellular signals like hormones (Gronemeyer et al., 2004; Moras and Gronemeyer, 1998). In the human genome, 48 members of this family were identified, which represent major targets for drug discovery. In contrast for invertebrates, only 21 NRs were found in the *Drosophila* genome sequence (Maglich et al., 2001). Among these, the EcR (Koelle et al., 1991) is the only one for which ligands have been identified. Nuclear receptors share the same modular domain structure, comprising a well-conserved DNA-binding domain (DBD) and a moderately-conserved LBD. The LBD mediates transcriptional repression, ligand-dependent activation (AF2), and homodimerization or heterodimerization. The AF2 is harbored in helix 12 at the carboxy-terminus end of the receptor. The AF2-mediated transcriptional activation requires a ligand-dependent repositioning of H12. The binding of an agonist ligand leads to the canonical active conformation allowing the formation of a hydrophobic cleft at the receptor surface to which co-activators can bind.

The ecdysone receptor that is absent from vertebrates serves as the key regulator involved in the development of most invertebrates (Riddiford et al., 2001). EcR is activated by ecdysteroids and in particular by 20-hydroxyecdysone (20E) the most representative active metabolite (Fig. 1). Ecdysteroids have been recognized as a class of steroidal molting hormones in arthropods and nonarthropod invertebrates (Kumar et al., 2002). These

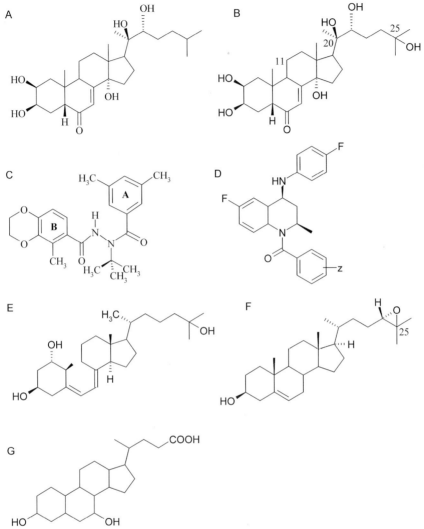

FIGURE 1. Schematic representation of NRs ligands. (A) ponasterone A (ponA), (B) 20-hydroxyecdysone (20E), (C) the dibenzoylhydrazine BYI06830, and (D) a tetrahydroquinoline with Z being a substituent, such as 3-F-4-CH$_3$ or 4-CH$_2$CH$_3$ (Kumar et al., 2004). (E) 1α, 25-Dihydroxyvitamin D$_3$ (vitamin D), (F) 24(S), 25-epoxycholesterol (eCH), and (G) 6α-ethylchenodeoxycholic acid (6ECDCA).

steroid hormones, together with juvenile hormones, regulate the major insect developmental transitions, notably, molting and metamorphosis, as well as cell differentiation and reproduction. This contrasts with the diversity of vertebrate hormones, which control a large variety of gene-regulatory pathways. Surprisingly, EcR binds its natural ligand 20E weakly and is by itself incapable of high-affinity DNA binding and transcriptional activation. In fact, the action of 20E relies upon the formation of a heterodimeric nuclear receptor complex composed of EcR and USP, the insect orthologue of the vertebrate RXR.

Ecdysone agonists are employed in several areas, as diverse as the regulation of gene expression or in agrochemical pest control. Ecdysteroids have the properties of being neither toxic nor teratogenic for vertebrates. These characteristics stimulated the development of gene switch systems based on EcR for the control of the expression of foreign genes with various applications in functional genomics or gene therapy (Christopherson *et al.*, 1992; No *et al.*, 1996; Palli *et al.*, 2003; Saez *et al.*, 2000; Suhr *et al.*, 1998). Besides, as the main regulator of gene expression in insects, EcR has been the obvious target for the development of new, environmentally safe insecticides against pest species, which cause billions of euros damage worldwide to agriculture every year. The first synthetic agonists were discovered by classical screening. The members of this class of compounds, the bisacylhydrazines, were reported to induce premature and lethal molting in different types of caterpillar pest larvae (Dhadialla *et al.*, 1998). Another class of nonsteroidal agonists was developed, known as the tetrahydroquinoline (THQ) ligand class (Smith *et al.*, 2003; Tice *et al.*, 2003a,b) (Fig. 1). These compounds show a weak activity against the lepidopteran *Heliothis virescens*, but bind to the *Aedes aegypti* EcR. One puzzling aspect of bisacylhydrazines (and THQ compounds) is the lack of structural similarity with ecdysteroids. Several attempts to understand the binding of bisacylhydrazines to EcR were reported (Kasuya *et al.*, 2003; Wurtz *et al.*, 2000). However, these studies could not satisfactorily explain the selectivity of these compounds for their cognate receptor. The explanation came with the publication of the crystallographic structures of the EcR in complex with ponA, an ecdysteroid, and with a bisacylhydrazine compound BYI06830 (Billas *et al.*, 2003) (Fig. 1). In this review, these two structures will be compared and discussed in terms of ligand interactions and specificity.

II. SEQUENCE COMPARISON OF INVERTEBRATE ECR LIGAND-BINDING DOMAINS

Figure 2 shows an alignment of EcR-LBD from different invertebrates, including a crab, a tick, and 17 insects spanning 4 distinct insect orders (one Coleoptera, two Orthoptera, eight Diptera, and seven Lepidoptera). The EcR-LBD sequences are well conserved across the different orders with more than 61% amino acid identities. This conservation is even higher for species within a given order (84% identity for Diptera, except for ctEcR where it is slightly lower (73%) and over 80% for Lepidoptera). The LBDs of the orthoptera EcR (lmEcR) and the coleopteran EcR (tmEcR) share about

88% amino acids in common, being closer to the dipteran (70% identity) than to the lepidopteran (60%). The 12 helices (H1–H12) composing the three layer α-helical sandwich fold are indicated according to the crystal structure of the EcR-ponA complex. Strong amino acid conservation is observed in the NR signature region encompassing the loop between H3 and H4 and characterized by the F(W)AKxxxxFxxLxxxDQxxLL motif (Wurtz *et al.*, 1996) and for residues belonging to helices H3, H4, H5, H8, and H12. A structure-based sequence analysis of all NR sequences known thus far partitions the NR superfamily into two classes (Brelivet *et al.*, 2004). The EcR belongs to class II characterized by NRs that form heterodimers. The signature motif of this class is a salt bridge between the differentially conserved arginine residue in loop H8–H9 (R449 for hvEcR) and the acidic residue at the H4–H5 kink (E378 for hvEcR). This salt bridge is present in the crystal structures of hvEcR-LBD. Furthermore, we notice the conservation of the glutamate in H12 (E524 in hvEcR) and the lysine residue in H4 (K362 in hvEcR) that form the charge clamp involved in co-factor binding (Darimont *et al.*, 1998; McInerney *et al.*, 1998; Nolte *et al.*, 1998). This suggests a similar co-factor binding mechanism between vertebrates and invertebrates.

III. CRYSTAL STRUCTURES OF THE ECR LIGAND-BINDING DOMAIN

A. INTRODUCTION

Our structural studies focused on the LBDs of the *H. virescens* EcR/USP heterodimer in complex with the ecdysteroid ponA or the nonsteroidal synthetic agonist BYI06830 (Billas *et al.*, 2003). The BYI06830 ligand is a lepidopteran-specific bisacylhydrazine compound (Fig. 1), similar to chromafenozide (Toya *et al.*, 2002). The crystal structure of the wild type EcR-LBD in complex with ponA was solved at 2.9 Å resolution. On the other hand, several mutants were also considered for their increased solubility. The mutations involve nonconserved hydrophobic surface residues (W303Y, A361S, L456S, and C483S), and do not affect the overall structure of the receptor as demonstrated by cotransfection assays and mass spectroscopic analysis (Fig. 3). Crystals of the mutant W303Y-A361S-L456S-C483S EcR-LBD in complex with BYI06830 were obtained and the corresponding structure was refined at 3.0 Å resolution (Billas *et al.*, 2003).

B. DIMERIC ARRANGEMENT OF THE ECR-LBD/USP-LBD COMPLEXES

The global structures of both EcR/USP heterodimers (i.e., when bound to ponA or to BYI06830) are comparable (Fig. 4A to D). The overall arrangement of the EcR/USP-LBD heterodimer is shaped like a butterfly when

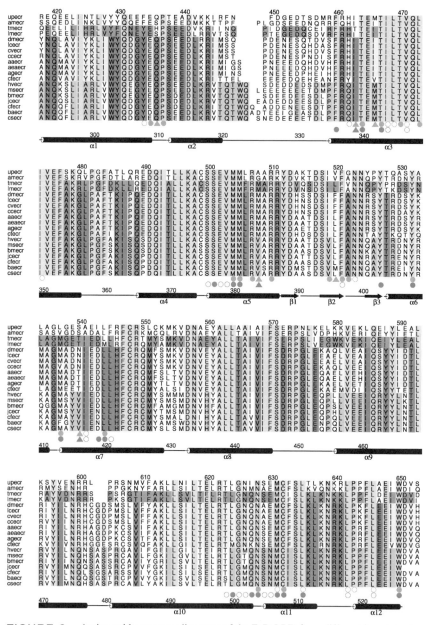

FIGURE 2. Amino acid sequence alignment of the EcR-LBD from different invertebrates. Shown are the EcR-LBD sequences of a crab (up: *Uca pugilator*), a tick (am: *Amblyomma americanu*), a beetle (tm: *Tenebrio molitor*), a grasshopper (lm: *Locusta migratoria*), and dipteran insects (dm: *Drosophila melanogaster*; lc: *Lucilia cuprina*, cv: *Calliphora vicina*, cc: *Ceratitis capitata*, aa: *A. aegypti*, aea: *Aedes albopictus*, ag: *Anopheles gambiae*, ct: *Chironomus tentans*) and

viewed from the front. This shape is similar to that seen for the vertebrate heterodimer complexes RARα/RXRα (Bourguet et al., 2000), PPARγ/RXRα (Gampe et al., 2000), LXRα/RXRβ (Svensson et al., 2003), and CAR/RXRα (Suino et al., 2004; Xu et al., 2004). Each LBD adopts the general NR fold consisting of a three-layer antiparallel α-helical sandwich and a β-sheet. Ponasterone A or BYI06830 and a phospholipid occupy the ligand-binding pockets of EcR and USP, respectively. The structure of USP-LBD in the heterodimeric complexes is virtually identical to those determined previously for the isolated USP-LBDs (Billas et al., 2001; Clayton et al., 2001). H12 of hvUSP-LBD in both complexes is in a so-called antagonist conformation. In contrast, EcR is in the active conformation with the AF2 helix packing against the main body of the LBD. These two EcR/USP heterodimeric complexes are thus the first crystal structures of nuclear receptor heterodimers with a dissymmetry in the H12 conformations. A feature characterizing vertebrate NR heterodimers is the asymmetrical interface, with the RXR partner rotated by ∼10° from the C2 symmetry axis of the RXR LBD (Bourguet et al., 2000; Gampe et al., 2000; Suino et al., 2004; Svensson et al., 2003; Xu et al., 2004). One of the consequences is the dissymmetrical distances separating H7 and the H8–H9 loop of the partner, with a clearly smaller distance between H7 of RXR and the H8–H9 loop of the dimerization partner than between the H8–H9 loop of RXR and H7 of the partner. In the case of EcR/USP, the asymmetry is less pronounced (Fig. 4B and D). As a matter of fact, the H8–H9 loop of USP moves by ∼3.5 Å towards EcR with respect to its position in the vertebrate heterodimers. As a consequence, the distances between H7 and the H8–H9 loop of the partner are very similar, and the EcR/USP heterodimer superposes well to the estrogen receptor (ER) homodimer. The EcR/USP heterodimer interface is composed of an intricate network of hydrophobic and polar interactions mediated by helices H7, H9, H10, and the loop between H8 and H9 of both receptors (Table I). In hvUSP a novel feature concerns the insertion loop between H5 and the β-sheet, which is partially stabilized. The electron density suggests possible contacts with the loop H8–H9 of the EcR-LBD

---

lepidopteran insects (ms: *Manduca sexta*, bm: *Bombyx mori*, cf: *C. fumiferana*, ba: *Bicyclus anynana*, jc: *Junonia coenia*, sf: *Spodoptera frugiperda*, hv: *H. virescens*, and cs: *Chilo suppressalis*). Residue numbering corresponding to the sequences of hvEcR and dmEcR are given at the bottom and at the top of the alignment, respectively. The secondary structure elements (α-helices and β-sheets) are indicated at the bottom of the alignment. Color code of the amino acid conservation: light blue: 100% conservation for all sequences; dark blue: 100% conservation for all insects; yellow: 100% conservation for dipteran insects; pink: 100% conservation for lepidopteran insects; grey: 100% conservation for the coleopteran and orthopteran insects. The residues of the ligand binding pocket at less than 4.5 and 6.5 Å from ponA are shown by red full and empty dots, respectively. Residues at less than 4.5 and 6.5 Å from BYI06830 are shown by blue full and empty dots, respectively. Colored arrows indicate residues located at less than 4.5 Å that are differentially conserved in the lepidopteran insect order. (See Color Insert.)

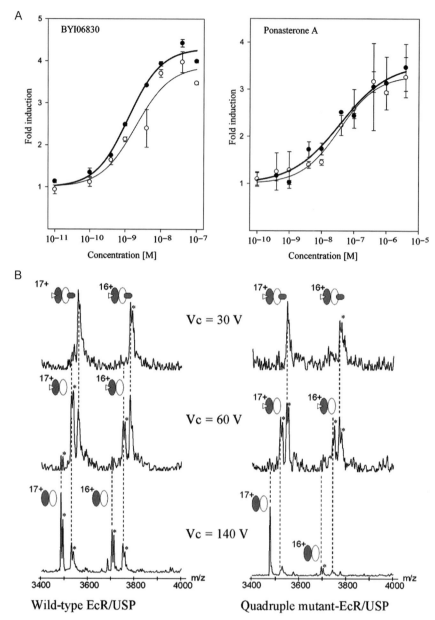

FIGURE 3. Four solubilizing point mutations (W303Y, A361S, L456S, and C483S) do not affect the function of the hvEcR-LBD. (A) Transcriptional activities of wild-type and quadruple mutant hvEcR in transient transfection assays. HEK-293 EBNA cells were transfected in 24-well plates with 250 ng per well of PAL1(5x)-TK-LUC reporter and 25 ng of wild type or mutant hvEcR and 25 ng hvUSP in pSG5 expression plasmids using standard calcium phosphate coprecipitation technique. Experiments were performed in the absence or presence of

partner. Since the H8–H9 loop represents an essential element of the heterodimer interface, the USP specific H5-β-sheet insertion might play an additional, yet unknown regulatory role.

C. THE ECR-LBD STRUCTURE

For the ponA-bound EcR-USP complex, the EcR-LBD is composed of 12 α-helices and three small β-strands. These secondary structural elements together with the hvEcR sequence are indicated in the EcR sequence alignment (Fig. 2). For the BYI06830-bound EcR-USP complex, marked structural differences are observed, which mostly concern the region composed of H6, H7, the β-sheet, and the loop between H1 and H3 (Fig. 5A). The N-terminal part of H7 is shifted by $\sim$2.5 Å compared to the ponA complex, being away to the USP partner H10–H11 helices. Helix H6 moves substantially concomitantly with the displacement of the amino-terminus of H7. Furthermore, the β-sheet observed for the ponA complex is drastically affected due to the disruption of interactions between the second and third strands. As a result, the three-stranded β-sheet observed for the ponA-bound EcR-LBD is replaced by a two-stranded β-sheet and a loop. Differences between the two complexes are best visualized in a surface representation of the receptor LBD colored according to the root mean square (rms) deviations between the ponA and the BYI06830 EcR-LBDs (Fig. 5B). These views show that the side of the receptor that comprises most of the dimerization interface as well as the AF2 helix remains identical, while the other side encompassing H1, H3, H6, H7, and the β-sheet differs more or less severely between the two EcR-LBD structures. Interestingly, this region is close to the C-terminus of H1, which was shown to be implicated in co-factor binding (Horlein *et al.*, 1995).

D. ADAPTABILITY IN THE LIGAND-BINDING POCKET OF ECR

Along with the differences observed in the receptor scaffold, the EcR-LBD in complex with the steroidal and nonsteroidal agonists exhibit different and only partially overlapping ligand-binding pockets (Fig. 6A). The EcR-LBD in complex with ponA possesses a long and thin L-shaped cavity extending towards H5 and the β-sheet, which is completely buried inside the

---

different concentrations of ponA or BYI06830. Data points represent the mean of assays performed in duplicate for at least three independent experiments. (B) Binding of ponA by wild-type and quadruple mutant EcR-LBDs as demonstrated by electrospray ionization mass-spectrometric measurements. Under nondenaturing conditions, noncovalent binding of ponA to EcR-LBD and of the phospholipids to USP-LBD is observed at low voltage (Vc = 30 V). Increasing the kinetic energy of the ions to Vc = 60 V led to the dissociation of ponA from EcR-LBD and at Vc = 140 V, the phospholipid dissociates from USP-LBD.

TABLE I. Interactions Between EcR and USP

Polar interactions		Nonpolar interactions	
EcR	USP	EcR	USP
(H7) D419	(L8–9) K380	(H7) H422	(L8–9) K380
(H7) H422	(L8–9) D378	(H7) C426	(L8–9) D378
(H7) S429	(L8–9) R385	(H7) C426	(L8–9) R385
(H9) E459	(H7) K355	(L8–9) P450	(H7) R347
(H9) R463	(L9–10) E411	(L8–9) P450	(H7) E351
(H10) K489	(H9) E389	(H9) Q462	(H10) L419
(H10) K489	(H9) E393	(H9) L466	(H10) A415
(H10) T495	(H10) R425	(H9) L466	(H10) L419
(H10) E496	(H10) R425	(H9) N467	(L9–10) E411
(H10) R498	(H10) S426	(H9) R470	(L9–10) E411
(H10) T499	(H10) R425	(H10) P481	(H9) L397
		(H10) A484	(H9) D400
		(H10) V485	(H9) E393
		(H10) V485	(H9) L397
		(H10) F487	(H10) F414
		(H10) G488	(H9) F396
		(H10) K489	(H9) E393
		(H10) K489	(H9) F396
		(H10) L491	(H10) A415
		(H10) L494	(H10) L419
		(H10) L494	(H10) P422
		(H10) T495	(H10) L421
		(H10) T495	(H10) P422
		(H10) E496	(L8–9) D378
		(H10) R498	(H10) P422
		(H10) R498	(H10) A423
		(H10) R498	(H10) R425
		(H10) R498	(H10) L429
		(H10) T499	(H10) L429
		(H10) M502	(H10) S426
		(H10) M502	(H10) L429
		(H10) M502	(H10) L430

Note: Residues involved in dimerization are listed along with the secondary structures. The intermolecular interactions are grouped into polar and nonpolar interactions (4.0 Å cutoff).

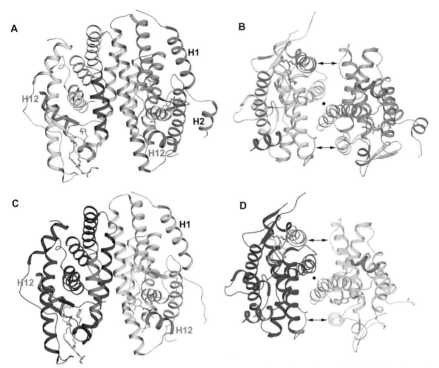

FIGURE 4. The two structures of the ligand-binding domains of EcR/USP. Overall structure of the EcR/USP heterodimer complexes in complex with the steroid hormone ponA (A and B) and with the nonsteroidal ligand BYI06830 (C and D). Figures B and D are 90° rotated views with the pseudo two-fold axis perpendicular to the plane of the sheet. USP-LBD is shown in cyan (A and B) and in blue (C and D). EcR-LBD is shown in orange (A and B) and in green (C and D). Helices H12 of both partners are shown as red ribbons. Helix H7 and the loop H8–H9 of USP are shown in magenta, helix H7 and the loop H8–H9 of EcR is shown in green for the ponA complex and in yellow for the BYI06830 complex. The EcR ligands and the phospholipid in USP are shown in stick representation with carbon colored in white, oxygen in red and nitrogen in blue.

receptor. In contrast, the ligand-binding pocket of the BYI06820-bound EcR-LBD consists of a rather bulky V-shaped cavity located close to H7, H11, and H12 with an open cleft between H7 and H10. This opening extends towards the H8–H9 loop of the heterodimer partner USP. The superimposition of the steroidal and nonsteroidal ligands as found in the EcR cavities indicates that the *tert*-butyl group together with the benzoyl A-ring of BYI06830 superimpose to the hydroxylated side chain of ponA at the C17-position (Fig. 6B). This superimposition contrasts with predictions of modeling studies, which considered a single binding niche (Kasuya *et al.*, 2003; Wurtz *et al.*, 2000). The accessible volumes of the ponA-bound and of the BYI06830-bound EcR cavity are 573 and 519 Å^3, respectively. This

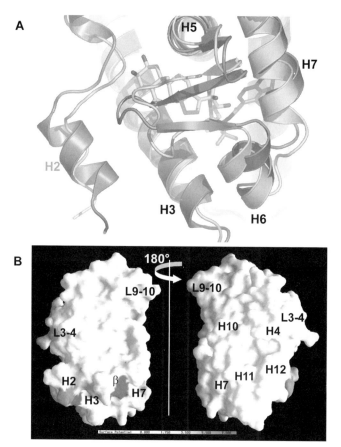

FIGURE 5. Flexible region in the EcR-LBD. (A) Ribbon diagram showing the superimposition of the structures of EcR-LBD in complex with ponA (orange ribbons) and with BYI06830 (green ribbons). The view is restricted to the regions differing the most between the two EcR-LBDs that includes H2, H6, H7, and the β-sheet. PonA and BYI06830 are shown in stick representation with carbon colored in cyan and light grey, respectively, oxygen in red and nitrogen in blue. (B) Surface representations of the ponA-bound EcR-LBD colored according to the root mean square (r.m.s.) deviations calculated residue by residue between the ponA- and the BYI06830-bound EcR-LBDs. The scale ranges from 0 (white) to 7.5 Å (red). The two views are related by a 180° rotation around the vertical axis. (See Color Insert.)

corresponds to occupancy of the cavity by ponA of ~73%, a value higher than typical values for other high affinity NRs (between 63 and 67%). On the other hand, the nonsteroidal ligand fills ~69% of the cavity.

1. The Ecdysteroid EcR Pocket

The binding mode of ponA is unambiguously determined from the well-defined electron density map (Fig. 7A). The steroid core of ponA adopts a chair conformation similar to that determined experimentally for crystals of

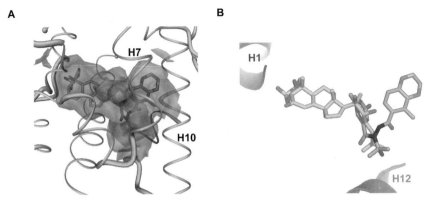

FIGURE 6. The LBDs of EcR complexed to a steroidal and to a nonsteroidal ligands exhibit different, and only partially overlapping ligand-binding cavities. (A) View of the two ligand-binding cavities of EcR-LBD, together with their respective ligand. The ponA-bound EcR cavity is shown in blue and the BYI06830-bound EcR cavity in green. (B) Superimposition of the steroidal and nonsteroidal EcR ligands bound to the EcR-LBD. Atom coloring is red for oxygen, blue for nitrogen, cyan for carbon in ponA and orange for carbon in BYI06830. (See Color Insert.)

the free 20E molecule with the 2-OH and the 3-OH groups in equatorial and axial orientations, respectively (Fabian *et al.*, 2002). The ponA is bound with the steroid A-ring oriented towards H1 and H2 and with the D-ring and the alkyl chain oriented towards the N-ter of H3 and H11 (Fig. 7B). The steroid skeleton makes hydrophobic contacts to Q310, P311, I339, T340, M342, L349, M380, M381, V384, R387, V395, and L396 and the alkyl chain makes van der Waals interactions to M413, V416, L420, N504, C508, L511, and W526. The three hydroxyl groups of the steroid core of ponA are hydrogen-bonded to EcR residues. The 2-βOH group makes a hydrogen bond to the R383 guanidinium moiety, the 3-βOH is hydrogen bonded to the backbone carbonyl group of E309 and the 14-αOH group is H-bonded to the side chains of T343 and T346. In addition, the C6 ketone moiety is H-bonded to the A398 amide group. For ponA, two additional hydroxyl groups, 20-OH and 22-OH, are present on the alkyl chain. However, only the 20-OH group is H-bonded to EcR through interaction with the side chain of Y408, the other hydroxyl group being in van der Waals contacts with V416 and Y408 and to a lesser extend with M413, L420, and N504.

The EcR residues participating in the binding of the steroid ligand are well conserved among invertebrate species (Fig. 2). In fact, only 5 out of 25 residues that are at less than 4.5 Å from the ligand differ among the different organisms. Residues whose side chains participate in H-bonds with the ligand (R383, T343, T346, and Y408) are strictly conserved for all EcRs. Of particular interest is Y408, which is H-bonded to the 20-OH group of ponA. In fact, ponA only differs from 20E by the lack of the hydroxyl group

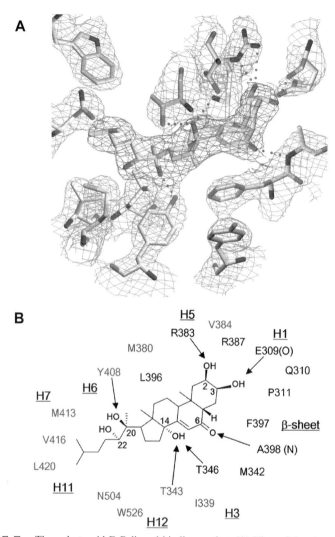

FIGURE 7. The ecdysteroid EcR ligand binding pocket. (A) View of the electron density for the ponA-bound EcR-LBD at 2.9 Å resolution. Two different maps are shown: a sigmaA weighted $2F_{obs} - F_{calc}$ omit map for the ligand (in blue) and a sigma A weighted $2F_{obs} - F_{calc}$ map (in magenta) for selected residues in the ligand-binding pocket. The map is contoured at 1σ and overlaid on the final refined models. Hydrogen bonds between ligand and residues are indicated by green dotted lines. (B) Schematic representation of the interactions of ponA with residues of the binding cavity. Arrows correspond to hydrogen bonds between ligand and amino acid residues. Residues in blue are common to both structures. (See Color Insert.)

at the C25 position. 20-Hydroxyecdysone that is the primary regulatory hormone of invertebrates is produced from ecdysone (E) by the action of a P450 monooxygenase that hydroxylates E at carbon 20 (Petryk *et al.*, 2003). 20-Hydroxyecdysone mediates the major developmental transitions in insects and other arthropods, although E itself and other ecdysteroids may also play a role at various stages of development (Gilbert *et al.*, 2002; Hiruma *et al.*, 1997). Given the similarity between ponA and 20E, the strong H-bond between the side chain of Y408 and the 20-OH group allows to rationalize the higher activity of 20E compared to E (Baker *et al.*, 2000) (Dinan *et al.*, 1999). In addition, the high conservation of the residues of the ponA binding pocket is consistent with the promiscuous character of 20E, the most active metabolite for species as diverse as flies, moths, ticks, or crabs (Fig. 2).

The two EcR-LBD structures mainly differ by the presence of a helix H2 and a three-stranded β-sheet for the steroidal ligand bound complex as compared to the nonsteroidal ligand bound EcR, where H2 is unwound and the interactions between the second and third strands of the β-sheet are disrupted. A detailed analysis shows that the steroidal ligand is directly involved in the stabilization of these structural elements. The 2-OH and the 3-OH groups allow scaffolding interactions to take place for the stabilization of H2 (Fig. 8A). The network of interactions involves hydrogen bonds between the E309 (H1) carbonyl group and the 3-OH (2.7 Å) and 2-OH (2.8 Å) groups. The carboxylate group of E309 makes a salt bridge to the R386 (H5) guanidinium group, which is also salt bridged to the carboxylate moiety of E306 (H1). Furthermore, the guanidinium moiety of R383 that is H-bonded to the ponA 3-OH group makes an additional hydrogen bond to the backbone carbonyl group of Y308 (H1). The small H2 helix packs its two hydrophobic residues, L316 and V319, against the core of the receptor, more precisely in the region close to the β-sheet tip and lined on the back side by H3 residues M342 and L345. The backbone carbonyl group of V319 located at the C-ter of H2 makes a hydrogen bond to the side chain of Q338, further helping the H2 positioning (Fig. 8B). The β-sheet region, and in particular residues of the second β-strand and of the β-sheet tip interact with ponA. In particular, its ketone moiety is hydrogen bonded to the backbone amide group of the β-turn residue A398. Furthermore, an extensive network of hydrophobic contacts is formed with the ponA steroid core. Of particular interest is the van der Waals interaction between the B-ring of ponA and the benzene ring of F397, which on the opposite side makes a stacking interaction to the side chain of Y403. As described in Section III.D.2, these two aromatic residues directly participate in the structural adaptation of the receptor to its ligand. We notice that the stabilization of the helical conformation of H2 and that of the β-sheet structure are intimately related, with the prerequisite that the EcR ligand can interact with these two structural elements.

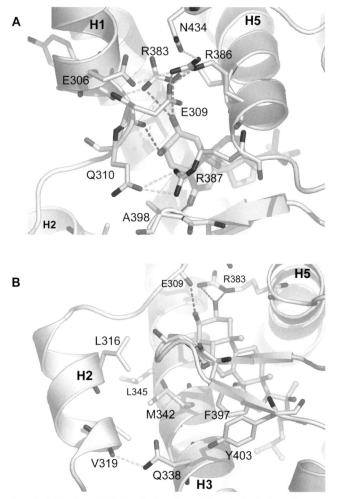

FIGURE 8. Stabilization of H2. PonA directly stabilizes the helical conformation of H2 via an intricate network of hydrogen bonds between the ligand and EcR residues (pink dashed lines) as well as between residues of neighboring structural elements (green dashed lines). Views of (A) the region comprising H1, H5, and the loop H1–H2 and (B) the region downwards encompassing helix H2 and the β-sheet. The amino acid residues and ponA are shown in stick representation with oxygen in red, nitrogen in blue, sulfur in green and carbon in white and cyan for EcR residues and ponA, respectively. (See Color Insert.)

2. The Bisacylhydrazine EcR Pocket

Examination of the receptor pocket reveals a V-shape electron density corresponding to the bound nonsteroidal ligand BYI06830 (Fig. 9A). Since the rings A and B of BYI06830 bear different substituents, their location inside the receptor could be clearly identified. The refined EcR-LBD structure indicates three hydrogen bonds between the ligand and EcR residues T343, Y408, and N504. In particular, the A-ring carbonyl group is H-bonded to T343, the B-ring carbonyl group interacts with Y408, and the unsubstituted amide group makes hydrogen bond with the carbonyl group of N504. However, at the resolution at which this structure was solved, experimental uncertainties remain on the exact ligand conformation (i.e., the positioning of the two ligand carbonyl groups and of the unsubstituted amide group). In fact, the B-ring carbonyl group could also point to the opposite direction and interact with the amide group of N504, and as a consequence the unsubstituted amide group would interact with Y408. This alternative conformation would better fit the chemical constraints. The determination of the EcR-LBD structure at higher resolution would be necessary to settle this issue.

Apart from the three H bonds discussed in earlier section, BYI06830 forms an extensive network of hydrophobic interactions with EcR (Fig. 9B). In particular, the A-ring is sandwiched between two adjacent methionine residues, M380 and M381, located at the carboxy-terminus of H5. The *tert*-butyl group is buried in a hydrophobic region of the cavity formed by residues belonging to H3 (F336, I339, and T340), H11 (M507, C508, and L511), H12 (L522), and to the loops H6–H7 (M413) and H11–H12 (L518). The presence of the *tert*-butyl group on the hydrazine is critical for the activity of the compound, since its replacement by a hydrogen atom results in an inactive compound (Mohammed-Ali *et al.*, 1995). The B-ring and its substituents point to the open cleft between H7 and H11 and make nonpolar interactions with EcR residues. The polar residues of this cleft, D419 and Q503, are too far from the benzodioxin substituent of the B-ring to make polar interactions, suggesting interesting possibilities for the optimization of dibenzoylhydrazine ligands. We further note the proximity of the USP K380 residue located in the H8–H9 loop, whose side chain is at ~4 Å from the carboxylate moiety of the EcR D416 residue.

In contrast to EcR-LBD in complex with ponA, residues of the BYI06830-bound EcR-LBD that belong to H1, to the loop H1–H2 and to the β-sheet no longer interact with the ligand. As a consequence, the scaffolding interactions that allow H2 to adopt a helical conformation in the ponA-bound EcR no longer exist. Furthermore, the binding of the synthetic ligand leads to the disruption of the interactions between the second and third β-sheet strands. The structural reorganization of the β-sheet is mainly due to the outer to inner motion of F397 and Y403, located on each side of the β-sheet tip. These two residues fill the region of the cavity occupied by the A-, B- and C-rings of the ecdysteroid but left empty by BYI06830 (Fig. 9C). The outer to inner switch of aromatic residues seems to be one of the key mechanisms in the structural moulding of the receptor to its ligand. Therefore, this adaptation phenomenon that involves desolvation processes is most likely entropy driven—a phenomenon that was already

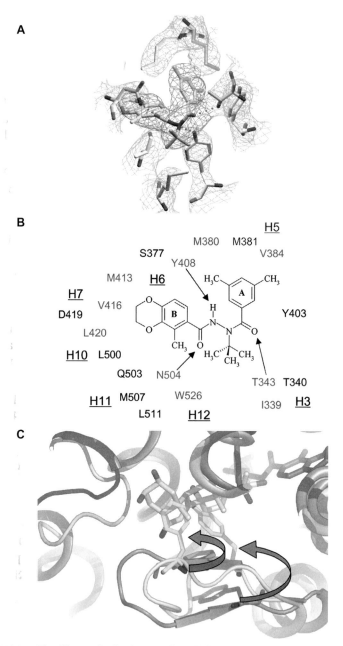

FIGURE 9. The dibenzoylhydrazine EcR ligand binding pocket. (A) View of the electron density for the BYI06830-bound EcR-LBD at 3.0 Å resolution. Two different maps are shown in each figure: a sigmaA weighted $2F_{obs} - F_{calc}$ omit map for the ligand (in blue) and a sigmaA weighted $2F_{obs} - F_{calc}$ map (in magenta) for selected residues in the ligand-binding pocket. The map is contoured at 1σ and overlaid on the final refined models. Hydrogen bonds between

observed in the structural transition from the apo to the holo forms of RXR (Egea et al., 2000).

The 21 residues of the cavity that are located at less than 4.5 Å from BYI06830 (Fig. 4, blue dots) are highly conserved (75% amino acid identity). Given this high residue conservation at the level of the binding pocket, the specificity of some synthetic agonists for lepidopteran insects is remarkable. The main characteristics of commercially available lepidopteran-specific dibenzoylhydrazine compounds is the presence on their A-ring of two methyl groups at positions C3 and C5 (Dhadialla et al., 1998). Examination of the local environment of these two CH_3 groups indicate van der Waals contacts with S377, V384, M380, M381, Y408, L420, N504, and W526. These residues are strictly conserved for all insects, except V384, which is conserved in Lepidoptera only and replaced by a methionine residue in other insects. Furthermore, the neighboring residue of V384, V395 is differentially conserved in Lepidoptera and replaced by an isoleucine residue in other insects. This suggests the possible implication of these two residues in the lepidopteran specificity of some of the dibenzoylhydrazine compounds. Docking studies indicate that the concerted replacement of V384 by Met and that of V395 by Ile would lead to steric interference with one of the methyl groups on the A-ring or with the protein itself (data not shown). In fact, despite the fact that methionine is a rather flexible residue, it is located in a very constraint environment, which does not allow any rotamer to be accommodated without steric clash. It would therefore be interesting to test whether this hypothesis on the lepidopteran specificity is correct using EcR mutants.

E. COMPARISON WITH OTHER NR LBD STRUCTURES

According to the nuclear receptor nomenclature (1999), the EcR belongs to the subfamily NR1H, which also comprises the liver X receptor (LXR) and the farnesoid X receptor (FXR), two receptors for steroid hormones (oxysterols and bile acids, respectively), which are absent in invertebrates. Together with the NRs of subfamily NR1I—vitamin D receptor (VDR),

---

ligand and residues are indicated by green dotted lines. (B) Schematic representation of the interactions of BYI06830 with the residues of the binding cavity. Arrows correspond to hydrogen bonds between ligand and amino acid residues. Residues in blue are common to both structures. (C) The structural adaptation of EcR upon binding of synthetic dibenzoylhydrazine compounds involves an inner to outer switch of two aromatic residues F397 and Y403 belonging to the β-sheet. In the BYI06830 EcR complex, these two residues fill the region of the pocket that is occupied by the ponA steroid core and left empty by the dibenzoylhydrazine ligand. The EcR structure and the carbon atom of the corresponding residues are shown in orange and in green for the ponA-bound and the BYI06830-bound complexes, respectively. The carbon atoms of ponA and BYI06830 are colored in grey and cyan, respectively. Oxygen and nitrogen atoms are shown in red and in blue, respectively. (See Color Insert.)

constitutive androstane receptor (CAR), and pregnane X receptor (PXR)—the NR1H receptors EcR, LXR, and FXR bind steroid-related compounds and heterodimerize with their ubiquitous partner RXR or its invertebrate orthologue USP. According to phylogenetic studies, the EcR protein is more closely related to LXR than to FXR (Bertrand et al., 2004). This relationship is also reflected in the similarity and identity in the LBD sequences of EcR and LXR (50% similarity and 35% identity for hsLXRβ) compared to FXR and VDR (43 and 41% similarity and 28 and 24% identity, respectively) (Fig. 10). It is therefore instructive to compare the structure of the ponA-bound EcR-LBD with the crystal structures of LXR (Farnegardh et al., 2003; Hoerer et al., 2003; Svensson et al., 2003; Williams et al., 2003), FXR (Downes et al., 2003; Mi et al., 2003) and VDR (Rochel et al., 2000). The FXR and VDR LBD crystal structures feature a helix H2 similar to that observed for EcR complexed to ponA, while LXR does not possess a helical structure at this place. On the other hand, the three stranded β-sheet is very similar to that seen in LXR and to a lesser extent to the β-sheet of VDR, while the β-sheet is replaced by a long helix H6 and a loop in FXR. The ligand-binding pockets of LXR bound to its endogenous activator 24(S),25-epoxycholesterol (eCH) (Williams et al., 2003) and of VDR bound to its endogenous ligand, the secosteroid 1α,25-dihydroxyvitamin D_3 (vitamin D) (Rochel et al., 2000) are similar in shape and size (685 and 707 Å^3, respectively) compared to the ponA bound EcR cavity (573 Å^3). The orientation of the respective steroid hormone inside the pocket is similar, with the A-ring of the steroid core oriented towards helix 1 and with the alkyl side chain oriented towards H12 (Fig. 11A and B). This orientation is similar to that of estradiol, progesterone, and dexamethasone in the complexes with the estrogen, progesterone, and glucocorticoid receptors, respectively. This is not the case for FXR bound to the bile acid 6α-ethyl-chenodeoxycholic acid (6ECDCA) (Mi et al., 2003), where the orientation of the ligand in the pocket is reversed (Fig. 11C). The vitamin D bound to its cognate receptor VDR adopts a curved shape. This conformation that is intrinsically possible due to the secosteroid nature of vitamin D cannot be adopted by the EcR ecdysteroid ponA and by the LXR agonist eCH. The latter adopt a more constraint geometry, with an almost planar configuration for eCH and a chair conformation for ponA. In addition, eCH superimposes very well to ponA, with the two angular methyl groups and the alkyl side chain roughly at the same positions.

The crystal structures of LXR (Farnegardh et al., 2003; Hoerer et al., 2003; Svensson et al., 2003; Williams et al., 2003) and VDR (Ciesielski et al., 2004; Rochel et al., 2000) complexed to different types of molecules indicate that local adaptations of the receptor LBD to its ligand can take place. In the case of VDR complexed to the synthetic Gemini agonist, the rearrangement of the beginning of H7 leads to a different pocket shape. For LXR, substantial variations in the size of the pocket are observed, with a

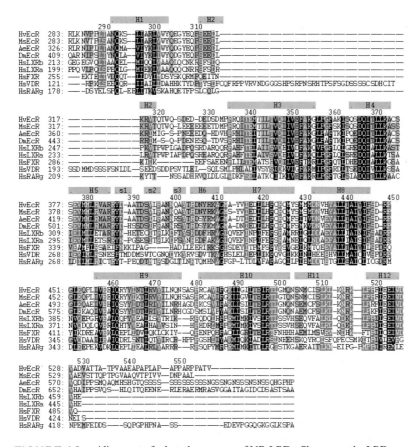

FIGURE 10. Alignment of selected sequences of NR LBDs. Shown are the LBD sequences of selected EcRs (hvEcR: *H. virescens*, msEcR: *M. sexta*, aeEcR: *A. aegypti*, dmEcR: *D. melanogaster*) and human NRs LXRβ, LXRα, FXR, VDR, and RARγ. Residue numbering corresponding to the hvEcR sequence is given at the top of the alignment. The first residue of each line in the alignment is numbered. The secondary structure elements are given for the hvEcR-LBD crystal structure complexed to ponA. Code of the amino acid conservation: white on black blue: 100% conservation, white on grey: 80% conservation, black on grey: 60% conservation.

volume going from 560 to 1090 Å^3 in the various complexes. The modifications of the pocket shape and size are not related to a change in the structure of the LBD but result from different rotational conformers of residues inside the ligand-binding cavity. The adaptation of the LXR pocket to the various ligands relies mainly in the flexibility of four Phe residues (F271 in H3, 329 in the β-sheet, F340 in H6, and F349 in H7) together with R319 (C-ter of H5) and the volume around the β-hairpin loop (Farnegardh *et al.*, 2003). For EcR, the adaptation of the protein domain to its ligand is also related to the

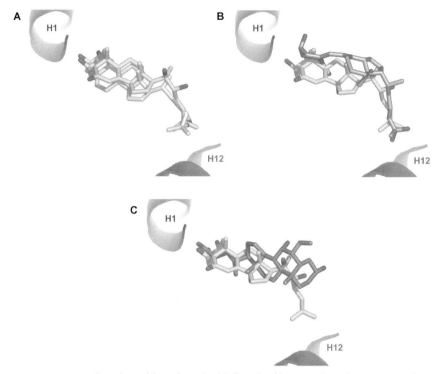

FIGURE 11. Superimposition of ponA with ligands of human NRs. The structures of the LBD of LXR bound to 24(S), 25-epoxycholesterol (eCH), VDR bound to 1α, 25-dihydroxyvitamin D_3 (vitamin D) and FXR bound to the bile acid 6α-ethyl-chenodeoxycholic acid (6ECDCA) were superimposed to the structure of ponA-bound EcR-LBD. Shown are the resulting superimpositions of ponA with (A) eCH, (B) vitamin D, and (C) 6ECDCA. Helices H1 and H12 are indicated. The orientation is identical to that of Fig. 6B. Carbon atoms are shown in cyan for ponA, grey for eCH, pink for vitamin D, and magenta for 6ECDCA. Oxygen atoms are shown in red. (See Color Insert.)

movement of aromatic residues F397 and Y403 located on each side of the β-sheet tip. However, in contrast to LXR and to the other NRs for which LBD structures are known, the EcR ligand-binding pocket is ligand-dependent due to different receptor structures, with distinct residues implicated in ligand recognition.

In both EcR complexes, no hydrogen bonds are formed between the ligand and residues of the AF2 helix. The two residues that face the interior of the pocket are L522 and W526. L522 is located in a hydrophobic patch of the receptor formed by residues of H3 and H4 rather far from the ligand (more than 5 Å) and adopts distinct rotamer conformations in the two structures. The Trp residue that is identically positioned in both complexes forms hydrophobic interactions with ponA and BYI06830 (the closest distance is 3.9 and 3.7 Å from ponA and BYI06830, respectively). The W526 indole is positioned so that its nitrogen atom makes a hydrogen bond to S376 (H5) (Fig. 12). Furthermore, W526 is at about 3.5 Å from the side chain of N504 (H11) with which it forms an electrostatic interaction of cation–π type (Ma and Dougherty, 1997). This Asn residue is hydrogen bonded to the nonsteroidal ligand BYI06830, while it makes hydrophobic contacts to the ponA side chain. The residues W526 and N504 are strictly conserved in all EcR sequences and are essential for steroidal and nonsteroidal ligand activation of EcR, as demonstrated by *in vitro* and *in vivo* experiments (see Section III.F.). The cation–π interaction between Trp and Asn is also observed for the other two receptors of the subfamily NR1H, LXR, and FXR, which also feature a Trp residue in H12, but where the Asn residue is replaced by a His residue (Downes *et al.*, 2003; Farnegardh *et al.*, 2003; Hoerer *et al.*, 2003; Mi *et al.*, 2003; Svensson *et al.*, 2003; Williams *et al.*, 2003). As pointed out in Williams *et al.* (2003), other receptors, such as VDR, TR, and RORα that possess a Phe residue instead of the Trp residue, could also use a cation–π mechanism through the Phe-His switch for the stabilization of the AF2 helix. Since only a subset of NRs use the cation–π mechanism for the stabilization of the active conformation of the AF2, its biological significance and the resulting differences compared to other NRs would justify a more thorough investigation.

F. VALIDATION OF THE STRUCTURAL DATA BY *IN VITRO* ASSAYS

Several mutants were generated to confirm the functional significance of our structural observations (Fig. 13). Residues belonging to the ligand-binding pocket that form key interactions with the ligand were mutated to alanine, and reporter activity was determined using a human embryonic kidney 293 EBNA cell assay. Mutation of the two residues that stabilize the AF2 helix by π–cation interaction, W526 and N504, results in a drastic reduction of transcriptional activity. These results agree with observations of the dominant negative phenotype of the W650A-dmEcR mutant (corresponding to W526A-hvEcR), for which neither ligand binding nor transcriptional activation are observed (Cherbas *et al.*, 2003; Hu *et al.*, 2003). Furthermore, our data are consistent with functional studies of the lepidopteran *Choristoneura fumiferana* (cf) EcR single mutants, in particular with the ecdysteroid-selective mutants R378A, F392A, and A393P (R383A, F397A, and A398P for hvEcR) (Kumar *et al.*, 2002). The A393P-cfEcR mutant is particularly interesting for its critical role in discriminating between steroidal and nonsteroidal ligands. For the ponA-bound hvEcR-LBD, A398 is located at the tip of the β-sheet and in close contact to the ligand, while for BYI06830-bound hvEcR-LBD, the β-sheet region including A398 is not involved in ligand recognition. Mutation of A398 to the very

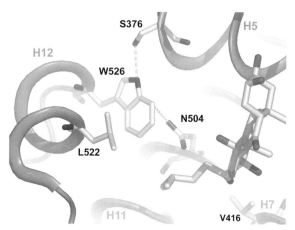

FIGURE 12. Stabilization of H12. The cation–π interaction between Trp526 and N504 (yellow dotted line) helps stabilizing the agonist conformation of the activation helix H12. These two residues are essential for steroidal and nonsteroidal ligand activation of EcR. In addition Trp526 is hydrogen bonded to S376 (green dotted line). The carbon atoms are colored in grey, oxygen atoms in red, and nitrogen atoms in blue. (See Color Insert.)

constraint proline residue affects the β-sheet structure by introducing local distortions of the protein backbone, which can explain the complete loss of activity for the steroidal and not for the nonsteroidal ligand. It is interesting to notice that mutation of V411 (H7) of cf EcR (corresponding to V416) to Phe or Tyr completely abolish the binding of ecdysteroids and bisacylhydrazine compounds (Kumar *et al.*, 2004). Both crystal structures show the proximity of this residue to ponA and BYI06830 and its mutation to a bulkier residue (F or Y) can readily explain the inability of both ligand types to bind to EcR. On the other hand, the mutants V411F-cf EcR, and V411Y-cf EcR, but not wild-type cf EcR, respond to the THQ class of compounds. However, the explanation for this behavior remains speculative and may involve either π-stacking interactions with the aromatic rings of Phe and Tyr, entropic effects or a different binding cavity of EcR in complex to THQ ligands.

IV. CONCLUSIONS

The structures of the EcR/USP heterodimer complexed to steroidal and nonsteroidal ligands highlight the extreme flexibility and adaptability of this protein domain, which allows the molding of the receptor around its ligand and consequently the formation of the proper ligand-binding pocket. The large plasticity of EcR to its ligand is so far unique for NRs and explains the ability of EcR to allow fundamentally different ligand types to bind and activate the receptor. Other NRs for which structures are determined, in

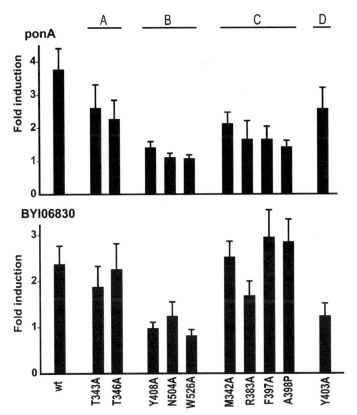

FIGURE 13. Differential effects of mutated residues in the binding cavities of hvEcR. Transcriptional activities of wild-type and point mutant hvEcR in transient transfection assays. HEK-293 EBNA cells were transfected in 24-well plates with 250 ng per well of PAL1(5×)-TK-LUC reporter and 25 ng of wild type or mutant hvEcR and 25 ng hvUSP in pSG5 expression plasmids using standard calcium phosphate coprecipitation technique. Experiments were performed in the absence or presence of 10^{-6} M ponA or 10^{-6} M BYI06830. Data points represent the mean of assays performed in duplicate for at least three independent experiments. Mutants (A) T343A and T346 show small reduction of the transcriptional activity for both ligands, (B) Y408A, N504A and W526A strongly affect the transcriptional activity for both ligands, (C) M342A, R383A, F397A, and A398P show specific reduction in ecdysteroid transcriptional activity, and (D) Y403A affects more specifically the transcriptional activity induced by BYI06830.

particular LXR, the NR phylogenetically closest to EcR, also show some degrees of plasticity. However, unlike EcR, their corresponding structures complexed to different ligands are essentially identical and never lead to ligand-binding pockets where different residues are implicated in ligand recognition. The extreme case of EcR, together with that of LXR and other NRs, support the unbiased experimental structure determination for NR ligand search.

ACKNOWLEDGMENTS

We thank all our colleagues at the Laboratoire de Génomique et Biologie Structurales of IGBMC (Illkirch, France) for fruitful exchanges and discussions. We thank Holger Greschik, Chris Browning and Natacha Rochel for critical reading of this manuscript. We thank the staff of the ESRF ID14 beamline (Grenoble, France) and the French beamline BM30A (Grenoble, France) for assistance during data collection. This work was supported by BayerCropScience and by grants from CNRS, INSERM, the Ministère de la Recherche et de la Technologie (Programme de Génomique Structurale) and EU-SPINE.

REFERENCES

Baker, K. D., Warren, J. T., Thummel, C. S., Gilbert, L. I., and Mangelsdorf, D. J. (2000). Transcriptional activation of the *Drosophila* ecdysone receptor by insect and plant ecdysteroids. *Insect Biochem. Mol. Biol.* **30**, 1037–1043.

Bertrand, S., Brunet, F. G., Escriva, H., Parmentier, G., Laudet, V., and Robinson-Rechavi, M. (2004). Evolutionary genomics of nuclear receptors: From twenty-five ancestral genes to derived endocrine systems. *Mol. Biol. Evol.* **21**, 1923–1937.

Billas, I. M. L., Iwema, T., Garnier, J. M., Mitschler, A., Rochel, N., and Moras, D. (2003). Structural adaptability in the ligand-binding pocket of the ecdysone hormone receptor. *Nature* **426**, 91–96.

Billas, I. M. L., Moulinier, L., Rochel, N., and Moras, D. (2001). Crystal structure of the ligand binding domain of the ultraspiracle protein USP, the ortholog of RXRs in insects. *J. Biol. Chem.* **276**, 7465–7474.

Bourguet, W., Vivat, V., Wurtz, J.-M., Chambon, P., Gronemeyer, H., and Moras, D. (2000). Crystal structure of a heterodimeric complex of RAR and RXR ligand-binding domains. *Mol. Cell* **5**, 289–298.

Brelivet, Y., Kammerer, S., Rochel, N., Poch, O., and Moras, D. (2004). Signature of the oligomeric behaviour of nuclear receptors at the sequence and structural level. *EMBO Rep.* **5**, 423–429.

Cherbas, L., Hu, X., Zhimulev, I., Belyaeva, E., and Cherbas, P. (2003). EcR isoforms in *Drosophila*: Testing tissue-specific requirements by targeted blockade and rescue. *Development* **130**, 271–284.

Christopherson, K. S., Mark, M. R., Bajaj, V., and Godowski, P. J. (1992). Ecdysteroid-dependent regulation of genes in mammalian cells by a *Drosophila* Ecdysone receptor and chimeric transactivators. *Proc. Natl. Acad. Sci. USA* **89**, 6314–6318.

Ciesielski, F., Rochel, N., Mitschler, A., Kouzmenko, A., and Moras, D. (2004). Structural investigation of the ligand binding domain of the zebrafish VDR in complexes with 1[alpha],25(OH)2D3 and Gemini: Purification, crystallization and preliminary X-ray diffraction analysis. *J. Steroid Biochem. Mol. Biol.* **89–90**, 55–59.

Clayton, G. M., Peak-Chew, S. Y., Evans, R. M., and Schwabe (2001). The structure of the ultraspiracle ligand-binding domain reveals a nuclear receptor locked in an inactive conformation. *Proc. Natl. Acad. Sci. USA* **98**, 1549–1554.

Darimont, B. D., Wagner, R. L., Apriletti, J. W., Stallcup, M. R., Kushner, P. J., Baxter, J. D., Fletterick, R. J., and Yamamoto, K. R. (1998). Structure and specificity of nuclear receptor-coactivator interactions. *Genes. Dev.* **12**, 3343–3356.

Dhadialla, T. S., Carlson, G. R., and Le, D. P. (1998). New insecticides with ecdysteroidal and juvenile hormone activity. *Annu. Rev. Entomol.* **43**, 545–569.

Dinan, L., Hormann, R. E., and Fujimoto, T. (1999). An extensive ecdysteroid CoMFA. *J. Comp. Aided Mol. Design* **13**, 185–207.

Downes, M., Verdecia, M. A., Roecker, A. J., Hughes, R., Hogenesch, J. B., Kast-Woelbern, H. R., Bowman, M. E., Ferrer, J. L., Anisfeld, A. M., Edwards, P. A., Rosenfeld, J. M., Alvarez, J. G., Noel, J. P., Nicolaou, K. C., and Evans, R. M. (2003). A chemical, genetic, and structural analysis of the nuclear bile acid receptor FXR. *Mol. Cell* **11,** 1079–1092.

Egea, P. F., Mitschler, A., Rochel, N., Ruff, M., Chambon, P., and Moras, D. (2000). Crystal structure of the human RXRalpha ligand-binding domain bound to its natural ligand 9-cis retinoic acid. *EMBO J.* **19,** 2592–2601.

Fabian, L., Argay, G., Kalman, A., and Bathori, M. (2002). Crystal structures of ecdysteroids: The role of solvent molecules in hydrogen bonding and isostructurality. *Acta Cryst. B* **58,** 710–720.

Farnegardh, M., Bonn, T., Sun, S., Ljunggren, J., Ahola, H., Wilhelmsson, A., Gustafsson, J. A., and Carlquist, M. (2003). The three dimensional structure of the liver X receptor beta reveals a flexible ligand binding pocket that can accommodate fundamentally different ligands. *J. Biol. Chem.* **278,** 38821–38828.

Gampe, R. T., Jr., Montana, V. G., Lambert, M. H., Miller, A. B., Bledsoe, R. K., Milburn, M. V., Kliewer, S. A., Willson, T. M., and Xu, H. E. (2000). Asymmetry in the PPARgamma/RXRalpha crystal structure reveals the molecular basis of heterodimerization among nuclear receptors. *Mol. Cell* **5,** 545–555.

Gilbert, L. I., Rybczynski, R., and Warren, J. T. (2002). Control and biochemical nature of the ecdysteroidogenic pathway. *Annu. Rev. Entomol.* **47,** 883–916.

Gronemeyer, H., Gustafsson, J. A., and Laudet, V. (2004). Principles for modulation of the nuclear receptor superfamily. *Nat. Rev. Drug Discov.* **3,** 950–964.

Hiruma, K., Bocking, D., Lafont, R., and Riddiford, L. M. (1997). Action of different ecdysteroids on the regulation of mRNAs for the ecdysone receptor, MHR3, dopa decarboxylase, and a larval cuticle protein in the larval epidermis of the tobacco hornworm, *Manduca sexta. Gen. Comp. Endocrinol.* **107,** 84–97.

Hoerer, S., Schmid, A., Heckel, A., Budzinski, R. M., and Nar, H. (2003). Crystal structure of the human liver X receptor [beta] ligand-binding domain in complex with a synthetic agonist. *J. Mol. Biol.* **334,** 853–861.

Horlein, A. J., Naar, A. M., Heinzel, T., Torchia, J., Gloss, B., Kurokawa, R., Ryan, A., Kamei, Y., Soderstrom, M., and Glass, C. K. (1995). Ligand-independent repression by the thyroid hormone receptor mediated by a nuclear receptor co-repressor. *Nature* **377,** 397–404.

Hu, X., Cherbas, L., and Cherbas, P. (2003). Transcription activation by the ecdysone receptor (EcR/USP): Identification of activation functions. *Mol. Endocrinol.* **17,** 716–731.

Kasuya, K., Sawada, Y., Tsukamoto, Y., Tanaka, K., Toya, T., and Yanagi, M. (2003). Binding mode of ecdysone agonists to the receptor: Comparative modeling and docking studies. *J. Mol. Model* **9,** 58–65.

Koelle, M. R., Talbot, W. S., Segraves, W. A., Bender, M., Cherbas, P., and Hogness, D. S. (1991). The *Drosophila* EcR gene encodes an ecdysone receptor, a new member of the steroid receptor superfamily. *Cell* **67,** 59–77.

Kumar, M. B., Fujimoto, T., Potter, D. W., Deng, Q., and Palli, S. R. (2002). A single point mutation in ecdysone receptor leads to increased ligand specificity: Implications for gene switch applications. *Proc. Natl. Acad. Sci. USA* **99,** 14710–14715.

Kumar, M. B., Potter, D. W., Horman, R. E., Edwards, A., Tice, C. M., Smith, H. C., Dipietro, M. A., Polley, M., Lawless, M., Wolohan, P. R. N., Kethidi, D. R., and Palli, S. R. (2004). Highly flexible ligand-binding pocket of ecdysone receptor: A single amino acid change leads to discrimination between two groups of nonsteroidal ecdysone agonists. *J. Biol. Chem.* **279,** 27211–27218.

Ma, J. C., and Dougherty, D. A. (1997). The Cationminus signpi Interaction. *Chem. Rev* **97,** 1303–1324.

Maglich, J. M., Sluder, A., Guan, X., Shi, Y., McKee, D. D., Carrick, K., Kamdar, K., Willson, T. M., and Moore, J. T. (2001). Comparison of complete nuclear receptor sets from the human, *Caenorhabditis elegans* and *Drosophila* genomes. *Genome Biol.* **2**, 0029.1–0029.7.

McInerney, E. M., Rose, D. W., Flynn, S. E., Westin, S., Mullen, T. M., Krones, A., Inostroza, J., Torchia, J., Nolte, R. T., Assa-Munt, N., Milburn, M. V., Glass, C. K., and Rosenfeld, M. G. (1998). Determinants of coactivator LXXLL motif specificity in nuclear receptor transcriptional activation. *Genes Dev.* **12**, 3357–3368.

Mi, L. Z., Devarakonda, S., Harp, J. M., Han, Q., Pellicciari, R., Willson, T. M., Khorasanizadeh, S., and Rastinejad, F. (2003). Structural basis for bile acid binding and activation of the nuclear receptor FXR. *Mol. Cell* **11**, 1093–1100.

Mohammed-Ali, A. K., Chan, T.-H., Thomas, A. W., Strunz, G. M., and Jewett, B. (1995). Structure-activity relationship study of synthetic hydrazines as ecdysone agonists in the control of spruce budworm (*Choristoneura fumiferana*). *Can. J. Chem.* **73**, 550–557.

Moras, D., and Gronemeyer, H. (1998). The nuclear receptor ligand-binding domain: Structure and function. *Curr. Opin. Cell Biol.* **10**, 384–391.

No, D., Yao, T.-P., and Evans, R. M. (1996). Ecdysone-inducible gene expression in mammalian cells and transgenic mice. *Proc. Natl. Acad. Sci. USA* **93**, 3346–3351.

Nolte, R. T., Wisely, G. B., Westin, S., Cobb, J. E., Lambert, M. H., Kurokawa, R., Rosenfeld, M. G., Willson, T. M., Glass, C. K., and Milburn, M. V. (1998). Ligand-binding and co-activator assembly of the peroxisome proliferator-activated receptor-gamma. *Nature* **395**, 137–143.

Nuclear Receptors Nomenclature committe. A Unified Nomenclature System for the Nuclear Receptor Superfamily (1999). *Cell* **97**, 161–163.

Palli, S. R., Kapitskaya, M. Z., Kumar, M. B., and Cress, D. E. (2003). Improved ecdysone receptor-based inducible gene regulation system. *Eur. J. Biochem.* **270**, 1308–1315.

Petryk, A., Warren, J. T., Marques, G., Jarcho, M. P., Gilbert, L. I., Kahler, J., Parvy, J. P., Li, Y., Dauphin-Villemant, C., and O'Connor, M. B. (2003). Shade is the *Drosophila* P450 enzyme that mediates the hydroxylation of ecdysone to the steroid insect molting hormone 20-hydroxyecdysone. *Proc. Natl. Acad. Sci. USA* **100**, 13773–13778.

Riddiford, L. M., Cherbas, P., and Truman, J. W. (2001). Ecdysone receptors and their biological actions. *Vitam. Horm.* **60**, 1–73.

Rochel, N., Wurtz, J.-M., Mitschler, A., Klaholz, B. P., and Moras, D. (2000). The crystal structure of the nuclear receptor of vitamin D bound to its natural ligand. *Mol. Cell* **5**, 173–179.

Saez, E., Nelson, M. C., Eshelman, B., Banayo, E., Koder, A., Cho, G. J., and Evans, R. M. (2000). Identification of ligands and coligands for the ecdysone-regulated gene switch. *Proc. Natl. Acad. Sci. USA* **97**, 14512–14517.

Smith, H. C., Cavanaugh, C. K., Friz, J. L., Thompson, C. S., Saggers, J. A., Michelotti, E. L., Garcia, J., and Tice, C. M. (2003). Synthesis and SAR of cis-1-Benzoyl-1,2,3,4-tetrahydroquinoline ligands for control of gene expression in ecdysone responsive systems. *Bioorg. Med. Chem. Lett.* **13**, 1943–1946.

Suhr, S. T., Gil, E. B., Senut, M. C., and Gage, F. H. (1998). High level transactivation by a modified Bombyx ecdysone receptor in mammalian cells without exogenous retinoid X receptor. *Proc. Natl. Acad. Sci. USA* **95**, 7999–8004.

Suino, K., Peng, L., Reynolds, R., Li, Y., Cha, J. Y., Repa, J. J., Kliewer, S. A., and Xu, H. E. (2004). The nuclear xenobiotic receptor CAR: Structural determinants of constitutive activation and heterodimerization. *Mol. Cell* **16**, 893–905.

Svensson, S., Ostberg, T., Jacobsson, M., Norstrom, C., Stefansson, K., Hallen, D., Johansson, I. C., Zachrisson, K., Ogg, D., and Jendeberg, L. (2003). Crystal structure of the heterodimeric complex of LXR{alpha} and RXR{beta} ligand-binding domains in a fully agonistic conformation. *EMBO J.* **22**, 4625–4633.

Tice, C. M., Hormann, R. E., Thompson, C. S., Friz, J. L., Cavanaugh, C. K., Michelotti, E. L., Garcia, J., Nicolas, E., and Albericio, F. (2003a). Synthesis and SAR of [alpha]-

Acylaminoketone ligands for control of gene expression. *Bioorg. Med. Chem. Lett.* **13**, 475–478.

Tice, C. M., Hormann, R. E., Thompson, C. S., Friz, J. L., Cavanaugh, C. K., and Saggers, J. A. (2003b). Optimization of [alpha]-acylaminoketone ecdysone agonists for control of gene expression. *Bioorg. Med. Chem. Lett.* **13**, 1883–1886.

Toya, T., Fukasawa, H., Masui, A., and Endo, Y. (2002). Potent and selective partial Ecdysone agonist activity of chromafenozide in Sf9 Cells. *Biochem. Biophys. Res. Comm.* **292**, 1087–1091.

Williams, S., Bledsoe, R. K., Collins, J. L., Boggs, S., Lambert, M. H., Miller, A. B., Moore, J., McKee, D. D., Moore, L., Nichols, J., Parks, D., Watson, M., Wisely, B., and Willson, T. M. (2003). X-ray crystal structure of the liver X receptor beta ligand binding domain: Regulation by a histidine-tryptophan switch. *J. Biol. Chem.* **278**, 27138–27143.

Wurtz, J.-M., Bourguet, W., Renaud, J.-P., Vivat, V., Chambon, P., Moras, D., and Gronemeyer, H. (1996). A canonical structure for the ligand-binding domain of nuclear receptors. *Nat. Struct. Biol.* **3**, 87–94.

Wurtz, J.-M., Guillot, B., Fagart, J., Moras, D., Tietjen, K., and Schindler, M. (2000). A new model for 20-hydroxyecdysone and dibenzoylhydrazine binding: A homology modeling and docking approach. *Protein Sci.* **9**, 1073–1084.

Xu, R. X., Lambert, M. H., Wisely, B. B., Warren, E. N., Weinert, E. E., Waitt, G. M., Williams, J. D., Collins, J. L., Moore, L. B., Willson, T. M., and Moore, J. T. (2004). A structural basis for constitutive activity in the human CAR/RXR[alpha] Heterodimer. *Mol. Cell* **16**, 919–928.

5

Nonsteroidal Ecdysone Agonists

Yoshiaki Nakagawa

*Division of Applied Life Sciences, Graduate School of Agriculture
Kyoto University, Kyoto 606–8502, Japan*

I. Introduction
II. Diacylhydrazines
   A. Chemistry
   B. Effects on In Vivo *Systems*
   C. Effects on In Vitro *Systems*
   D. Effects on Receptors
III. Other Nonsteroidal Ecdysone Agonists
IV. Conclusions
    References

Nonsteroidal ecdysone agonists are novel compounds that have become attractive candidates not only as pest control agents in agriculture but also as tools for research. Their narrow spectrum of activity makes them relatively safe as pesticides, and their mode of action as ligands for gene expression has found application in gene therapy and inducing transgenic gene expression in plants. These diacylhydrazines (DAHs) are potent nonsteroidal ecdysone agonists, and four of them, tebufenozide, methoxyfenozide, chromafenozide, and halofenozide, have been developed as insecticides. Although these compounds are very toxic to insects, they are safe for mammals and are environmentally benign. Their action on insects is also selective, the first three are effective against Lepidoptera but weakly active or inactive on Diptera and Coleoptera. On the other hand, halofenozide is effective on Coleoptera but mildly active on Lepidoptera. Previous reviews on ecdysone agonists have concentrated on the biological response of some DAHs and their effects on pests. In this review, the chemistry, biological effects and their modes of action at the molecular level will be covered. In addition, a few studies on other nonsteroidal ecdysone agonists, such as 3,5-*di-tert*-butyl-4-hydroxy-*N-iso*-butylbenzamide, acylaminoketones, and benzoyl-1,2,3,4-tetrahydroquinolines, will be briefly reviewed. © 2005 Elsevier Inc.

I. INTRODUCTION

Molting and metamorphosis in insects are regulated primarily by the molting hormone, 20-hydroxyecdysone (20E) and juvenile hormones (JHs). The chemical structure of 20E was elucidated in the 1960s, and its three-dimensional (3-D) structure was solved in 1965 (Huber and Hoppe, 1965; Karlson, 1980). A few years later the chemical structure of JH-I was resolved (Roller *et al.*, 1967), and subsequently the structures of the other JHs—JH-0, JH-II, JH-III, and 4-methyl JH-I—were isolated from various insects and characterized. Potent analogues of JH, such as phenoxycarb (Dorn *et al.*, 1981) and pyriproxyfen (Kawada *et al.*, 1989), were developed as insecticides before ecdysone mimics were commercialized.

To date a number of ecdysteroids have been isolated from plants (phytoecdysteroids) and animals (zooecdysteroids) and first reviewed by Hetru and Horn (1980). Later, *The Ecdysone Handbook* was edited by Lafont and Wilson. More than 300 compounds were listed at the onset of 2000, and the ecdysteroid database is available on the internet (http://ecdysbase.org), based on the third edition of *The Ecdysone Handbook*. Among the phytoecdysteroids, ponasterone A (PonA), isolated from *Podocarpus nakaii* Hay, was found to be very active (Kobayashi *et al.*, 1967; Nakanishi *et al.*, 1966). Although a number of ecdysteroids have been isolated to date, they have never been used as insecticides because they were not easy to synthesize, had no topical effect, and were subject to oxidative degradation.

The synthesis of a nonsteroidal ecdysone agonist, *N-tert*-butyl-*N,N'*-dibenzoylhydrazine (RH-5849; Fig. 1) (Hsu, 1991; Wing, 1988), stimulated the development of a new class of insect growth regulator (IGR), based on ecdysteroid activity. This dibenzoylhydrazine possessed not only molting hormonal activity but also had insecticidal properties (Wing *et al.*, 1988). After the structural optimization of RH-5849, two new compounds, tebufenozide (RH-5992; Fig. 1) and methoxyfenozide (RH-2485; Fig. 1), were introduced into the market as lepidopteran specific insecticides. Subsequently, a chlorinated analogue, halofenozide (RH-0345; Fig. 1), was

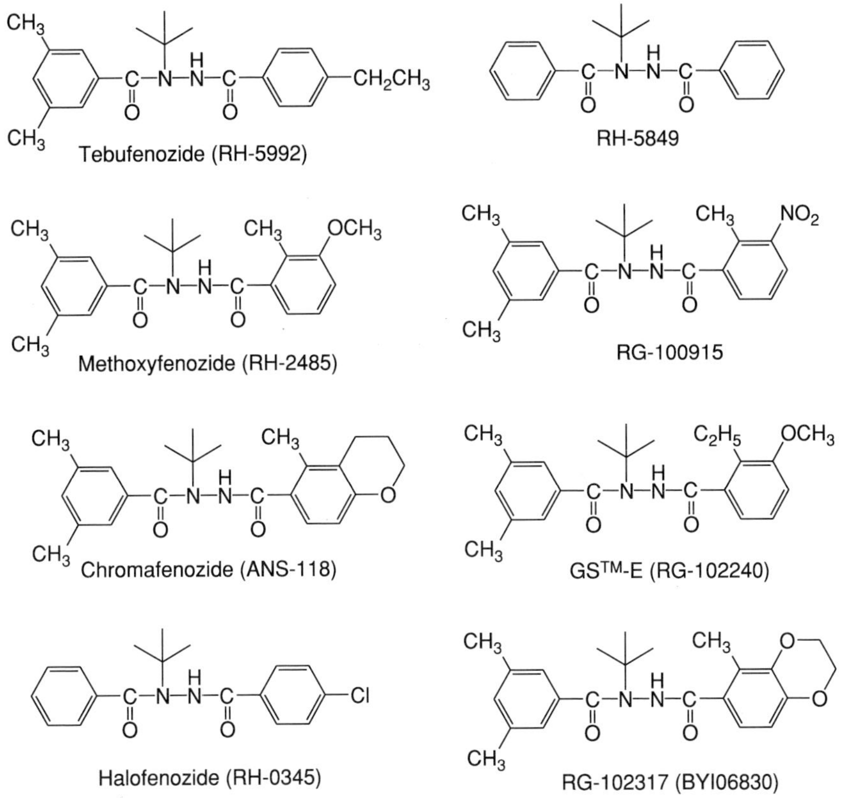

FIGURE 1. Structures of diacylhydrazine congeners.

introduced as an insecticide against primarily Coleoptera and certain Lepidoptera (Dhadialla et al., 1998). In 2000, another diacylhydrazine (DAH)-type compound containing a chromane ring, chromafenozide (ANS-118; Fig. 1), was developed in Japan against Lepidoptera (Sawada et al., 2003c; Tanaka et al., 2001). Interest in developing ecdysteroid-based insecticides date back as early as 1996 when 3,5-*di-tert*-butyl-4-hydroxy-*N-iso*-butylbenzamide was reported as an ecdysone agonist against *Drosophila* (Mikitani, 1996). It turned out that the binding and hormonal activities of these compounds were very low compared to 20E and DAHs. In the meantime, new chemical structures, *cis*-1-benzoyl-1,2,3,4-tetrahydroquinoline (Smith et al., 2003) and α-acylaminoketone derivatives (Tice et al., 2003a,b), were reported as ecdysone agonists. Interestingly, tetrahydroquinoline analogues were effective against dipterans, such as the mosquito *Aedes aegypti* L. However, the activity of these new chemicals has not been fully evaluated.

20-Hydroxyecdysone is a steroid hormone that regulates molting, metamorphosis, reproduction, and various other developmental processes in insects. Ecdysone functions through a heterodimeric receptor complex composed of ecdysone receptor (EcR) and ultraspiracle (USP) (Yao *et al.*, 1993). The DAH analogues also specifically bind to this heterodimeric receptor complex. These EcR and USP proteins have been shown to be members of the steroid hormone receptor superfamily. Members of this superfamily are characterized by the presence of five modular domains, A/B (transactivation), C (DNA-binding/heterodimerization or DBD), D (hinge, heterodimerization), E (ligand-binding domain, heterodimeriziation, transactivation, or LBD), and F (transactivation). Crystallographic studies on the E domain structures of several nuclear receptors showed a conserved fold composed of 11 helices (H1 and H3–H12) and 2 short strands (s1 and s2). The crystal structures of USP were shown to have a long H1–H3 loop and an insert between H5 and H6. These structures appear to lock USP in an inactive conformation by displacing H12 from agonist conformation. In both crystal structures, USP had a large hydrophobic cavity, which contained phospholipids ligands (Palli *et al.*, 2003). To date, primary sequences of EcRs had been identified not only from insects (Cho *et al.*, 1995; Dhadialla and Tzertzinis, 1997; Fujiwara *et al.*, 1995; Hannan and Hill, 1997; Imhof *et al.*, 1993; Kothapalli *et al.*, 1995; Martinez *et al.*, 1999; Minakuchi *et al.*, 2002; Mouillet *et al.*, 1997; Ogura *et al.*, 2005b; Saleh *et al.*, 1998; Swevers *et al.*, 1995; Verras *et al.*, 1999) but also from tick (Guo *et al.*, 1997) and crab (Chung *et al.*, 1998). Furthermore, the crystal structures have been solved for LBD of both EcR (Billas *et al.*, 2003; Carmichael *et al.*, 2005) and USP (Billas *et al.*, 2001; Clayton *et al.*, 2001). EcR-LBDs were crystallized with either PonA or a DAH congener, BYI06830, although four amino acids that do not participate in the ligand binding were mutated in EcR-LBD/BYI06830 complex (Billas *et al.*, 2003).

Nonsteroidal ecdysone agonists, such as DAHs, are not only used as insecticides but also as ligands for gene expression and gene therapy. It is imperative therefore that the biological and pharmacological responses, as well as structure–activity relationships both *in vivo* and *in vitro*, have to be fully understood for the rational and safe use of these analogues. It is important to understand the dose-response, effective concentrations, and physicochemical properties of these compounds. In order to understand their target specificity and degradation, their metabolic and environmental fates should be elucidated.

II. DIACYLHYDRAZINES

A. CHEMISTRY

Hsu (1991) reported that synthesis of DAH is relatively straight forward as illustrated in Fig. 2. Basically the different benzoic acids are converted to their corresponding benzoyl chlorides, and the products reacted with

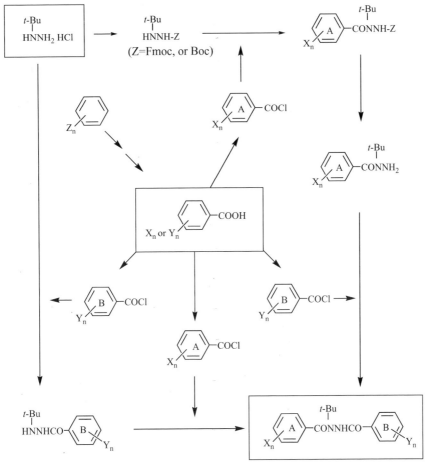

FIGURE 2. Representative synthetic scheme of DAHs.

N-alkylhydrazine hydrochlorides, such as *tert*-butylhydrazine hydrochloride, followed by coupling with other benzoyl chlorides. Where benzoic acids not commercially available, benzoic acids were derived from substituted benzenes, such as aniline and benzonitrile, as shown in Fig. 2. To synthesize alkanoyl analogues, corresponding acyl chlorides were used instead of benzoyl chloride (Nakagawa *et al.*, 1998; Shimizu *et al.*, 1997). Using this procedure, a number of DAHs have been synthesized in our laboratory (Nakagawa *et al.*, 1998, 1999, 2000a; Oikawa *et al.*, 1994a,b; Shimizu *et al.*, 1997). In this review, the benzene ring of dibenzoylhydrazine close to the *tert*-butyl group will be referred to as A-ring, and the other ring will be called the B-ring as indicated in Fig. 2. The structure–activity relationships and essential physicochemical properties were quantitatively analyzed for both *in vivo* and *in vitro* activity (Nakagawa *et al.*, 1995a, 1998, 1999, 2000a, 2001, 2002b, 2005; Ogura *et al.*, 2005a; Oikawa *et al.*, 1994a,b; Smagghe *et al.*, 1999) using Hansch–Fujita's classical quantitative structure-activity relationship (QSAR) method (Hansch and Fujita, 1964) and a comparative molecular field analysis (CoMFA) method (Cramer III *et al.*, 1988). The modifications of the hydrazine bridge of RH-5849 by dialkylation and cyclization resulted in the formation of biologically inactive compounds (Hsu, 1991; Toya *et al.*, 2002b).

A new type of DAH, chromafenozide (Fig. 1), was developed in Japan to control lepidopteran pests (Tanaka *et al.*, 2001). Sawada *et al.* (2003b) synthesized a series of DAHs in which one of benzoyl moieties (B-ring) of dibenzoylhydrazines was replaced with benzoheterocyclecarbonyl substructure, and the resulting compound with 2,3-dihydro-1,4-benzodioxine structure was found to be comparable in insecticidal activity to tebufenozide against *Spodoptera litura*. In a further study, the effects of the heterocyclic ring size as well as the substituent effects on the benzene ring on the insecticidal activity were examined. Although chromafenozide containing 5-methylchromane-6-carbonyl moiety was registered as a commercial insecticide, other compounds carrying different heterocyclic systems, such as benzodiole and benzodioxan, were also found to have insecticidal effects and were more potent than tebufenozide (Sawada *et al.*, 2003c). The effects of modifying the bridge moiety was also investigated (Sawada *et al.*, 2003a). The conversion of *tert*-butyl group of chromafenozide to $CH(CH_3)C(CH_3)_3$ group did not decrease the insecticidal activity. In addition, the substitution of NH hydrogen of chromafenozide with CN, $SCCl_3$, and CHO exhibited good and moderate insecticidal activity, whereas the substitutions with CH_3, $COCH_3$, and $CH_2OC_2H_5$ were detrimental to the activity.

The conformations of RH-5849 and tebufenozide were solved by X-ray crystal structure analyses (Cao *et al.*, 2001; Chan *et al.*, 1990; Hsu *et al.*, 1997; Nakagawa *et al.*, 1995a). These conformations were the same as that of the crystal structure of BYI06833 bound to EcR-LBD (Billas *et al.*, 2003). The conformational analysis of RH-5849 derived two energetically equivalent conformers (Wurtz *et al.*, 2000).

B. EFFECTS ON *IN VIVO* SYSTEMS

1. Larvae

Wing *et al.* (1988) discovered that RH-5849 caused the premature initiation of larval molting and had larvicidal effects against the tobacco hornworm, *Manduca sexta*. They also showed a positive correlation between induction of premature molting and Kc cell ecdysteroid receptor affinity for various ecdysone analogues, although it was later demonstrated that some compounds have selective toxicity among insect orders (Nakagawa *et al.*, 1999, 2001, 2002b; Smagghe *et al.*, 1999b). Topical application of 20E at

52 nmol resulted in no mortality of the rice stem borer, *Chilo suppressalis* Walker (Minakuchi *et al.*, 2003a). In addition, no mortality was observed in the application of 20E to 4th-instar larvae of *Chironomus tentans* by dipping in 10 mg/liter of 10% ethanol–water (Smagghe *et al.*, 2002b). However, it may be due to the unfavorable solvent in the topical application method. According to Retnakaran *et al.* (1997b), topical application was effective only when nonaqueous carriers, such as acetone and dimethyl sulfoxide, were used.

In the early 1990s the insecticidal activity of RH-5849 was measured against various insects and reported to be very potent to some Coleoptera (*Leptinotarsa decemlineata* Say, *Phaedon cochleariae* F.) and Lepidoptera (*Plutella xylostella* L., *Mamestra brassicae* L., *Pieris brassicae* Hübner), and moderately potent to Heteroptera (*Oncopertus fasciatus* Dallas) and Homoptera (*Acyrthosiphon pisum* Harris), but inactive against Dictyoptera (*Periplaneta americana* L., *Blatella germanica* L.) (Darvas *et al.*, 1992). They also reported that RH-5849 has no detectable effect on the pupariation process of Diptera (*Neobellieria bullata* Parker), although it was effective during larval molts (Darvas *et al.*, 1995). Even susceptible larvae, such as the cotton leafworms *Spodoptera littoralis* Boisd, were insensitive to RH-5849 just before the onset of wandering as well as 14–20 h earlier (Pszczolkowski and Kuszczak, 1996).

In the mid-1990s tebufenozide and methoxyfenozide were registered as insecticides to control Lepidoptera. Smagghe and Degheele (1994a) examined the susceptibility of tebufenozide to the last-instar larvae of Lepidoptera and concluded that the susceptibility to these insecticides decreased in the order *Spodoptera exempta* Walker > *M. brassicae* > *S. litoralis* > *Spodoptera exigua* Hübner > *Galleria mellonera* L., when insects were topically treated. African armyworm *S. exempta* was 85-fold more sensitive than the greater wax moth, *G. mellonera*. Although the difference of LD_{50} of tebufenozide between *S. exempta* and *S. exigua* is 8 times in the topical application, it was 74 times when the compound was orally applied. The LD_{50} values of tebufenozide against a beneficial insect, *B. mori*, were also determined to be 16.2 and 12.0 μg/larva for 72 and 96 h, respectively (Kumar *et al.*, 2000). However, tebufenozide had no activity on larvae of *L. decemlineata*, *Diabrotica virgifera virgifera* LeConte, and *Locusta migratoria migratorioides* R & F and nymphs of *Podisus sagitta* (Smagghe and Degheele, 1994a). They also found that RH-5849 and tebufenozide had no toxic effects on *Podisus nigrispinus* Dallus nymphs and on 3rd-instar and last-instar nymphs of *Podisus maculiventris* Say (Smagghe and Degheele, 1995b). Methoxyfenozide exhibited high insecticidal efficacy against a wide range of caterpillar pests, including many members of families Pyralidae, Pieridae, Tortricidae, and Noctuidae (Carlson *et al.*, 2001). In the larvicidal test to 4th-instar latvae of dipteran insect, *A. aegypti*, methoxyfenozide was approximately 10 times more active than RH-5849 and tebufenozide (Darvas *et al.*, 1998). Insecticidal activity of ecdysone agonists against *C. tentans* was in the order methoxyfenozide > tebufenozide > RH-5849 (Smagghe *et al.*, 2002b). The larvicidal activity of these representative 3 nonsteroidal ecdysone agonists against 5th-instar larvae of spruce budworms *Choristoneura fumiferana* Clem was measured, with the result that methoxyfenozide was the most toxic (1.05 ng/larva) followed by tebufenozide (8.8 ng/larva) and RH-5849 (236.9 ng/larva) (Sundaram *et al.*, 1998a).

As described earlier, tebufenozide was highly toxic to Lepidoptera but not to Coleoptera. Another DAH congener, halofenozide is, however, registered as an insecticide to control scarabids (Coleoptera) (Dhadialla *et al.*, 1998). Halofenozide was found to be toxic to various coleopterans, such as the Japanese beetle, *Popillia japonica* Newman; the European chafer, *Rhizotrogus (Amphimallon) majalis* Razoumowsky; and the oriental beetle, *Anomala orientals* Waterhouse. Among the coleopterans tested, the Japanese beetle was the most sensitive and the European chafer the least sensitive (Cowles and Villani, 1996). It was also tested against the oriental beetle, *Exomala orientalis* (Waterhouse), and the Asiatic garden beetle, *Maladera castanea* Arrow, by application to turf plots. Asiatic garden beetles were insensitive to halofenozide at all tested doses (Cowles *et al.*, 1999). According to Sundaram *et al.* (1998a), halofenozide (41 ng/larva) is more potent than RH-5849 against *C. fumiferana* but less potent than tebufenozide and methoxyfenozide.

The structure–activity relationships for the insecticidal toxicity of DAHs are very different among insect orders. In our laboratory, we conducted QSAR analyses to understand the physicochemical requirement for insecticidal potency. Oikawa *et al.* (1994a,b) synthesized two series of dibenzoylhydrazine congeners with various substituents at A-rings and B-rings (Fig. 2) and measured their larvicidal activity against *C. suppressalis*. In their bioassay procedure, compounds were topically applied in DMSO solution under the synergistic condition with an inhibitor of oxidation, piperonyl butoxide (PB). Using similar topical application methods, Nakagawa *et al.* tested substituted dibenzoylhydrazines for larvicidal activity against a lepidopteran species, *S. exigua*, and a coleopteran species, *L. decemlineata* (Nakagawa *et al.*, 1999, 2001, 2002b; Smagghe *et al.*, 1999b). In these studies, larvicidal activity was determined as the reciprocal logarithm of 50% lethal dose (LD_{50} mmol/insect), pLD_{50}, which is shown in Table I, with respect to 4 representative dibenzoylhydrazines. Structure–activity relationships for the larvicidal activities were similar between two lepidopteran insects, *C. suppressalis* and *S. exigua,* but were very different for *L. decemlineata*. Substituent effects on the larvicidal activity were quantitatively analyzed by the Hansch–Fujita method (Fujita, 1990; Hansch and Fujita, 1964). A part of the QSAR study for mono-substituted compounds is shown in Eqs. (1) and (2), which were derived for B-ring moiety against *C. suppressalis* (Oikawa *et al.*, 1994b) and *L. decemlineata,* (Nakagawa *et al.*, 1999) respectively. The QSAR equation for *S. exigua* was similar to Eq. (1) for *C. suppressalis* (Smagghe *et al.*, 1999b).

TABLE I. Larvicidal Activity of Representative Dibenzoylhydrazines

Compounds	pLD$_{50}$ (mmol/insect)			pLC$_{50}$ (mole/kg)^a
	C. suppressalis	S. exigua	L. decemlineata	A. aegypti
RH-5849	6.27	4.92	5.38	5.14
Tebufenozide	7.32	7.50	3.38	5.30
Methoxyfenozide	7.37^b	8.18	3.29	6.22
Halofenozide	6.61^b	6.10	6.09	5.34

^{a}Original data are listed in terms of LC$_{50}$ (mg/kg).
^{b}Unpublished data.

C. suppressalis:

$$pLD_{50}(B-ring) = 0.72 \log P - 0.88\Delta L^{ortho} - 0.98\Delta V_w^{meta} \\ -0.59\Delta L^{para} + 4.93 \quad n=30, \; s=0.254, \; r=0.912, \; F_{4,25}=30.91 \quad (1)$$

L. decemlineata:

$$pLD_{50}(B-ring) = 0.89 \log P + 0.40\Delta B_5^{ortho} + 0.44 E_s^{meta} + 0.65 ES^{para} \\ +1.09 HB + 2.97 \quad n=28, \; s=0.328, \; r=0.842, \; F_{5,22}=10.68 \quad (2)$$

In these equations, log P (P: partition coefficient in *n*-octanol and water) is the molecular hydrophobicity of compounds (Fujita et al., 1964). V$_w$ is the van der Waals volume of substituents (Bondi, 1964), and L and B$_5$ are STERIMOL parameters representing the length and maximum width of substituents (Verloop, 1983; Verloop et al., 1976). E$_s$ is a steric parameter derived from the difference in the rates of hydrolysis between CH$_3$COOC$_2$H$_5$ and XCH$_2$COOC$_2$H$_5$ (Kutter and Hansch, 1969). For hydrogen-bond acceptor substituents at the *para* position, a site-specific effect represented by the indicator variable HB was significant. To substituents with the hydrogen-accepting site at the β-position from the benzene ring, HB values were assigned so that HB = 0 for OCH$_3$, HB = 1 for CN and COCH$_3$, HB = 2 for NO$_2$ and SO$_2$CH$_3$. Multi-substituted compounds are also combinable with mono-substituted compounds in QSAR analyses by using either indicator variables or without counting one of the substituent effects for vicinal substitutions. Equations (1) and (2) indicate that hydrophobicity is the important parameter for the insecticidal activity. The significance of the hydrophobicity parameter for the QSAR of insecticidal activity against *L. decemlineata* is only valid for compounds containing meta-substituents on the A-ring moiety (Nakagawa et al., 2001). Steric effects were very different between *L. decemlineata* and *C. suppressalis*, suggesting that the reception pockets are different between them. Although the molecular mechanism can be roughly predicted from QSARs for *in vivo* activity, QSAR analyses for the binding affinity to receptors will help in understanding the ligand–receptor interactions.

Ecdysteroid titer profiles during the last larval stage of various insects have been compared, and the results indicated a common ecdysteroid profile with an initial small metamorphic peak followed by a large molt peak (Dean *et al.*, 1980; Nijhout, 1994). Smagghe *et al.* (1995c) investigated the ecdysteroid titer in last-instar larvae of *S. exigua* and *L. decemlineata* and examined the effects of nonsteroidal ecdysone agonists on the hormone titer. In the last-instar larvae of *S. exigua*, treated with either RH-5849 (100 mg/ml) or tebufenozide (3 mg/ml), precocious larval molting was induced within 24 h of treatment, and the hormone concentration remained at a basal level of 20–30 ng/ml without any elevation. Treatment of *L. decemlineata* larvae with tebufenozide caused no hormonal or physiological abnormalities, but the content of 20E remained at approximately 30 ng/ml. RH-5849 induced the supernumerary larval molt in diapause and nondiapause larvae when the compound was applied very early in the instar (Gadenne *et al.*, 1990). According to Retnakaran *et al.* (1997a,b), the newly molted 6th-instar spruce budworm larvae fed with tebufenozide (100 ng) went into a precocious molt and showed apolysis in 48 h, but no ecdysis or melanization (white new head capsule) was observed. Expression of a gene cascade was sequentially upregulated and downregulated by a 20E pulse during each instar. As a result of the expression of these upregulated genes, the larvae underwent apolysis, head capsule slippage, and other physiological events. Although 20E was cleared at this juncture, allowing the downregulatory genes to be expressed, ecdysone agonists strongly bound to receptors and remained in place, repressing all the down-regulatory genes. As a result, the treated larvae went into a precocious incomplete molt that was lethal (Retnakaran *et al.*, 2003). Tebufenozide is effective only if it is ingested prior to the appearance of the ecdysteroid peak in hemolymph. If the larva ingests the material after the peak, the effect was manifested in the succeeding larval instar (Palli *et al.*, 1995). Tebufenozide induced the expression of early genes, such as transcription factors *Choristoneura* hormone receptor 75 (CHR75) and CHR3, but since the compound bound strongly to the receptor and persisted in the epidermal cells, the genes that were expressed in the absence of ecdysone (downregulated genes) were not expressed (Palli *et al.*, 1995; Retnakaran *et al.*, 1995). Effect of DAH on the expression of transcription factors is described in a later section.

2. Pupae, Adults, and Eggs

Several studies on the effects of DAH on adults and eggs have been reported. The sexual attractiveness and responsiveness of the adult codling moth, *Cydia pomonella* L., exposed to surfaces treated with methoxyfenozide were investigated in wind tunnel and orientation-tube assays by Hoelscher and Barrett (2003). They concluded that a male's ability to respond to a calling female was more negatively affected by the ecdysone agonist than a female's ability to call and attract males. Smagghe and Degheele (1994b) stated that RH-5849 has a chemosterilizing activity resulting in total inhibition of oviposition within 2 days of continuous treatment in Lepidoptera, such as *S. exempta* and *S. exigua*, but when they did deposit eggs, they were viable similar to the controls. A considerable increase in fecundity (threefold) and no effect on egg viability was observed when 6th-instar larvae were fed on cotton leaves treated with 0.25 mg a.i./liter tebufenozide (Ishaaya *et al.*, 1995). RH-5849 at 10 μg/tick increased the wet weight of the ovary slightly, but tebufenozide failed to induce a similar increase in weight. Diacylhydrazines did not have any effect on the tick salivary gland ecdysteroid system (Charrois *et al.*, 1996; Smagghe and Degheele, 1994b). Similar chemosterilizing effects on reproduction occurred in *L. decemlineata* but at higher concentrations (Smagghe and Degheele, 1994a). According to Farinos *et al.*, adult *L. decemlineata* treated with halofenozide resulted in a rapid cessation of oviposition due to distorted ovaries, detrimental oocyte growth, and loss of oviposition with reduction of yolk protein synthesis. In the sugar beet weevil, *Aubeonymus mariaefranciscae* Roudier, the development of the progeny was strongly affected by halofenozide, and more than 80% of 1st-instar larvae died after hatching in the first 25 days after the treatment, which was due to premature molting and inhibition of ecdysis (Farinos *et al.*, 1999). Ultrastructural observations of the ovaries at the end of vitellogenesis in treated females with halofenozide showed no evident differences with the fine structure of follicular cells or oocytes in controls (Taibi *et al.*, 2003). Both RH-5849 and tebufenozide had no effect or chemosterilizing effect on predatory soldier bugs, *P. nigrispinusi* and *P. maculiventris*, both by oral treatment and topical application (Smagghe and Degheele, 1995b). Degeneration of the epithelial cells, reduction in the number of veins, precocious cuticle formation, and inhibition of growth of normal wing scales were observed in insects treated with 50–100 ng/pupa. At the higher dose (200 ng/pupa), there was significant mortality of pupae of *C. fumiferana*. Injection of tebufenozide into pupa resulted in a dose-dependent induction of mRNA for an ecdysone-induced transcription factor (CHR3), in a manner similar to that of larvae. However, the effects are not expressed overtly and are camouflaged by the pharmacological effects (Sundaram *et al.*, 2002). RH-5849 and tebufenozide adversely affected the mating success of *S. litura* when the surviving treated males were crossed with normal females and decreased the longevity of treated males and of untreated females when crossed with treated males (Seth *et al.*, 2004). The number of eggs laid by untreated females mated to treated males decreased, and the fertility of the eggs was reduced. Tebufenozide was active at lower concentrations than RH-5849. Injections of RH-5849 given to pharate adults or newly emerged adult *S. litura* also caused drastic reduction in the number of sperm in the upper regions of the male tract. Ovicidal activity was observed when eggs of the sugarcane borer, *Diatraesacchavlis* F, were exposed to tebufenozide. Concentrations of 100 ppm or more reduced survival to less than 10% (Rodriguez *et al.*, 2001). Trisyono and Chippendale (1997) examined the effects of methoxyfenozide and tebufenozide on eggs and larvae of the European corn borer, *Ostrinia nubilalis* Hübner. More than 90% of eggs died when egg masses were dipped in a solution of 100 ppm of methoxyfenozide and tebufenozide in acetone:water (1:1). Even at lower concentrations the survival rate was low (Trisyono and Chippendale, 1997). They also reported that both tebufenozide and methoxyfenozide exhibited a concentration-dependent ovicidal activity against the southern corn borer *Diatraea grandiosella* Dyar, and more than 95% of eggs died when egg masses were dipped in solutions of 100 or 200 mg/liter of 50% acetone–water compounds (Trisyono and Chippendale, 1998).

3. Resistance

The selective toxicity of DAH-type ecdysone agonists is primarily determined by the different binding affinity of ligands to receptors, which in turn is due to the difference of the primary sequence of the target receptor site of receptors. The difference in detoxifying (metabolizing the DAH) ability between species is another factor responsible for the selective toxicity. To date, many resistant strains to insecticides, such as organophosphate and pyrethroids, have emerged around the world. The green-headed leafroller, *Planotortrix octo* Dugdale, from New Zealand that was known to be resistant to an organophosphate insecticide, azinphosmethyl, has been shown to be cross-resistant to tebufenozide (Wearing, 1998). Waldstein and Reissig (2000) also found that the organophosphate-resistant oblique-banded leafroller, *Choristoneura rosaceama* Harris was resistant to tebufenozide with the resistance ratio of 12.8 but only 1.1–3.2 for other neurotoxic compounds. Another experiment was executed using populations of *C. rosaceama* in British Columbia by Smirle *et al.* (2002). Resistance to methoxyfenozide and tebufenozide was highly correlated with resistance to azinphosmethyl across populations, indicating the existence of cross-resistance between these compounds. They concluded that a resistance management strategy, such as the rotation of azinphosmethyl and dibenzoylhydrazines, is unlikely to be successful for *C. rosaceama*, even though the rotation of these compounds, which have different modes of actions, are generally favored to minimize the appearance of a tebufenozide-resistant population. Susceptibility to tebufenozide and methoxyfenozide of *S. exigua* from the southern United States and Thailand was determined through exposure to dipped leaves (Moulton *et al.*, 2002). Thailand strains were 45–68 times and 150–1500 times less susceptible to tebufenozide and 320–340 times and 67–120 times less susceptible to methoxyfenozide as first and third instars, respectively, when compared with the laboratory reference strain. Even among U.S. field populations, their susceptibility to both methoxyfenozide and tebufenozide varied from area to area.

Larvicidal activity of methoxyfenozide, tebufenozide, halofenozide, and RH-5849 was evaluated in a multi-resistant population of *S. littoria* that originated from Israeli fields (Ishaaya *et al.*, 1995), and resistance ratios were calculated to be 2.99, 1.00, and 1.18 for methoxyfenozide, tebufenozide, and halofenozide, respectively (Smagghe *et al.*, 2001). Larvae of *S. littoralis* collected in cotton fields in Israel had been sprayed heavily with pyrethroids and organophosphates and were >100-fold more resistant to a pyrethroid than a well-defined susceptible strain. Methoxyfenozide was threefold to sevenfold more potent than tebufenozide on the susceptible strain, and sevenfold to 14-fold more potent on the field strain (Ishaaya *et al.*, 1995). Cross-resistance was also found between insect growth regulators. Two populations of *C. pomonella* that were resistant to teflubenzuron (sevenfold resistance) and triflumuron (102-fold resistance) in France were cross-resistant to tebufenozide (26-fold) (Sauphanor and Bouvier, 1995).

Rates of penetration and excretion as well as absorption of compounds also affect their potency. Farinos *et al.* (1999) examined the rate of penetration of labeled halofenozide through the cuticle of *L. decemlineata* and *A. mariaefranciscae*, obtaining a similar pattern in these insects, whereas the rate of excretion was much more rapid in *A. mariaefranciscae*. The extremely slow excretion of halofenozide in *L. decemlineata* must contribute to its higher larvicidal activity in *L. decemlineata*. Retention of halofenozide in the reproductive system of males of *A. mariaefranciscae* and *L. decemlineata* was low, whereas radioactivity was recovered in females at a high level in both ovaries and eggs. Comparing females of both Coleoptera insects, *L. decemlineata* retained a higher amount of halofenozide than *A. mariaefranciscae* (Farinos *et al.*, 1999). The fact that absorption of halofenozide in multi-resistant strains is 25% lower than in susceptible strains was consistent with a 20% reduction in toxicity (Smagghe *et al.*, 2001). It was also shown that Cf-203 cells accumulated [^{14}C]tebufenozide, but Dm-2 cells excluded the material. The fact that tebufenozide is lepidopteran-specific and RH-5849 is Dipteran-specific is reflected in the respective effects of the two compounds on lepidopteran Cf-203 and dipteran Dm-2 cell lines (Retnakaran *et al.*, 2001).

Accumulation and active exclusion of PonA was similar between lepidopteran and dipteran cell lines. But lepidopteran cell lines (Cf-203 and Md-66) retained more of this compound within the cells than the dipteran cell lines (Dm-2 and Kc). The clearance was blocked by 10^{-5}M of oubain, Na$^+$, K$^+$ATPase inhibitor, suggesting that tebufenozide enters into the cells passively (influx) and is accumulated in Cf-203 cells, but it is actively excluded (efflux) from Dm-2 cells (Sundaram *et al.*, 1998b). In a further study, it was reported that tebufenozide was actively excluded from resistant cells by ATP-binding cassette (ABC) transporters (Retnakaran *et al.*, 2001). Among various transporters, Pdr5p was responsible for the active exclusion of [^{14}C]tebufenozide in yeast. This exclusion was temperature-dependent and blocked by ATPase inhibitors, such as oligomycin and vanadate. Mutants with the pleiotropic drug resistance 5 (*PDR5*) deletion can also selectively accumulate [^{14}C]halofenozide and [^{14}C]methoxyfenozide, but not [^{14}C]RH-5849, indicating that these three compounds (tebufenozide, methoxyfenozide, and halofenozide) share the same transporter Pdr5p for efflux (Hu et al., 2001). They discovered that older instars of the white-marked tussock moth, *Orgyia leucostigma*, are resistant to tebufenozide, which is probably due to the presence of such an exclusion system (Retnakaran et al., 2001).

Grebe et al. (2000) successfully selected clones with defects in ecdysteroid receptor function by treating an epithelial cell line from *C. tentans* with tebufenbozide at 0.1 nM to 0.1 μM. With regard to hormone binding, several types of hormone resistance were distinguished: (1) the same two high-affinity hormone recognition sites are present as in wild-type cells (K_{D1} = 0.31 nM, K_{D2} = 6.5 nM), but the number of binding sites is reduced. (2) The binding site with the lower affinity (K_{D2}) is missing. (3) The binding site with the higher affinity (K_{D1}) is missing. (4) No specific binding is observed. Ponasterone A binding can be rescued by addition of EcR but not USP. (5) Ligand specificity is altered. Tebufenozide cannot compete efficiently with [^{3}H]PonA in wild-type cells. Enhancement of the metabolism of 20E is also associated with hormone resistance in clones of the epithelial cell line from *C. tentans*, selected under the continuous presence of 20E (Kayser et al., 1997; Spindler-Barth and Spindler, 1998).

4. Metabolism

Nakagawa et al. (1985, 1987, 1989) reported that PB and *S,S,S*-tributylphosphorotrithioate (DEF), an inhibitor of hydrolysis, significantly enhanced the larvicidal activity of some benzoylphenylureas. Even in *in vitro* cultured integument systems, the activity of the compounds with electron-donating substituents at phenyl moiety was enhanced 20–30 times under the synergistic activity of PB (Nakagawa et al., 1992). *S,S,S*-tributylphosphorotrithioate was effective for compounds with a sterically small and electron-withdrawing substituent at the benzoyl moiety of benzoylphenylureas (Nakagawa et al., 1985). However, most dibenzoylhydrazines are only marginally synergized *in vivo* with these metabolic inhibitors, while no synergistic effect was observed in *in vitro* system (Nakagawa et al., 1995b). This indicates that dibenzoylhydrazines show greater resistance to oxidation and hydrolysis compared to benzoylphenylureas. In a susceptible strain of *C. rosaceama*, both PB and DEF did not show any significant level of synergism with most of insecticides, such as indoxacarb, cypermethrin, chlorpyrifos, azinphosmethyl, tebufenozide, and chlorfenapyr (Ahmad and Hollingworth, 2004). In a resistant strain of this species, tebufenozide was synergized with DEF but not PB. Another synergist, diethyl maleate (DEM), was synergistic to some extent in both the susceptible and resistant strains equally. However, when the concentration was increased to 20 ppm, both PB and DEM significantly synergized the toxicity of tebufenozide in the resistant as well as the susceptible colonies (threefold to fourfold) of *C. rosaceama* (Waldstein and Reissig, 2000).

According to Darvas *et al.* (1998) the inhibitor of cytochrome P-450, verbutin, did not have any significant synergistic effect with halofenozide and methoxyfenozide but showed mild synergism with RH-5849 and tebufenozide. Smagghe and Degheele (1993, 1994c) studied the pharmacokinetics and metabolic detoxication of RH-5849 and tebufenozide using radiolabeled compounds in several species of insect species, such as *S. exempta*, *S. exigua*, and *L. decemlineata*, and they concluded that the difference in their toxicity were not due to pharmacokinetic and metabolic differences.

5. Environmental Fate and Toxicity

Dibenzoylhydrazines show very low toxicity to a wide range of mammals, birds, amphibians, and fish. They are also well tolerated at high doses in both sub-chronic and chronic dietary exposure studies, and are nonirritating, nonmutagenic, and noncarcinogenic and do not cause adverse reproductive or growth effects (Carlson, 2000). In particular, acute oral LD_{50} values of commercial insecticides, such as tebufenozide, methoxyfenozide and chromafenozide, are very low (>5000 mg/kg for rats and mice). Halofenozide is slightly more toxic for rats (2850 mg/kg) and mice (2214 mg/kg), but the toxicity of halofenozide is still very low compared to other insecticides, such as imidacloprid (450 mg/kg) and fipronil (95–97 mg/kg) (see *The Pesticide Manual*, 2003, BCPC). Methoxyfenozide is slightly systemic through the root system but has negligible systemic effects in leaf systems. Evidence collected to date indicates that methoxyfenozide has an excellent margin of safety to nontarget organisms, including a wide range of nontarget and beneficial insects (Carlson *et al.*, 2001).

It is reported that the mosquito species, *Aedes taeniorhynchut* and *A. aegypti*, are susceptible to tebufenozide, whereas aquatic crustaceans, such as *Artemia* sp. and *Daphnia magna* Straus, are tolerant, indicating that tebufenozide is a selective insecticide with little or no adverse effects on nontarget crustacean organisms (Song *et al.*, 1997). Kunkel *et al.* (1999) evaluated the impact of halofenozide on earthworms and beneficial arthropods and found no reduction in abundance of any group of beneficial invertebrates. Application of halofenozide followed by irrigation will have relatively little impact on beneficial invertebrates, although halofenozide is persistent enough to control *P. japonica*, and *Cyclocephara* spp. grubs in the soil that will emerge several months later.

Sundaram *et al.* (1997a) investigated the persistence and metabolic fate of tebufenozide in clay loam soil, forest litter, spruce needles, and shoots after a spray application of an aqueous formulation of tebufenozide. Tebufenozide was lost very rapidly from the sandy substrates than from the clay substrates in the laboratory microcosm studies. They concluded that clay substrates have a greater adsorptive capacity, and clay-type soil provides a greater protection against downward mobility (Sundaram, 1995, 1997a). The persistence characteristics observed in laboratory and field studies were due to the fluctuating environmental conditions and water pH encountered in the field study, compared with the constant environmental conditions and water pH utilized in the laboratory study (Sundaram, 1997b). They also investigated the persistence of tebufenozide in white spruce foliage under an actual forest environment. Foliage was collected at different intervals of time up to 64 h, and tebufenozide residues were measured by HPLC. Foliage was also fed to laboratory-reared 4th-instar and 6th-instar *C. fumiferana*. Tebufenozide residues in foliage declined with time linearly following first-order kinetics, and half-life was 20–45 days. Larval mortality declined gradually in keeping with the residue levels, but the effect lasted even when larvae were fed with foliage collected 64 days after treatment (Sundaram *et al.*, 1996a,b). The field trial was conducted to determine the foliar deposit of tebufenozide applied aerially against *C. fumiferana* using a commercial formulation, Mimic 240LV, containing 250 g active material per liter of formulation. Tebufenozide recovered from foliage ranged from 2.5 to 5.9 μg/g foliage when 1 liter of formulation was sprayed per hactare, and 5.8–6.8 l/g foliage after 2 liter/ha were sprayed. The mean percentage of population reduction ranged from 61.9 to 93.6% for 1 liter/ha spray and 85.6–98.3% for 2 liter/ha application (Cadogan *et al.*, 1998).

C. EFFECTS ON *IN VITRO* SYSTEMS

1. Tissues and Cells

Molting hormonal activity can be evaluated *in vivo* by observing the morphological change induced in the ligated abdomen of the insect (behind the prothoracic glands that secrete the native hormone) and treating with hormone agonists. They can also be studied by the partial rescue of wild-type phenotypic expression in ecdysone deficient mutants. In the ligated *Galleria* larvae, ED_{50} values of 20E, RH-5849, and tebufenozide were 1.3, 1.75 and 0.9 μg, respectively (Slama, 1995). ED_{50} values of 20E and RH-5849 were 1.9 and 1.2 μg in *Manduca*, and 0.1 and 0.013 μg in *Pieris*. However, both RH-5849 and tebufenozide were completely ineffective in the standard assays on ligated larvae of *Sarcophaga* and *Calliphora*. Since the molting hormone agonists induce various biological responses even on isolated organs, tissues, and cells, various *in vitro* systems have been used to evaluate the molting hormone activity: (1) induction of process evaluation and inhibition of proliferation in cells (Wing, 1988); (2) induction of early ecdysone-specific puffs on the polytene chromosomes of the salivary glands; and (3) evagination of imaginal discs (Frisrom and Yund, 1976; Smagghe *et al.*, 1996a, 2000).

Wing (1988) evaluated the hormonal effect of DAHs by measuring the cellular changes in Kc cells and estimated the EC_{50} values of 20E (0.035 μM) and RH-5849 (4.8 μM). Sohi et al. (1995) investigated the effects of 20E, RH-5849, and tebufenozide on three insect cell lines, Md-66 cell line from the forest tent caterpillar, *Malacosoma disstria*, and two cell lines Cf-1 and Cf-70 from *C. fumiferana*. Md-66 cells responded to all three compounds by forming clumps and by producing filamentous extensions, whereas cell attachment was induced and cell proliferation reduced by ecdysone agonists. Although Cf-70 cells responded to all three compounds to a lesser extent, Cf-1 cells showed little or no morphological response. Because tebufenozide responded to Md-66 cells both dose and time dependently and induced the expression of the *M. disstria* hormone receptor (MdHR3) as was the case with 20E, the Md-66 cell line was thought to be an excellent *in vitro* model system to study the action of ecdysone agonists. Quack et al. (1995) reported that RH-5849 and tebufenozide induced differentiation in the epithelial cell line from *C. tentans*, arresting cell growth and regulating chitin metabolism, which are elicited by 20E.

The molting hormonal activity of 20E in inducing the evagination of imaginal discs was 0.031 μM in terms of EC_{50} (Frisrom and Yund, 1976), which was similar to that evaluated in Kc cells by Wing (1988). Smagghe et al. (2000) determined the potency of RH-5849, tebufenozide, methoxyfenozide, halofenozide, and 20E in the cultured imaginal discs of *S. littoralis*. The resulting order of potency of these compounds was methoxyfenozide (0.011 μM) > 20E (0.291 μM) > tebufenozide (0.403 μM) > halofenozide (0.472 μM) > RH-5849 (44.5 μM). They also reported the hormonal activity of RH-5849, tebufenozide, and 20E on the cultured mesothoracial imaginal wing discs of *L. decemlineata* and *G. mellonella* (Smagghe et al., 1996a). The order of receptor binding in terms of IC_{50} in the wing discs of *G. mellonella* was tebufenozide (0.022 μM), 20E (0.106 μM), and RH-5849 (0.911 μM), and this was in agreement with the order of the biological activity for wing discs in terms of EC_{50}—tebufenozide (0.009 μM) > 20E (0.321 μM) > RH-5849 (0.865 μM). EC_{50} values of tebufenozide, 20E, and RH-5849 in the isolated imaginal discs of *L. decemlineata* were 0.757, 0.061, and 0.461 μM, respectively. The potency of tebufenozide was very different between *L. decemlineata* and *G. mellonella,* although the potency of 20E was similar between them (Smagghe et al., 1996a). The EC_{50} values of RH-5849, 20E, tebufenozide, and PonA to elicit evagination were 0.870, 0.090, 0.012, and 0.003 μM, respectively, in the cultured imaginal wing discs of last-instar larvae of *S. exigua*. These values are linearly correlated to the inhibition of receptor binding evaluated in imaginal discs (Smagghe and Degheele, 1995a). Farkas and Slama (1999) examined the dose-response relationships of 20E, RH-5849, and tebufenozide for the evagination response of the explanted imaginal discs of *D. melanogaster* and concluded that 20E was approximately 100 times more potent than RH-5849 and tebufenozide, although there was no difference in the activity between RH-5849 and tebufenozide.

Farkas and Slama (1999) reported that RH-5849 and tebufenozide exhibited a similar dose-response relationship in the induction of early chromosomal puffs (74EF and 75B) or regression of the preexisting puffs (25AC and 68C) as did 20E. 20-Hydroxyecdysone was two orders of magnitude more active than RH-5849 and tebufenozide in assays related to glycoprotein glue secretion, evagination of imaginal discs, and rescue of phenotypic expression in ecdysone deficient mutants, but the difference in the activity was small between RH-5849 and tebufenozide. Ecdysteroids caused salivary gland degeneration in female ixodid ticks, *Amblyomma hebraeum* Koch, but no ecdysone mimicking effect was observed when treated with levels up to 51 μM of RH-5849 and up to 28 μM of tebufenozide, although both DHAs had an ecdysone-mimicking effect *in vivo* (Charrois *et al.*, 1996). New cuticle was incomplete in last-instar and 3rd-instar larvae of *S. exigua* treated with tebufenozide, but in cultured discs depositition a new cuticle was stimulated within 12 h after cultivation in a medium containing tebufenozide. These observations indicated a hyperecdysteroid action and confirm that the molt accelerating mode of action of tebufenozide resulted in a forced, untimely synthesis of cuticle by activation of epidermal and epithelial cells, and that its ecdysis inhibitory activity is mediated by its effect on post-apolysis process (Smagghe *et al.*, 1996b).

Reporter gene assay has been used to measure hormonal activity (Mikitani, 1995; Swevers *et al.*, 2004; Toya *et al.*, 2002a). Both PonA and 20E bound to EcRs and activated the luciferase gene constructed in the ecdysone responsive reporter plasmid in a dose-response manner in both Sf-9 and Kc cells. Chromafenozide activated reporter transcription with comparable potency to PonA only in Sf-9 cells, but its maximum activity was fourfold lower than that of PonA (Toya *et al.*, 2002a). It appears to bind lepidopteran EcR with comparable affinity to PonA but may activate the EcR in a different manner. Swevers *et al.* (2004) constructed a cell-based high throughput screening system using *Bombyx mori* cell lines with a reporter gene construct encoding green fluorescent protein and measured the molting hormonal activity of seven DAHs, including RH-5849, tebufenozide, methoxyfenozide, and halofenozide. The EC_{50} values of tebufenozide and methoxyfenozide were determined to be 0.5 and 0.9 nM, respectively (Swevers *et al.*, 2004). The molting hormonal activity of various DHAs has been measured in this *in vitro* system, and a QSAR study is in progress.

2. Enzymes and Proteins

Nakagawa *et al.* (1995b) developed a method to evaluate the activity of chitin synthesis inhibitors in cultured integument segments treated for 24 h with 20E. In this integument system 20E worked as an inhibitor of chitin synthesis and IC_{50} was determined to be approximately 1 μM, indicating that the chitin synthesis was not initiated without reduction of 20E titer in the hemolymph. IC_{50} values of five DAHs varied from 0.08 to 7.8 μM, whereas that of a potent chitin synthesis inhibitor, diflubenzuron, was 0.02 μM (Nakagawa et al., 1992). Insect cuticle is made of chitin and proteins, and the inhibition of cuticle protein synthesis is detrimental to growth. There are several reports of the effects of RH-5849 on the deposition of cuticle proteins and enzyme activity. Smagghe and Degheele (1992) examined the differences in protein pattern of the cuticle between treated and untreated 6th-instar larvae by using electrophoresis and demonstrated that some proteins were missing in the newly induced larval cuticle or expressed with less intensity. The lack of several bands in the pattern of cuticular and hemolymph proteins of treated versus untreated larvae, probably from proteins specific for the pupal instar, was suggested as a cause of unsuccessful pupation in the treatment (Smagghe and Degheele, 1992). Retnakaran et al. (1995) showed that tebufenozide and 20E inhibited the expression of 14-kDa larval cuticle protein transcript levels in cultured segments of *Manduca*. Larvae of *L. decemlineata* treated with RH-5849 (10 and 50 mg/liter), tebufenozide (2 g/liter), halofenozide (0.1 mg/liter), and 20E (2 g/liter) by oral administration showed symptoms of premature molting and failed to undergo ecdysis. The effects of compounds on hemolymphal and cuticular proteins were probably linked to premature and new epicuticle deposition (Smagghe et al., 1999a).

20-Hydroxyecdysone appears as a peak during each instar with a rise and a fall, which results in the upregulation and downregulation of genes involved among other things in the molting process. Interestingly, molting hormone stimulates at least one universal route for its own inactivation by inducing ecdysteroid 26-hydroxylase, which is analogous to vitamin D inactivation in vertebrates (Bergman and Postlind, 1991). Inactivation of ecdysteroids was accomplished by: (1) 26-hydroxylation and further oxidation to the 26-oic acid; (2) formation of various conjugates; and (3) conversion into 3-epiecdysteroid (Takeuchi et al., 2000, 2001; Williams et al., 1997, 2000). Since ecdysteroids have some effects on the induction of metabolic enzymes, and consequently DAHs may have a similar effects as ecdysteroids on enzymes. Keogh and Smith (1991) found that RH-5849 inhibited the cytochrome P-450–dependent ecdysone-20-monooxygenase in the midgut of wandering last-instar larvae of *M. sexta*. RH-5849 was also found to elicit a dramatic increase in midgut steroid hydroxylase activity when injected into competent head or thoracic ligated prewandering last-instar larvae. Williams et al. (1997) examined the effect of RH-5849 on the induction of oxidase. Administration of RH-5849 but not 20E to *M. sexta* resulted in induction of midgut cytosolic ecdysone oxidase and ecdysteroid phosphotransferase activities. In addition, both RH-5849 and 20E caused induction of ecdysteroid 26-hydroxylase activity in midgut mitochondria and microsomes, whereas

20-hydroxylase was induced to a lesser extent by 20E in mitochondria and by either RH-5849 or 20E in microsomes. According to Taibi et al. (2003), halofenozide did not affect the amount of ecdsone, but significantly reduced the amount of 20E. Lower amounts of 20E may be due to a negative effect on ecdysone-20-monooxygenase, which is necessary for conversion of ecdysone to 20E. It is also reported that the activity of ecdysone-20-monooxygenase of the Chinese oak silkmoth *Antheraea pernyi* was strongly reduced by treatment with halofenozide and significantly reduced by RH-5849 and tebufenozide (Williams et al., 2002). Dopa decarboxylase (DDC) transcription normally occurs at the end of the molt, and the epidermis cultured *in vitro* requires initial exposure to 20E for 17 h followed by its removal. Both *in vivo* and *in vitro* treatments with tebufenozide prevented DDC expression up to 48 h after the removal of the compound (Retnakaran et al., 1995). A putative RNA helicase cDNA (*CfrHlc64*) with 1998 nucleotides was isolated from *C. fumiferana*. Tebufenozide enhanced the expression of *CfrHlc64* in a dose-dependent manner (Zhang et al., 2004). The induction of acetylcholinesterase is known to be a characteristic effect of ecdysone agonists, and EC_{50} values of 20E and RH-5849 for the increase in the specific activity of acetylcholinesterase were determined to be 0.007 and 1.05 μM, being consistent with their hormonal activity (Wing, 1988).

Tebufenozide and methoxyfenozide caused vitellogenin accumulation in hemolymph of *C. pomonella*. The expression of 65-kDa EcR and 60- and 64-kDa USP proteins in the ovaries of *C. pomonella* was enhanced by tebufenozide and methoxyfenozide at both the transcription and translation levels. Northern hybridization analysis indicated that this EcR protein was encoded by EcRB1 transcript and that two USP bands were the products of USP-1 transcript. The data suggest that these compounds regulate the expression of these specific proteins, which might eventually lead to the inhibition of fecundity in *C. pomonella* (Sun et al., 2003b). Tebufenozide and methoxyfenozide also significantly reduced the 47-kDa EcR expression after 12 h exposure and greatly inhibited this protein expression after 24 h in the female fat bodies from *C. pomonella*. At 48 h, the level of the 47-kDa EcR was significantly increased in only the tebufenozide-treated sample and not in methoxyfenozide. Moreover, these ecdysone agonists significantly induced the expression of USP proteins in female fat bodies after 12 h exposure, and then the pattern was followed by a dramatic inhibition following 24 h exposure. When the exposure length reached 48 h, the USP expression was significantly enhanced by up to 3.8-fold by tebufenozide but still inhibited completely by methoxyfenozide (Sun et al., 2003a). Injection of tebufenozide into developmentally arrested pupal abdomens of *B. mori* initiated vitellogenesis in ovarian follicles, but further development toward the eggshell (chorion) production is not sustained. The developmental arrest occurs during mid-vitellogenesis, prior to the initiation of a cascade of changes in the expression of regulatory factors (Swevers and Iatrou, 1999).

RH-5849 stimulated N-acetylglucosamine (Glc-NAc) uptake and inhibited cellular proliferation in IAL-PID2 cells at similar concentrations (Shilhacek et al., 1990). The cell line usually forms multi-cellular vesicles consisting of a squamous monolayer, but treatment with ecdysteroids and DAHs led them to the formation of a stratified columnar epithelium. This differentiation accompanied by changes in protein pattern was observed in the epithelial cell line from C. tentans. Ecdysteroids and DAHs not only inhibit chitin synthesis but stimulate the degradation and secretion of chitin (Quack et al., 1995). Cultured discs were stimulated to deposit a new cuticle within 12 h following cultivation in a medium containing tebufenozide. Inhibitory activity of ecdysis is mediated by its effect on postapolysis processes (Smagghe et al., 1996b). Based on the role of 20E in the induction of chitin synthesis in vitro, Oikawa et al. (1993) constructed the in vitro assay method using cultured integument to evaluate the molting hormonal activity. In their in vitro system ecdysone agonists afforded the typical hormonal response (bell-shaped dose-response relationship) for the chitin deposition, although the transfer of integument fragments to the hormone-free condition is necessary for the chitin synthesis. The 50% effective concentration (EC_{50}, M) was determined for various ecdysone agonists from each concentration–response curve, and its reciprocal logarithm value (pEC_{50}) was used as the index of the hormonal activity. With this activity index, Nakagawa et al. (1995b) quantitatively analyzed the structure–activity relationship of dibenzoylhydrazines with various substituents at A-rings and B-rings, which was similar to the QSAR equation for larvicidal activity (Oikawa et al., 1994a,b). They also quantitatively analyzed DAHs, including alkanoyl analogues, to derive Eq. (3) (Nakagawa et al., 2000a).

$$pEC_{50} = 7.78 \log P - 0.78 (\log P)^2 - 0.893 D_A - 0.609 D_B - 3.227 \quad (3)$$
$$n = 23, \ s = 0.557, \ r = 0.929, \ F_{4,18} = 28.291, \ \log P_{opt} = 5.15$$

As shown in Eq. (3) an optimum hydrophobicity ($\log P_{opt}$) is required for the hormonal activity. In Eq. (3), D_A and D_B correspond to respective length of the A-ring and B-ring moieties of the DAHs, indicating that long-substituted benzoyl and alkanoyl groups are not favorable for the activity. Furthermore, the structure–activity relationship was examined using CoMFA (Cramer et al., 1988) to visualize the physicochemical properties, which increase the activity. In CoMFA, sterically favorable and unfavorable fields as well as electrostatically negative and positive fields necessary for the hormonal activity were visualized (Nakagawa et al., 1995a, 1998).

The cell line established from the midgut of C. fumiferana (Cf-203) responded to 20E and tebufenozide, increasing expression of the transcription factor, CHR3, a homologue of Manduca hormone receptor 3 (MHR3) (Hu et al., 2004). This expression was similar to that seen in the M. disstria cell line (Md-66) (Sohi et al., 1995). No significant difference in morphological changes was observed between 20E and tebufenozide treatments, but the cells appeared to be more sensitive to 20E than to tebufenozide in terms of cell appearance. Cell numbers of the adapted cell lines in 20E and tebufenozide containing media were not significantly decreased, but both Cf-203 and Md-66 cell lines accumulated less [^{14}C]tebufenozide and lost the ability to express CHR3 in response to these compounds (Hu et al., 2004). When tebufenozide was administered after the peak of ecdysteroids, the larva molted normally into the next stage, but an incomplete molt in the subsequent larval stage was observed, due to persistence and carryover of tebufenozide. *Choristoneura* hormone receptor 3 was induced in the epidermis, fat body, and midgut of 6th-instar larvae treated with tebufenozide during all days of the 6th stadium (Palli et al., 1995). The relative potencies of RH-5849, tebufenozide, halofenozide, and methoxyfenozide to induce CHR3 in the 5th-instar larvae of *C. fumiferana* were studied by Sundaram et al. (1998a). Detectable levels of CHR3 mRNA were induced by methoxyfenozide at a concentration as low as 1.5 ng/larva, but similar inductions required concentrations as high as 7.5 ng of tebufenozide, 37.5 ng of halofenozide, and 187.5 ng of RH-5849. The order of potency of these four ecdysone agonists for inducing CHR3 mRNA was the same as that of their larval toxicity. A number of ecdysteroids were assayed for their ability to activate *Drosophila* nuclear receptors in transfected tissue culture cells. None of the compounds had a significant effect on the activity of three *Drosophila* hormone receptors, DHR38, DHR78, or DHR96 (Baker et al., 2000). On the other hand, similar to 20E, tebufenozide induced MHR3 mRNA and suppressed 14-kDa larval cuticular protein (LCP-14) transcript levels in the epidermis of *Manduca*, which was cultured *in vitro*. The ED_{50} (0.14 μM) of tebufenozide for the induction of MHR3 was 10 times less than that of 20E. When the epidermis was exposed to 20E for 17 h and then cultured in hormone-free medium for a further 48 h, the LCP-14 mRNA level went up to nearly 60% of the maximum level reached in the untreated control. However, when treated with tebufenozide instead of 20E, the LCP-14 mRNA level remained low even after 48 h (Retnakaran et al., 1995). It was reported that hormone receptor 3 of the African bollworm *Helicoverpa armigera* (HHR3) could be induced from the epidermis of newly molted 6th-instar larvae by methoxyfenozide (Zhao et al., 2004).

3. Nonmolt Related Effects

RH-5849 is moderately toxic to rats (LD_{50} = 435 mg/kg), but methoxyfenozide and tebufenozide are harmless to mammals (>5000 mg/kg) (*The Pesticide Manual*, 2003, BCPC). In order to investigate the neurotoxic effects of DAHs, Salgado injected RH-5849 into cockroaches and other insects at 50 μg/g, which caused a rapid onset of hyperactivity, leading to prostration and constant movement of all appendages within a few minutes, then paralysis after several hours (Salgado, 1992). RH-5849 at 100 μM induced strong contractions in muscles of larvae of the housefly *Musca domestica*.

Maintained voltage-gated K^+ current in muscles was reduced by 70% by treatment with 100 μM of RH-5849 (Salgado, 1998). At cercal nerve–giant axon synapses of *P. americana*, RH-5849 greatly prolonged excitatory postsynaptic potentials, but the transient K^+ current and Ca^{2+}-dependent–maintained K^+ current were not affected. RH-5849 and its analogues blocked K^+ channels in neurons as well as muscle in *M. domestica*. Nishimura *et al.* (1996) demonstrated that the rising and falling phases of an action potential, electrically induced in the cockroach giant axon, was decelerated by varying DAHs at moderately high concentrations. These compounds caused convulsions in *P. americana* when injected. However, the convulsive activity against *P. americana* did not correlate with the larvicidal activity against *C. suppressalis*. Very potent larvicides against *C. suppressalis* were either moderately active or inactive in convulsion.

Shilhacek *et al.* (1990) reported that RH-5849 inhibited the growth of larvae of the Indian meal moth *Plodia interpunctella* in a dose-response manner, but the deleterious effects of RH-5849 could be prevented by the simultaneous application of the juvenoid, methoprene. Larvae simultaneously treated with both RH-5849 and methoprene continued to grow until they attained a size about three times normal. According to Muszynska-Pytel *et al.* (1992) RH-5849 significantly increased allatotropic activity of the brain and also activated synthesis of JH by the corpora cardiaca/corpora allata complex. Simultaneous application of RH-5849 and fluoromevalonate (FMeV), a potent inhibitor of JH synthesis, to young final instar larvae lowered the incidence of perfect supernumerary larval molts. They concluded that the effect of RH-5849 on the developmental program in *G. mellonella* is mediated by the corpora allata.

D. EFFECTS ON RECEPTORS

1. Binding to *In Vitro* Translated Proteins

After the characterization of the *Drosophila* gene encoding EcR (Koelle *et al.*, 1991), various cDNAs coding EcRs of other insects (Cho *et al.*, 1995; Dhadialla and Tzertzinis, 1997; Fujiwara *et al.*, 1995; Hannan and Hill, 1997; Imhof *et al.*, 1993; Kamimura *et al.*, 1996; Kothapalli *et al.*, 1995; Martinez *et al.*, 1999; Minakuchi *et al.*, 2002; Mouillet *et al.*, 1997; Ogura *et al.*, 2005b; Saleh *et al.*, 1998; Swevers *et al.*, 1995; Verras *et al.*, 1999) as well as tick (Guo *et al.*, 1997) and crab (Chung *et al.*, 1998) have been successfully cloned. It is also reported that the heterodimeridization between EcR and USP was required for binding to the response element in the DNA as well as the binding of 20E (Yao *et al.*, 1993). The 3-D structure of the EcR-LBD of the tobacco burworm *Heliothis virescens*, was resolved by x-ray crystallography in 2003 (Billas *et al.*, 2003), whereas the 3-D structure of USP-LBD was resolved 2 years earlier (Billas *et al.*, 2001; Clayton *et al.*, 2001). Billas *et al.* (2003) crystallized two EcR-LBD/ligand complexes with 20E and a diacylhydrazine compound (BYI06830) and showed the ligand–receptor interaction as well as the superposition of 20E and BYI06830. Interestingly, ecdysteroid and nonsteroidal ecdysone only partially overlap ligand-binding pockets as shown in Fig. 3. The *t*-butyl group is placed to the terminal *i*-propyl moiety of the side chain, and 3,5-dimethylbenzoyl moiety matches with the main side chain of PonA. This superposition could not be predicted by molecular modeling and docking studies, which were done earlier (Hormann *et al.*, 2003; Kasuya *et al.*, 2003; Nakagawa *et al.*, 1995a, 1998; Wurtz *et al.*, 2000).

Wurtz *et al.* constructed the EcR-LBD complex based on the crystal structure of receptor proteins for vitamin D and retinoic acid and built an analogous docking model between LBD and either 20E or RH-5849. The proposed superposition was, however, fairly different from that obtained from x-ray crystal structure analyses. Another superposition proposed by Kasuya *et al.* (2003) in which chromafenozide was matched to the steroid nucleus of 20E was also different from the actual structure. Nakagawa *et al.* superposed ecdysteroids and DAHs by fitting the side chain moiety of steroids to the benzoyl moiety of DAHs (Nakagawa *et al.*, 1995a, 1998; Shimizu *et al.*, 1997). This idea was consistent with the actual structure, even though the benzoyl (B-ring) moiety that is not matched with any structural moiety of PonA in the crystal structure analysis was fitted on the side chain of PonA. The binding of all DAHs to EcR may not be same as that of BYI06830. In fact, a few DAHs could not be included in QSAR equations (Nakagawa *et al.*, 2005; Ogura *et al.*, 2005a). Before the publication of the

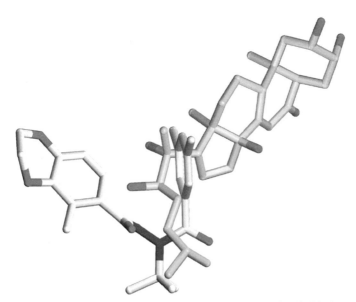

FIGURE 3. Superposition between PonA and BYI06830 (Reproduced with the permission of Nature Publishing Group). (See Color Insert.)

crystal structure of HvEcR-LBD, a CfEcR-LBD was constructed by homology modeling techniques, and 17 amino acid residues were identified as critical for 20E binding. In the mutation experiments of these amino acids, one particular mutant (A110P) failed to respond to steroids, but its response to RG-102240 (GS^{TM}E; Fig. 1) and RG-102317 (Fig. 1; BYI06830) was unaffected. One particular mutant A110P displayed a drastic reduction in the PonA binding but not in RG-102240 (70% of the WT activity). The R95A mutant also failed to respond to PonA, but unlike the A110P mutant, the R95A mutant showed a significant reduction in RG102240 activity. Interestingly, both A110 and R95A were predicted to be in the vicinity of the tail end of the 20E ligand (Kumar et al., 2002).

Minakuchi et al. succeeded in the cloning of EcR (Minakuchi et al., 2002) and USP (Minakuchi et al., 2003b) genes of *C. suppressalis* and performed the binding assay using *in vitro* translated proteins (Minakuchi et al., 2003a). They also performed the binding assay using DmEcR/DmUSP (Minakuchi et al., 2005). A cloning study of EcR and USP of *L. decemlineata* and a binding study using *in vitro* translated EcR/USP were performed by Ogura et al. (2005b). The potency in terms of pIC_{50} (IC_{50}: concentration [M] to inhibit 50% of PonA binding) of representative ecdysone agonists for the binding to *in vitro* translated EcR/USP is listed in Table II. The binding affinity of both steroidal and nonsteroidal ecdysone agonists toward the EcR/USP heterodimer of *C. suppressalis* was linearly correlated with that against the natural receptor proteins prepared from tissues of *C. suppressalis*

TABLE II. Binding of Ecydsone Agonists to *In Vitro* Translated EcR/USP Heterodimers

	pIC_{50} (M)		
Compounds	*C. suppressalis*[a]	*D. melanogaster*[b]	*L. decemliniata*[c]
PonasteroneA	8.08	8.27	8.13
20-Hydroxyecdysone	6.66	7.03	6.36
Cyasterone	6.65	7.07	6.29
Makisterone A	6.33	6.87	5.76
Ecdysone	4.70	5.24	4.98
RH-5849	6.50	5.16	4.97
Tebufenozide	8.85	6.01	5.18
Methoxyfenozide	8.87	6.49	5.99
Halofenozide	6.92	5.95	5.23
Chromafenozide	9.13	6.54	5.77

[a] Minakuchi et al., 2003a.
[b] Minakuchi et al., 2005.
[c] Ogura et al., 2005b.

(Minakuchi et al., 2003a), as well as the molting hormonal activity evaluated in the cultured integument of C. suppressalis (Nakagawa et al., 1998, 2000a). Carlson et al. (2001) compared the potency of RH-5849, methoxyfenozide, tebufenozide, and 20E by comparing their Kd values between *Plodia* EcR (methoxyfenozide [0.0005 μM] > tebufenozide [0.003 μM] > halofenozide [0.129 μM] > 20E [0.210 μM]) and *Drosophila* EcR (20E [0.060 μM]) > methoxyfenozide [0.124 μM] > tebufenozide [0.192 μM] > halofenozide [0.493 μM]).

2. Binding to Whole Tissue or Cultured Cells

Smagghe et al. (1996a) measured the binding affinity of ecdysone agonists using a whole imaginal disc of *S. exigua*. The IC_{50} values of 20E, PonA, RH-5849, and tebufenozide were 0.29 μM (pIC_{50} = 6.54), 0.007 μM (pIC_{50} = 8.15), 1.1 μM (pIC_{50} = 5.96), and 0.033 μM (pIC_{50} = 7.48), respectively, which is the same order as that for the biological response for the evagination. They also determined IC_{50} values of 20E, RH-5849, and tebufenozide to be 0.425 μM (pIC_{50} = 6.37), 0.74 μM (pIC_{50} = 6.13), and 1.32 μM (pIC_{50} = 5.88), respectively, in imaginal discs of *L. decemlineata* (Smagghe et al., 1996a). The binding of PonA to LmEcR/LmUSP of *L. migratoria* was significantly more frequent than for RH-5849, tebufenozide, methoxyfenozide, and halofenozide (Hayward et al., 2003). IC_{50} values of 20E, RH-5849, and tebufenozide were also evaluated in wing discs of *G. mellonella* to be 0.106 μM (pIC_{50} = 6.97), 0.911 μM (pIC_{50} = 6.04), and 0.022 μM (pIC_{50} = 7.66), respectively (Smagghe et al., 1996a). The order of toxicity to *S. littoralis* of ecdysone agonists corresponded with the binding competition with whole imaginal discs (methoxyfenozide > tebufenozide > halofenozide > RH-5849) (Smagghe et al., 2000). In *A. hebraeum* RH-5849 displaced PonA binding by only 30% even at high concentrations (230 μM) and tebufenozide by 43% at 79 μM. Spindler-Barth et al. (1991) demonstrated that ecdysone agonists were specific to EcR and did not replace dexamethasone or estradiol from the corresponding receptors of vertebrate. Ecdysone agonists inhibited the [^{3}H]PonA-binding dose dependently in *C. tentans* cell line, and pIC_{50} values of 20E and RH-5849 were calculated to be approximately 6.5 and 6.1, respectively.

Nakagawa et al. (2000a, 2002b) evaluated the binding affinity of various ecdysone agonists by measuring the competitive inhibition of [^{3}H]PonA uptake to intact insect cells, Sf9 and Kc. It was later demonstrated that the activity measured in intact cells was equivalent to the binding affinity using cell free receptor preparations from both Sf-9 and Kc (Minakuchi et al., 2003c). Since a binding assay using intact cells is convenient and economical, they measured the activity of compounds in intact cells and discussed the structure–activity relationship. The binding affinity of compounds to Sf-9 cells was linearly correlated with the hormonal activity measured using a cultured integument of *C. suppressalis* with a high correlation coefficient

($r = 0.97$) (Minakuchi *et al.*, 2003a; Nakagawa *et al.*, 2000b), although there is no correlation between activities against Sf-9 and Kc cells (Nakagawa *et al.*, 2002a). These results suggest that the ligand–receptor binding modes are very similar within Lepidoptera but different between Lepidoptera and Diptera. Ogura *et al.* (2005a) measured the binding affinity of a series of dibenzoylhydrazines with various substituents at the *para*-position of the B-ring against Sf-9 cells and quantitatively analyzed the substituent effects. Equation (4) was derived for 17 congeners with various substituents, such as alkyls (C1–C4), halogens (F, Cl, Br, I), OCH_3, NO_2, CN, CF_3, $COCH_3$, and SO_2CH_3. Only the compound containing C_6H_5 group was omitted from the correlation.

$$\text{pIC}_{50} = 0.61 \log P - 0.82\sigma - 0.37B_1 + 5.50$$
$$n = 17, \ s = 0.243, \ r = 0.914, \ F_{3,13} = 46.026 \quad (4)$$

Equation (4) indicates that ligand-binding increased with the introduction of the electron-donating hydrophobic group but decreased with the introduction of the electron-withdrawing bulky group. Probably the electronic effects are related to the hydrogen-bonding interaction between C=O and the acidic site of the receptor, which is coupled with the electron transfer from the basic site to the NH-moiety. An electronic effect was significant in the QSAR analysis, which suggests an electronic interaction between the ligand and the binding site of receptor, although no electronic effect was significant in the QSAR analysis for larvicidal activity. Nakagawa *et al.* (2005) used CoMFA to visualize the physicochemical properties of ligand binding in Sf-9 cells, such as steric and electrostatic interactions of compounds required for binding to the receptor.

In Nakagawa's research group, the steroid moiety of PonA was modified and the receptor binding assay was performed in the insect cell lines. Since brassinosteroids have similar structures to ecdysteroids, it is assumed that brassinosteroids may have ecdysteroid activity. In fact 24-epibrassinolide and 24-epicastasterone showed weak binding to the cultured imaginal discs of *S. littoralis*, and the order of activity was tebufenozide (0.087 μM) > 20E (0.158 μM) > 24-epibrassinolide (3.65 μM) > 24-epicastasterone (2.03 μM) (Smagghe *et al.*, 2002a). Sobek *et al.* (1993) performed the binding assay using cytosol or the nuclear extract of *G. mellonella* and determined the Kd values with the various compounds, such as PonA (0.004 μM) > 20E (0.12 μM) > RH-5849 (0.41 μM) > 22S,23S-homocastasterone (7.8 μM) > ecdysone (10.6 μM). According to these data, brassinosteroids were equipotent to ecdysone in the receptor binding but were approximately 10 times lower in activity than 20E and RH-5849. In the binding assay using Sf-9 cells, castasterone, one of the brassinosteroids, was reported to be inactive (Nakagawa *et al.*, 2000b). Watanabe *et al.* (2003) synthesized a hybrid compound (Cas/PonA) in which the side chain of castasterone was replaced with the side chain of PonA with racemic 22-OH and found that it showed ecdysone-like activity. In a further study, they stereoselectively synthesized 22R and 22S enantiomers and demonstrated that the 22R-form was 100 times more potent than the corresponding 22S-form in the receptor binding. The order of hormonal activity was PonA (0.03 μM) > 20E (0.18 μM) > Cas/PonA-22R (2.69 μM) > ecdysone (8.91 μM) ≫ Cas/PonA-22S (inactive) (Watanabe et al., 2004). On the other hand, these hybrid compounds were inactive as brassinosteroids in the rice lamina inclination assay (Fujioka et al., 1998).

3. Binding to Nongenomic Receptors

Nongenomic actions have been demonstrated for various steroid hormones, such as progesterone, estradiol, aldosterone, glucocorticoids, as well as thyroid hormones and vitamine D (Losel et al., 2003; Losel and Wehling, 2003; Wehling, 1997). For example, aldosterone induced a cell volume increase in less than 10 min in living endothelial cells (Oberleithner et al., 2000). Elmogy et al. (2004) demonstrated that 20E bound to a putative membrane receptor (mEcR) located in the plasma membrane of the anterior silk glands of B. mori. The mEcR exhibited saturable binding for [^{3}H]PonA (Kd = 0.017 μM). They measured the binding affinities of steroidal and nonsteroidal ecdysone agonists against mEcR of B. mori. The order of the binding affinity was 20E > PonA > methoxyfenozide > tebufenozide > RH-5849, which was different from that for the binding affinity to nuclear receptor proteins of C. suppressalis (tebufenozide ≈ methoxyfenozide > PonA > 20E ≈ RH-5849).

4. Application to Gene Switch

Precise control of gene expression is an invaluable tool in studying development and other physiological processes for gene therapy (Christopherson et al., 1992; Jaenisch, 1988). In the past, tetracycline-, steroid-, and rapamycin-regulated systems have been utilized in transgenic mice and cultured cells (Furth et al., 1994; Shockett et al., 1995). Scientists are interested in ecdysone-inducible systems because ecdysones do not affect mammalian physiology like glucocorticoids or progesterone and are not toxic like tetracycline. No et al. (1996) reported that ecdysone is an efficient and potent inducer of gene expression in cultured mammalian cells and transgenic mice. Thus, nonsteroidal ecdysone agonists are thought to be feasible ligands for a gene switch. Hoppe et al. (2000) reported that an dibenzoylhydrazine analogue (GS-E; Fig. 1) could be utilized as a ligand to chimeric receptor Drosophila/Bombyx-EcR in vivo for activation of gene expression.

By homology modeling of CfEcR, 17 amino acid residues were identified as critical for 20E binding by Kumar et al. (2002). They found that one particular mutant (A110P) that failed to respond to steroids responded to RG-102240 (GS^{TM}E) and RG-10237 by analyzing the transactivation effects of mutants and performing ligand-binding assays. They believe this steroid-insensitive EcR mutant is possibly a potential gene switch in insects and plants, that have endogenous ecdysteroids. In addition, this mutant would also be useful for developing orthogonal EcR-ligand pairs for simultaneous regulation of multiple genes in the same cell. Palli *et al.* (2003) reported another EcR-based inducible gene-regulation system. They prepared the constructs in which DEF domains of CfEcR, CfUSP, and MmRXR were fused to either the GAL4 DNA-binding domain or VP16-activation domain and tested in mammalian cells to evaluate their ability to transactivate the luciferase gene. A two-hybrid format switch, where the GAL4-DNA-binding domain was fused to CfEcR (DEF) and Vp16 activation domain was fused to MmRXR (EF), was found to be the best combination. They used RG-102240 (GS^{TM}-E) and RG-102317 (Fig. 1) as synthetic ligands. The reporter gene induction using GAL4:CfE (DEF) + VP16:MmR(DEF) switch was dose dependent, and significant levels of reporter gene induction were observed at 1 μM or higher concentration of RG-102240, 0.04 μM or higher concentration of RG-102317, 5 μM or higher concentration of PonA, and 25 μM or higher concentration of the ecdysteroid, muristerone A. This new CfEcR-based switch had most of the desirable properties of an optimal gene-regulation system and is currently being evaluated for *in vivo* efficacy. Tice *et al.* (2003a) synthesized a new ligand, α-acylaminoketone (Fig. 4), to control gene expression.

Gene regulation in plants was also successfully controlled by methoxyfenozide (Koo *et al.*, 2004). Constitutive expression of the gene-encoding tobacco mosaic virus (TMV) coat protein (CP) in transgenic plants confers resistance to infection by TMV and related tobamoviruses. Resistance to TMV was examined by observing temporal and quantitative control of TMV Cg CP (CgCP) gene expression in a methoxyfenozide-inducible *Arabidopsis* system. By soil drenching with a commercial methoxyfenozide, most transgenic lines were induced from undetectable levels to 0.05–0.8% protein levels. This induction was about four times that of CP produced by the constitutive cauliflower mosaic virus (CaMV) double 35S promoter.

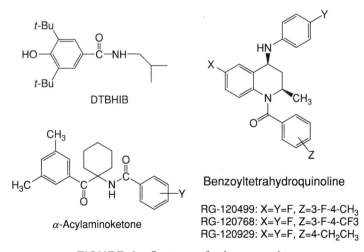

FIGURE 4. Structures of ecdysone agonists.

III. OTHER NONSTEROIDAL ECDYSONE AGONISTS

3,5-Di-t-butyl-N-i-butyl-4-hydroxybenzamide (DTBHIB) was reported to be an ecdysone agonist in 1996, whereas the corresponding N-iso-propyl analogue was reported to be inactive (Mikitani, 1996). Nakagawa and co-workers replaced the i-Bu group with other alkyl groups, such as t-Bu, i-Pent and n-Hex, and measured their binding potencies against EcRs in Kc and Sf-9 cells as well as the molting hormonal activity in the cultured integument of *C. suppressalis* (unpublished). As shown in Table III, no compounds are as potent as the original DTBHIB in the competitive inhibition of binding of PonA. Only i-Pr derivative was equipotent to DTBHIB against Sf-9 cell but

TABLE III. Inhibition of the Incorporation of ^{3}[H]PoA to the Insect Cell Lines and the Molting Hormonal Activity of Ecdysone Agonists

| No. | Compounds | pIC$_{50}$ (M) | | pEC$_{50}$ (M)^c |
		Kc^a	Sf-9^b	*C. suppressalis*
1.	RH-5849	5.24	6.44	6.40
2.	Tebufenozide	6.39	8.81	8.94
3.	Methoxyfenozide	6.55	8.46	8.95
4.	Halofenozide	6.17^c	6.48	7.10^d
5.	Chromafenozide	6.83^d	8.78	8.83
6.	Ponesterone A	8.89	8.05	7.53
7.	20-Hydroxyecdysone	7.34	6.78	6.75
8.	Ecdysone	5.59	5.63^d	5.50^e
9.	R = i-Pr	<4.06 (43%)^d	5.23^d	<4.48 (19%)^d
10.	i-Bu	5.30^d	5.28^d	<4.00 (0%)^d
11.	t-Bu	<3.60 (42%)^d	<3.60 (16%)^d	<4.00 (41%)^d
12.	i-Pentyl	4.15 ± 0.12^d	<3.60 (18%)^d	<4.00 (2%)^d
13.	n-Heptyl	<3.60 (42%)^d	<3.60 (30%)^d	<4.00 (6%)^d

Structure (between rows 8 and 9): HO–(3,5-di-t-Bu-phenyl)–C(=O)–R

^{a}Unless noted from Nakagawa *et al.*, 2002a.
^{b}Unless noted from Nakagawa *et al.*, 2000b.
^{c}Unless noted from Nakagawa *et al.*, 1998, 2000.
^{d}Unpublished data.
^{e}Watanabe *et al.*, 2003.

was inactive against Kc cells, being consistent with Mikitani's result (Mikitani, 1996). The *i*-pentyl analogue showed weak activity against Kc cells. Even in the most potent compound, the activity was 1/20 to that of ecdysone to both Kc and Sf-9 cells. In addition, no compounds showed molting hormonal activity in the cultured integument. 8-*O*-Acetylhalpagide isolated from the ecdysteroids-rich plant *Ajuga reptans*, as an ecdysone agonist (Elbrecht *et al*., 1996), has been removed from the list of ecdysone agonists (Dinan *et al*., 2001). Dinan *et al*. (2001) demonstrated that the purified 8-*O*-acetylhalpagide was not active in the *D. melanogaster* BII cell bioassay (Clement *et al*., 1993), either as an agonist or as an antagonist. In addition, this compound did not inhibit the [^{3}H]PonA binding in dipteran and lepidopteran receptor complexes.

Tice *et al*. (2003b) reported that an α-acylaminoketone derivative (Fig. 4) has molting hormonal activity equal to that of RG-102240 (GS^{TM}-E, Fig. 1), which was designed as a gene switch. The α-acylaminoketone is the derivative of dibenzoylhydrazine in which the *t*-butyl amino part was converted to cyclohexyl substructure, indicating that an *N* atom is not necessary in ecdysone agonists. As described in an earlier section, the cyclic RH-5849 analogue was inactive, even though the compound was conformationally similar to RH-5849 (Toya *et al*., 2002b), suggesting that NH is essential in ecdysone agonists.

Smith *et al*. (2003) succeeded in synthesizing new nonsteroidal ecdysone agonists, benzoyltetrahydroquinoline (BTHQ) analogues, whose structures are very different from dibenzoylhydrazines and α-acylaminoketone. Interestingly, BTHQs were reported to be potent against dipteran *A. aegypti* in the reporter gene assay, although they were relatively weak against Lepiaoptera. However, since the activity has not been evaluated as pIC$_{50}$ and pEC$_{50}$, it is difficult to compare the activity of these BTHQs with that of previously reported compounds, such as DHAs and DTBHIB congeners. An interesting study was performed regarding BTHQ compounds by Kumar *et al*. (2004). Mutants of EcR (CfEcR) ligand-binding domain (V128F or V128Y) responded well to BTHQ ligands but poorly to both ecdysteroid and diacylhydrazine ligands, although the original receptor binds to both ecdysteroids and DAHs. These mutants were further improved by introducing a second mutation, A110P, which was previously reported to cause ecdysteroid insensitivity (Kumar *et al*., 2002).

IV. CONCLUSIONS

To date, four DAHs have been developed as insecticides and their insecticidal activity and various biological effects have been reported. Advances in molecular biology techniques have aided in the molecular analysis of the mode of action of ecdysone agonists. As indicated in this review, tebufenozide, methoxyfenozide, and chromafenozide bind lepidopteran EcRs with high affinity, whereas they bind with low affinity to the EcRs of Diptera or Coleoptera. These compounds also have strong larvicidal activity against Lepidoptera, although they are inactive or weakly active against Diptera and Coleoptera. Halofenozide is more potent than tebufenozide or methoxyfenozide against *L. decemlineata* and is registered as the insecticide to control scarabids. However, the binding affinity of halofenozide to the EcR of *L. decemlineata* was not high, being slightly lower than those of tebufenozide and methoxyfenozide. The compounds with strong binding affinity to receptors have strong insecticidal activity against Lepidoptera but there was poor correlation between them in *L. decemlineata*. The prolonged persistence of halofenozide in the body of *L. decemlineata* is thought to cause the high insecticidal activity. It was also reported that one of the ABC transporters is related to the extrusion of tebufenozide, methoxyfenozide, and halofenozide. It is conceivable that other transporters might be responsible for other congeners and structures. The fact that the receptor binding pocket of a diacylhydrazine compound was only partly occupied by the side chain moiety of 20E suggested that it may be possible to develop new structures as ecdysone agonists. In fact, BTHQ analogues have been developed as new ecdysone agonists against Diptera. Although neither *in vitro* potency (IC_{50}) nor *in vivo* activity (LD_{50}) has been reported, the discovery of the new chemistry of ecdysone agonists is fascinating. In particular, as mentioned in this review, ecdysone agonists are not only suitable as insecticides but also as pharmaceuticals for gene therapy and inducers of virus resistance in plants.

ACKNOWLEDGMENTS

I am grateful to Dr. Arthur Retnakaran and Dr. Chieka Minakuchi for their careful attention and helpful suggestions in the preparation of this manuscript. I also wish to thank my colleagues who have devoted their efforts to various experiments.

REFERENCES

Ahmad, M., and Hollingworth, R. M. (2004). Synergism of insecticides provides evidence of metabolic mechanisms of resistance in the obliquebanded leafroller *Choristoneura rosaceama* (Lepidoptera: Tortricidae). *Pest Manag. Sci.* **60**, 465–473.

Baker, K. D., Warren, J. T., Thummel, C. S., Gilbert, L. I., and Mangelsdorf, D. J. (2000). Transcriptional activitation of the *Drosophila* ecdsone receptor by insect and plant ecdysteroids. *Insect Biochem. Mol. Biol.* **30**, 1037–1043.

Bergman, T., and Postlind, H. (1991). Characterization of mitochondrial cytochromes P-450 from pig kidney and liver catalysing 26-hydroxylation of 25-hydroxyvitamin D_3 and C_{27} steroids. *Biochem. J.* **276**, 427–432.

Billas, I. M. L., Moulinier, L., Rochel, N., and Moras, D. (2001). Crystal structure of the ligand-binding domain of the ultraspiracle protein USP, the ortholog of retinoid X receptors in insects. *J. Biol. Chem.* **276**, 7465–7474.

Billas, I. M. L., Iwema, T., Garnier, J. M., Mitschler, A., Rochel, N., and Moras, D. (2003). Structural adaptability in the ligand-binding pocket of the ecdysone hormone receptor. *Nature* **426**, 91–96.
Bondi, A. (1964). van der Waals volumes and radii. *J. Phys. Chem.* **68**, 441–451.
Cadogan, B. L., Thompson, D., Retnakaran, A., Scharbach, R. D., Robinson, A., and Staznik, B. (1998). Deposition of aerially applied tebufenozide (RH5992) on balsam fir (*Abies balsamea*) and its control of spruce budworm (*Choristoneura fumiferana* [Clem.]). *Pestic. Sci.* **53**, 80–90.
Cao, S., Qian, X., and Song, G. (2001). N'-*tert*-Butyl-N'-aroyl-N-(alkoxycarbonylmethyl)-N-aroylhydrazines, a novel nonsteroidal ecdysone agonists: Syntheses, insecticidal activity, conformational, and crystal structure analysis. *Can. J. Chem.* **79**, 272–278.
Carlson, G. R. (2000). Tebufenozide: A novel caterpillar control agent with unusually high target selectivity. *In* "Green Chemical Syntheses and Processes" (P. T. Anastas, L. G. Heine, and T. C. Williamson, Eds.), vol. 767, pp. 8–17. American Chemical Society, Washington, DC.
Carlson, G. R., Dhadialla, T. S., Hunter, R., Jansson, R. K., Jany, C. S., Lidert, Z., and Slawecki, R. A. (2001). The chemical and biological properties of methoxyfenozide, a new insecticidal ecdysteroid agonist. *Pest Manag. Sci.* **57**, 115–119.
Carmichael, J. A., Lawrence, M. C., Graham, L. D., Pilling, P. A., Epa, V. C., Noyce, L., Lovrecz, G., Winkler, D. A., Pawlak-Skrzecz, A., Eaton, R. E., Hannan, G. N., and Hill, R. J. (2005). The X-ray structure of a hemipteran ecdysone receptor ligand-binding domain: comparison with a lepidopteran ecdysone receptor ligand-binding domain and implications for insecticide design. *J. Biol. Chem.* **280**, 22258–22269.
Chan, T. H., Ali, A., Britten, J. F., Thomas, A. W., Strunz, G. M., and Salonius, A. (1990). The crystal structure of 1,2-dibenzoyl-1-*tert*-butylhydrazine, a nonsteroidal ecdysone agonist, and its effects on spruce budworm (*Choristoneura fumiferana*). *Can. J. Chem.* **68**, 1178–1181.
Charrois, G. J. R., Mao, H., and Kaufman, W. R. (1996). Impact on salivary gland degeneration by putative ecdysteroid antagonists and agonists in the ixodid tick *Amblyomma hebraeum*. *Pestic. Biochem. Physiol.* **55**, 140–149.
Cho, W. L., Kapitskaya, M. Z., and Raikhel, A. S. (1995). Mosquito ecdysteroid receptor: Analysis of the cDNA and expression during vitellogenesis. *Insect Biochem. Mol. Biol.* **25**, 19–27.
Christopherson, K. S., Mark, M. R., Bajaj, V., and Godowski, P. J. (1992). Ecdysteroid-dependent regulation of genes in mammalian cells by a *Drosophila* ecdysone receptor and chimeric transactivators. *Proc. Natl. Acad. Sci. USA* **89**, 6314–6318.
Chung, A. C.-K., Durica, D. S., Clifton, S. W., Roe, B. A., and Hopkins, P. M. (1998). Cloning of crustacean ecdysteroid receptor and retinoid-X receptor gene homologs and elevation of retinoid-X receptor mRNA by retinoic acid. *Mol. Cell. Endocrinol.* **139**, 209–227.
Clayton, G. M., Peak-Chew, S. Y., Evans, R. M., and Schwabe, J. W. R. (2001). The structure of the ultraspiracle ligand-binding domain reveals a nuclear receptor locked in an inactive conformation. *Proc. Natl. Acad. Sci. USA* **98**, 1549–1554.
Clement, C. Y., Bradbrook, D. A., Lafont, R., and Dinan, L. (1993). Assessment of a microplate-based bioassay for the detection of ecdysteroid-like or antiecdysteroid activities. *Insect Biochem. Mol. Biol.* **23**, 187–193.
Cowles, R. S., and Villani, M. G. (1996). Susceptibility of Japanese beetle, oriental beetle, and European chafer (Coleoptera: Scarabaeidae) to halofenozide, an insect growth regulator. *J. Econ. Entomol.* **89**, 1556–1565.
Cowles, R. S., Alm, S. R., and Villani, M. G. (1999). Selective toxicity of halofenozide to exotic white grubs (Coleoptera: Scarabaeidae). *J. Econ. Entomol.* **92**, 427–434.
Cramer, R. D., III, Patterson, D. E., and Bunce, J. D. (1988). Comparative molecular field analysis (CoMFA). 1. Effect of shape on binding of steroids to carrier proteins. *J. Am. Chem. Soc.* **110**, 5959–5967.
Darvas, B., Polgar, L., Tag El-Din, M. H., Eross, K., and Wing, K. D. (1992). Developmental disturbances in different insect orders caused by an ecdysteroid agonist, RH-5849. *J. Econ. Entomol.* **85**, 2107–2112.

Darvas, B., Rees, H. H., Hoggard, N., Farag, A. I., O'Hanlon, G., and Mercer, J. (1995). Effects of wet environment on ecdysone 20-mono-oxygenase and ecdysteroid levels during wandering behavior of *Neobellieria bullata* and *Parasarcophaga argyrostoma* larvae. *Comp. Biochem. Physiol.* **110B,** 57–63.

Darvas, B., Pap, L., Kelemen, M., and Polgar, L. A. (1998). Synergistic effects of verbutin with dibenzoylhydrazine-type ecdysteroid agonists on larvae of *Aedes aegypti* (Diptera: Culicidae). *J. Econ. Entomol.* **91,** 1260–1264.

Dean, R. L., Bollenbacher, W. E., Locke, M., Smith, S. L., and Gilbert, L. I. (1980). Haemolymph ecdysteroid levels and cellular events in the intermoult/moult sequence of *Calpodes ethlius*. *J. Insect Physiol.* **26,** 267–280.

Dhadialla, T. S., and Tzertzinis, G. (1997). Characterization and partial cloning of ecdysteroid receptor from a cotton boll weevil embryonic cell line. *Arch. Insect Biochem. Physiol.* **35,** 45–57.

Dhadialla, T. S., Carlson, G. R., and Le, D. P. (1998). New insecticides with ecdysteroidal and juvenile hormone activity. *Annu. Rev. Entomol.* **43,** 545–569.

Dinan, L., Whiting, P., Bourne, P., and Coll, J. (2001). 8-*O*-Acetylharpagide is not an ecdysteroid agonist. *Insect Biochem. Mol. Biol.* **31,** 1077–1082.

Dorn, S., Frischknecht, M. L., Martinez, V., Zurfluh, R., and Fischer, U. (1981). A novel non-neurotoxic insecticide with a broad activity. *Z. Pflanzenkr Pflanzenschutz* **88,** 269–275.

Elbrecht, A., Chen, Y., Jurgens, T., Hensens, O. D., Zink, D. L., Beck, H. T., Balick, M. J., and Borris, R. (1996). 8-*O*-Acetylharpagide is a nonsteroidal ecdysteroid agonist. *Insect Biochem. Mol. Biol.* **26,** 519–523.

Elmogy, M., Iwami, M., and Sakurai, S. (2004). Presence of membrane ecdysone receptor in the anterior silk gland of the silkworm *Bombyx mori*. *Eur. J. Biochem.* **271,** 3171–3179.

Farinos, G. P., Smagghe, G., Tirry, L., and Castanera, P. (1999). Action and pharmacokinetics of a novel insect growth regulator, halofenozide, in adult beetles of *Aubeonymus mariaefranciscae* and *Leptinotarsa decemlineata*. *Arch. Insect Biochem. Physiol.* **41,** 201–213.

Farkas, R., and Slama, K. (1999). Effect of bisacylhydrazine ecdysteroid mimics (RH-5849 and RH-5992) on chromosomal puffing, imaginal disc proliferation and pupariation in larvae of *Drosophila melanogaster*. *Insect Biochem. Mol. Biol.* **29,** 1015–1027.

Frisrom, J. W., and Yund, M. A. (1976). Characteristics of the action of ecdysones on *Drosophila* imaginal discs cultured *in vitro*. *In* "Invertebrate Tissue Culture Research Application" (K. Maramorosch, Ed.), pp. 161–178. Academic Press, New York.

Fujioka, S., Noguchi, T., Takatsuto, S., and Yoshida, S. (1998). Activity of brassinosteroids in the dwarf rice lamina inclination bioassay. *Phytochemistry* **49,** 1841–1848.

Fujita, T., Iwasa, J., and Hansch, C. (1964). A new substituent constant, p, derived from partition coefficients. *J. Am. Chem. Soc.* **86,** 5175–5180.

Fujita, T. (1990). The extrathermodynamic approach to drug design. *In* "Comprehensive Medicinal Chemistry" (C. A. Ramsden, Ed.), pp. 497–560. Pergamon Press, Oxford.

Fujiwara, H., Jindra, M., Newitt, R., Palli, S. R., Hiruma, K., and Riddiford, L. M. (1995). Cloning of an ecdysone receptor homolog from *Manduca sexta* and the developmental profile of its mRNA in wings. *Insect Biochem. Mol. Biol.* **25,** 845–856.

Furth, P. A., St. Onge, L., Boger, H., Gruss, P., Gossen, M., Kistner, A., Bujard, H., and Hennighausen, L. (1994). Temporal control of gene expression in transgenic mice by a tetracycline-responsive promoter. *Proc. Natl. Acad. Sci. USA* **91,** 9302–9306.

Gadenne, C., Varjas, L., and Mauchamp, B. (1990). Effects of the non-steroidal ecdysone mimic, RH-5849, on diapause and non-diapause larvae of the European corn borer, *Ostrinia nubilalis* Hbn. *J. Insect Physiol.* **36,** 555–559.

Grebe, M., Rauch, P., and Spindler-Barth, M. (2000). Characterization of subclones of the epithelial cell line from *Chironomus tentans* resistant to the insecticide RH 5992, a non-steroidal moulting hormone agonist. *Insect Biochem. Mol. Biol.* **30,** 591–600.

Guo, X., Harmon, M. A., Laudet, V., Mangelsdorf, D. J., and Palmer, M. J. (1997). Isolation of a functional ecdysteroid receptor homologue from the ixodid tick *Amblyomma americanum* (L.). *Insect Biochem. Mol. Biol.* **27,** 945–962.

Hannan, G. N., and Hill, R. J. (1997). Cloning and characterization of LcEcR: A functional ecdysone receptor from the sheep blowfly *Lucilia cuprina*. *Insect Biochem. Mol. Biol.* **27**, 479–488.

Hansch, C., and Fujita, T. (1964). r-s-p Analysis. A method for the correlation of biological activity and chemical structure. *J. Am. Chem. Soc.* **86**, 1616–1626.

Hayward, D. C., Dhadialla, T. S., Zhou, S., Kuiper, M. J., Ball, E. E., Wyatt, G. R., and Walker, V. K. (2003). Ligand specificity and developmental expression of RXR and ecdysone receptor in the migratory locust. *J. Insect Physiol.* **49**, 1135–1144.

Hetru, C., and Horn, D. H. S. (1980). Pytoecdysteroids and zooecdysteroids. *In* "Progress in Ecdysone Research" (J. A. Hoffmann, Ed.), pp. 13–28. Elsevier, Netherlands.

Hoelscher, J. A., and Barrett, B. A. (2003). Effects of methoxyfenozide-treated surfaces on the attractiveness and responsiveness of adult codling moth (Lepidoptera: Tortricidae). *J. Econ. Entomol.* **96**, 623–628.

Hoppe, U. C., Marban, E., and Johns, D. C. (2000). Adenovirus-mediated inducible gene expression *in vivo* by a hybrid ecdysone receptor. *Mol. Ther.* **1**, 159–164.

Hormann, R. E., Dinan, L., and Whiting, P. (2003). Superimposition evaluation of ecdysteroid agonist chemotypes through multi-dimensional QSAR. *J. Comput. Aided Mol. Des.* **17**, 135–153.

Hsu, A. C.-T. (1991). 1,2-Diacyl-1-alkylhydrazines, a new class of insect growth regulators. *In* "Synthesis and Chemistry of Agrochemicals II" (D. R. Baker, J. G. Fenyes, and W. K. Moberg, Eds.), vol. 443, pp. 478–490. American Chemical Society, Washington, DC.

Hsu, A. C.-T., Fujimoto, T. T., and Dhadialla, T. S. (1997). Structure-activity study and conformational analysis of RH-5992, the first commercialized nonsteroidal ecdysone agonist. *In* "Phytochemicals for Pest Control" (P. A. Hedin, R. M. Hollingworth, E. P. Masler, J. Miyamoto, and D. G. Thompson, Eds.), vol. 658, pp. 206–219. American Chemical Society, Washington, DC.

Hu, W., Feng, Q., Palli, S. R., Krell, P. J., Arif, B. M., and Retnakaran, A. (2001). The ABC transporter Pdr5p mediates the efflux of nonsteroidal ecdysone agonists in *Saccharomyces cerevisiae*. *Eur. J. Biochem.* **268**, 3416–3422.

Hu, W., Cook, B. J., Ampasala, D. R., Zheng, S., Caputo, G., Krell, P. J., Retnakaran, A., Arif, B. M., and Feng, Q. (2004). Morphological and molecular effects of 20-hydroxyecdysone and its agonist tebufenozide on CF-203, a midgut-derived cell line from the spruce budworm, *Choristoneura fumiferana*. *Arch. Insect Biochem. Physiol.* **55**, 68–78.

Huber, R., and Hoppe, W. (1965). Zur Chemie des Ecdysons, VII: Die Kristall- und Molekulstructuranalyse des Insektenverpuppungshormons Ecdyson mit der automatisierten Faltmolekulmethode. *Chem. Ber.* **98**, 2403–2424.

Imhof, M. O., Rusconi, S., and Lezzi, M. (1993). Cloning of a *Chironomus tentans* cDNA encoding a protein (cEcRH) homologous to the *Drosophila melanogaster* ecdysteroid receptor (dEcR). *Insect Biochem. Mol. Biol.* **23**, 115–124.

Ishaaya, I., Yablonski, S., and Horowitz, A. R. (1995). Comparative toxicity of two ecdysteroid agonists, RH-2485 and RH-5992, on susceptible and pyrethroid-resistant strains of the Egyptian cotton Leaf worm, *Spodoptera littoralis*. *Phytoparasitica* **23**, 139–145.

Jaenisch, R. (1988). Transgenic animals. *Science* **240**, 1468–1474.

Kamimura, M., Tomita, S., and Fujiwara, H. (1996). Molecular cloning of an ecdysone receptor (B1 isoform) homologue from the silkworm, *Bombyx mori*, and its mRNA expression during wing disc development. *Comp. Biochem. Physiol.* **113B**, 341–347.

Karlson, P. (1980). Ecdysone in retrospect and prospect. *In* "Progress in Ecdysone Research" (J. A. Hoffmann, Ed.), pp. 1–11. Elsevier, Netherlands.

Kasuya, A., Sawada, Y., Tsukamoto, Y., Tanaka, K., Toya, T., and Yanagi, M. (2003). Binding mode of ecdysone agonists to the receptor: Comparative modeling and docking studies. *J. Mol. Model.* **9**, 58–65.

Kawada, H., Kojima, I., and Shinjo, G. (1989). Laboratory evaluation of a new insect growth regulator pyriproxyfen, as a cockroach control agent. *Jpn. J. Sanit. Zool.* **40**, 195–201.

Kayser, H., Winkler, T., and Spindler-Barth, M. (1997). 26-Hydroxylation of ecdysteroids is catalyzed by a typical cytochrome P-450-dependent oxidase and related to ecdysteroid resistance in an insect cell line. *Eur. J. Biochem.* **248**, 707–716.

Keogh, D. P., and Smith, S. L. (1991). Regulation of cytochrome P-450 dependent steroid hydroxylase activity in *Manduca sexta*: Effects of the ecdysone agonist RH 5849 on ecdysone 20-monooxygenase activity. *Biochem. Biophys. Res. Commun.* **176**, 522–527.

Kobayashi, M., Nakanishi, K., and Koreeda, M. (1967). The moulting hormone activity of ponasterones on *Musca domestica* (Diptera) and *Bombyx mori* (Lepidoptera). *Steroids* **9**, 529–536.

Koelle, M. R., Talbot, W. S., Segraves, W. A., Bender, M. T., Cherbas, P., and Hogness, D. S. (1991). The *Drosophila EcR* gene encodes an ecdysone receptor, a new member of the steroid receptor superfamily. *Cell* **67**, 59–77.

Koo, J. C., Asurmendi, S., Bick, J., Woodford-Thomas, T., and Beachy, R. N. (2004). Ecdysone agonist-inducible expression of a coat protein gene from tobacco mosaic virus confers viral resistance in transgenic *Arabidopsis*. *Plant J.* **37**, 439–448.

Kothapalli, R., Palli, S. R., Ladd, T. R., Sohi, S. S., Cress, D., Dhadialla, T. S., Tzertzinis, G., and Retnakaran, A. (1995). Cloning and developmental expression of the ecdysone receptor gene from the spruce budworm, *Choristoneura fumiferana*. *Dev. Genet.* **17**, 319–330.

Kumar, V. S., Santhi, M., and Krishnan, M. (2000). RH-5992–an ecdysone agonist on model system of the silkworm *Bombyx mori*. *Indian J. Exp. Biol.* **38**, 137–144.

Kumar, M. B., Fujimoto, T., Potter, D. W., Deng, Q., and Palli, S. R. (2002). A single point mutation in ecdysone receptor leads to increased ligand specificity: Implications for gene switch applications. *Proc. Natl. Acad. Sci. USA* **99**, 14710–14715.

Kumar, M. B., Potter, D. W., Hormann, R. E., Edwards, A., Tice, C. M., Smith, H. C., Dipietro, M. A., Polley, M., Lawless, M., Wolohan, P. R. N., Kethidi, D. R., and Palli, S. R. (2004). Highly flexible ligand binding pocket of ecdysone receptor: A single amino acid change leads to discrimination between two groups of nonsteroidal ecdysone agonists. *J. Biol. Chem.* **279**, 27211–27218.

Kunkel, B. A., Held, D. W., and Potter, D. A. (1999). Impact of halofenozide, imidacloprid, and bendiocarb on beneficial invertebrates and predatory activity in turfgrass. *J. Econ. Entomol.* **92**, 922–930.

Kutter, E., and Hansch, C. (1969). Steric parameters in drug design. Monoamine oxidase inhibitors and antihistamines. *J. Med. Chem.* **64**, 647–652.

Losel, R., and Wehling, M. (2003). Nongenomic actions of steroid hormones. *Nat. Rev. Mol. Cell. Biol.* **4**, 46–56.

Losel, R. M., Falkenstein, E., Feuring, M., Schultz, A., Tillmann, H. C., Rossol-Haseroth, K., and Wehling, M. (2003). Nongenomic steroid action: Controversies, questions, and answers. *Physiol. Rev.* **83**, 965–1016.

Martinez, A., Scanlon, D., Gross, B., Perara, S. C., Palli, S. R., Greenland, A. J., Windass, J., Pongs, O., Broad, P., and Jepson, I. (1999). Transcriptional activation of the cloned *Heliothis virescens* (Lepidoptera) ecdysone receptor (HvEcR) by Muristerone A. *Insect Biochem. Mol. Biol.* **29**, 915–930.

Mikitani, K. (1995). Sensitive, rapid and simple method for evaluation of ecdysteroid agonist activity based on the mode of action of the hormone. *J. Seric. Sci. Jpn.* **64**, 534–539.

Mikitani, K. (1996). A new nonsteroidal chemical class of ligand for the ecdysteroid receptor 3,5-di-*tert*-butyl-4-hydroxy-*N*-isobutyl-benzamide shows apparent insect molting hormone activities at molecular and cellular levels. *Biochem. Biophys. Res. Commun.* **227**, 427–432.

Minakuchi, C., Nakagawa, Y., Kiuchi, M., Tomita, S., and Kamimura, M. (2002). Molecular cloning, expression analysis and functional confirmation of two ecdysone receptor isoforms from the rice stem borer *Chilo suppressalis*. *Insect Biochem. Mol. Biol.* **32**, 999–1008.

Minakuchi, C., Nakagawa, Y., Kamimura, M., and Miyagawa, H. (2003a). Binding affinity of nonsteroidal ecdysone agonists against the ecdysone receptor complex determines the strength of their molting hormonal activity. *Eur. J. Biochem.* **270**, 4095–4104.

Minakuchi, C., Nakagawa, Y., Kiuchi, M., Seino, A., Tomita, S., and Kamimura, M. (2003b). Molecular cloning and expression analysis of ultraspiracle (USP) from the rice stem borer *Chilo suppressalis*. *Insect Biochem. Mol. Biol.* **33**, 41–49.

Minakuchi, C., Nakagawa, Y., and Miyagawa, H. (2003c). Validity analysis of a receptor binding assay for ecdysone agonists using cultured intact insect cells. *J. Pestic. Sci.* **28**, 55–57.

Minakuchi, C., Nakagawa, Y., Kamimura, M., and Miyagawa, H. (2005). Measurement of receptor-binding activity of non-steroidal ecdysone agonists using *in vitro* expressed receptor proteins (EcR/USP complex) of *Chilo suppressalis* and *Drosophila melanogaster*. *In* "New Discoveries in Agrochemicals" (J. M. Clark and H. Ohkawa, Eds.), Vol. 892, pp. 191–200. American Chemical Society, Washington, DC.

Mouillet, J.-F., Delbecque, J.-P., Quennedey, B., and Delachambre, J. (1997). Cloning of two putative ecdysteroid receptor isoforms from *Tenebrio molitor* and their developmental expression in the epidermis during metamorphosis. *Eur. J. Biochem.* **248**, 856–863.

Moulton, J. K., Pepper, D. A., Jansson, R. K., and Dennehy, T. J. (2002). Pro-active management of beet armyworm (Lepidoptera: Noctuidae) resistance to tebufenozide and methoxyfenozide: Baseline monitoring risk assessment, and isolation of resistance. *J. Econ. Entomol.* **95**, 414–424.

Muszynska-Pytel, M., Mikolajczyk, P., Pszczolkowski, M. A., and Cymborowski, B. (1992). Juvenilizing effect of ecdysone mimic RH-5849 in *Galleria mellonella* larvae. *Experientia* **48**, 1013–1017.

Nakagawa, Y., Iwamura, H., and Fujita, T. (1985). Quantitative structure-activity studies of benzoylphenylurea larvicides. II. Effect of benzyloxy substituents at aniline moiety against *Chilo suppressalis* Walker. *Pestic. Biochem. Physiol.* **23**, 7–12.

Nakagawa, Y., Sotomatsu, T., Irie, K., Kitahara, K., Iwamura, H., and Fujita, T. (1987). Quantitative structure-activity studies of benzoylphenylurea larvicides. III. Effects of substituents at the benzoyl moiety. *Pestic. Biochem. Physiol.* **27**, 143–155.

Nakagawa, Y., Akagi, T., Iwamura, H., and Fujita, T. (1989). Quantitative structure-activity studies of benzoylphenylurea larvicides. VI. Comparison of substituent effects among activities against different insect species. *Pestic. Biochem. Physiol.* **33**, 144–157.

Nakagawa, Y., Matsutani, M., Kurihara, N., Nishimura, K., and Fujita, T. (1992). Quantitative strucutre-activity studies of benzoylphenylurea larvicides. VIII. Inhibition of *N*-acetylglucosamine incorporation into the cultured integument of *Chilo suppressalis* Walker. *Pestic. Biochem. Physiol.* **43**, 141–151.

Nakagawa, Y., Shimizu, B., Oikawa, N., Akamatsu, M., Nishimura, K., Kurihara, N., Ueno, T., and Fujita, T. (1995a). Three-dimensional quantitative structure-activity analysis of steroidal and dibenzoylhydrazine-type ecdysone agonists. *In* "Classical and Three-Dimensional QSAR in Agrochemistry" (C. Hansch and T. Fujita, Eds.), Vol. 606, pp. 288–301. American Chemical Society, Washington, DC.

Nakagawa, Y., Soya, Y., Nakai, K., Oikawa, N., Nishimura, K., Ueno, T., Fujita, T., and Kurihara, N. (1995b). Quantitative structure-activity studies of insect growth regulators. XI. Stimulation and inhibition of *N*-acetylglucosamine incorporation in a cultured integument system by substituted *N-tert*-butyl-*N,N'*-dibenzoylhydrazines. *Pestic. Sci.* **43**, 339–345.

Nakagawa, Y., Hattori, K., Shimizu, B., Akamatsu, M., Miyagawa, H., and Ueno, T. (1998). Quantitative structure-activity studies of insect growth regulators XIV. Three dimensional quantitative structure-activity relationship of ecdysone agonists including dibenzoylhydrazine analogs. *Pestic. Sci.* **53**, 267–277.

Nakagawa, Y., Smagghe, G., Kugimiya, S., Hattori, K., Ueno, T., Tirry, L., and Fujita, T. (1999). Quantitative structure-activity studies of insect growth regulators: XVI. Substituent effects of dibenzoylhydrazines on the insecticidal activity to Colorado potato beetle *Leptinotarsa decemlineata. Pestic. Sci.* **55,** 909–918.

Nakagawa, Y., Hattori, K., Minakuchi, C., Kugimiya, S., and Ueno, T. (2000a). Relationships between structure and molting hormonal activity of tebufenozide, methoxyfenozide, and their analogs in cultured integument system of *Chilo suppressalis* Walker. *Steroids* **65,** 117–123.

Nakagawa, Y., Minakuchi, C., and Ueno, T. (2000b). Inhibition of [^{3}H]ponasterone A binding by ecdysone agonists in the intact Sf-9 cell line. *Steroids* **65,** 537–542.

Nakagawa, Y., Smagghe, G., Van Paemel, M., Tirry, L., and Fujita, T. (2001). Quantitative structure-activity studies of insect growth regulators: XVIII. Effects of substituents on the aromatic moiety of dibenzoylhydrazines on larvicidal activity against the Colorado potato beetle *Leptinotarsa decemlineata. Pest Manag. Sci.* **57,** 858–865.

Nakagawa, Y., Minakuchi, C., Takahashi, K., and Ueno, T. (2002a). Inhibition of [^{3}H] ponasterone A binding by ecdysone agonists in the intact Kc cell line. *Insect Biochem. Mol. Biol.* **32,** 175–180.

Nakagawa, Y., Smagghe, G., Tirry, L., and Fujita, T. (2002b). Quantitative structure-activity studies of insect growth regulators: XIX. Effects of substituents on the aromatic moiety of dibenzoylhydrazines on larvicidal activity against the beet armyworm *Spodoptera exigua. Pest Manag. Sci.* **58,** 131–138.

Nakagawa, Y., Takahashi, K., Kishikawa, H., Ogura, T., Minakuchi, C., and Miyagawa, H. (2005). Classical and three-dimensional QSAR for the Inhibition of [^{3}H]ponasterone A binding by diacylhydrazine-type ecdysone agonists to insect Sf-9 cells. *Bioorg. Med. Chem.* **13,** 1333–1340.

Nakanishi, K., Koreeda, M., Sasaki, L., Chang, M. L., and Hsu, H. Y. (1966). Insect hormones I. the structure of ponasterone A, an insect molting hormone from the leaves of *Podocarpus makaii* H. *J. Chem. Soc., Chem. Commun.* **91,** 915–917.

Nijhout, H. F. (1994). "Insect Hormones," Chapter 5, pp. 89–141. Princeton University Press, New Jersey.

Nishimura, K., Tada, T., and Nakagawa, Y. (1996). Effect of the insect growth regulators, *N-tert*-butyl-*N,N'*-dibenzoylhydrazines, on neural activity of the American cockroach. *Comp. Biochem. Physiol.* **114C,** 141–144.

No, D., Yao, T.-P., and Evans, R. M. (1996). Ecdysone-inducible gene expression in mammalian cells and transgenic mice. *Proc. Natl. Acad. Sci. USA* **93,** 3346–3351.

Oberleithner, H., Reinhardt, J., Schillers, H., Pagel, P., and Schneider, S. W. (2000). Aldosterone and nuclear volume cycling. *Cell. Physiol. Biochem.* **10,** 429–434.

Ogura, T., Nakagawa, Y., Minakuchi, C., and Miyagawa, H. (2005a). QSAR for binding activity of substituted dibenzoylhydrazines to intact Sf-9 cells. *J. Pestic. Sci.* **30,** 1–6.

Ogura, T., Minakuchi, C., Nakagawa, Y., Smagge, G., and Miyagawa, H. (2005b). Molecular cloning, expression analysis and functional confirmation of ecdysone receptor and ultraspiracle from the Colorado potato beetle *Leptinotarsa decemlineata. FEBS J.* **272,** 4114–4128.

Oikawa, N., Nakagawa, Y., Soya, Y., Nishimura, K., Kurihara, N., Ueno, T., and Fujita, T. (1993). Enhancement of *N*-acetylglucosamine incorporation into the cultured integument of *Chilo suppressalis* by molting hormone and dibenzoylhydrazine insecticides. *Pestic. Biochem. Physiol.* **47,** 165–170.

Oikawa, N., Nakagawa, Y., Nishimura, K., Ueno, T., and Fujita, T. (1994a). Quantitative structure-activity analysis of larvicidal 1-(substituted benzoyl)-2-benzoyl-1-*tert*-butylhydrazines against *Chilo suppressalis. Pestic. Sci.* **41,** 139–148.

Oikawa, N., Nakagawa, Y., Nishimura, K., Ueno, T., and Fujita, T. (1994b). Quantitative structure-activity studies of insect growth regulators X. Substituent effects on larvicidal activity of 1-*tert*-butyl-1-(2-chlorobenzoyl)-2-(substituted benzoyl)hydrazines against *Chilo suppressalis* and design synthesis of potent derivatives. *Pestic. Biochem. Physiol.* **48,** 135–144.

Palli, S. R., Primavera, M., Tomkins, W., Lambert, D., and Retnakaran, A. (1995). Age-specific effects of a non-steroidal ecdysteroid agonist, RH-5992, on the spruce budworm, *Choristoneura fumiferana* (Lepidoptera: Tortricidae). *Eur. J. Entomol.* **92**, 325–332.

Palli, S. R., Kapitskaya, M. Z., Kumar, M. B., and Cress, D. E. (2003). Improved ecdysone receptor-based inducible gene regulation system. *Eur. J. Biochem.* **270**, 1308–1315.

Pszczolkowski, M. A., and Kuszczak, B. (1996). Effect of an ecdsyone agonist, RH-5849, on wandering behaviour in *Spodoptera littoralis*. *Comp. Biochem. Physiol.* **113C**, 359–367.

Quack, S., Fretz, A., Spindler-Barth, M., and Spindler, K. D. (1995). Receptor affinities and biological responses of nonsteroidal ecdysteroid agonists on the epithlial cell line from *Chironomus tentans* (Diptera: Chironomidae). *Eur. J. Entomol.* **92**, 341–347.

Retnakaran, A., Hiruma, K., Palli, S. R., and Riddiford, L. M. (1995). Molecular analysis of the mode of action of RH-5992, a lepidopteran-specific,. non-steroidal ecdysteroid agonist. *Insect Biochem. Mol. Biol.* **25**, 109–117.

Retnakaran, A., Mac Donald, A., Tomkins, W., Davis, C., Brownwright, A. J., and Palli, S. R. (1997a). Ultrastructural effects of a non-steroidal ecdysone agonist, RH-5992, on the sixth instar larva of the spruce budworm, *Choristoneura fumiferana*. *J. Insect. Physiol.* **43**, 55–68.

Retnakaran, A., Smith, L. F. R., Tomkins, W. L., Primavera, M., Palli, S. R., Payne, N., and Jobin, L. (1997b). Effect of RH-5992, a nonsteroidal ecdysone agonist, on the spruce budworm, *Choristoneura fumiferana* (Lepidoptera: Tortricidae): Laboratory, greenhouse and ground spray trials. *Can. Ent.* **129**, 871–885.

Retnakaran, A., Gelbic, I., Sundaram, M., Tomkins, W., Ladd, T., Primavera, M., Feng, Q., Arif, B., Palli, R., and Krell, P. (2001). Mode of action of the ecdysone agonist tebufenozide (RH-5992), and an exclusion mechanism to explain resistance to it. *Pest Manag. Sci.* **57**, 951–957.

Retnakaran, A., Krell, P., Feng, Q., and Arif, B. (2003). Ecdysone agonists: Mechanism and importance in controlling insect pests of agriculture and forestry. *Arch. Insect Biochem. Physiol.* **54**, 187–199.

Rodriguez, L. M., Ottea, J. A., and Reagan, T. E. (2001). Selection, egg viability, and fecundity of the sugarcane borer (Lepidoptera: Crambidae) with tebufenozide. *J. Econ. Entomol.* **94**, 1553–1557.

Roller, H., Dahm, K. H., Sweeley, C. C., and Trost, B. M. (1967). Die Struktur des Juvenilhormons. *Angew. Chem.* **79**, 190–191.

Saleh, D. S., Zhang, J., Wyatt, G. R., and Walker, V. K. (1998). Cloning and characterization of an ecdysone receptor cDNA from *Locusta migratoria*. *Mol. Cell. Endocrinol.* **143**, 91–99.

Salgado, V. L. (1992). The neurotoxic insecticidal mechanism of the nonsteroidal ecdysone agonist RH-5849: K^+ channel block in nerve and muscle. *Pestic. Biochem. Physiol.* **43**, 1–13.

Salgado, V. L. (1998). Block of neuronal voltage-dependent K+ channels by diacylhydrazine insecticides. *Neurotoxicology* **19**, 245–252.

Sauphanor, B., and Bouvier, J. C. (1995). Cross-resistance between benzoylureas and benzoylhydrazines in the codling moth, *Cydia pomonella* L. *Pestic. Sci.* **45**, 369–375.

Sawada, Y., Yanai, T., Nakagawa, H., Tsukamoto, Y., Tamagawa, Y., Yokoi, S., Yanagi, M., Toya, T., Sugizaki, H., Kato, Y., Shirakura, H., Watanabe, T., Yajima, Y., Kodama, S., and Masui, A. (2003a). Synthesis and insecticidal activity of benzoheterocyclic analogues of N'-benzoyl-N-($tert$-butyl)benzohydrazide: Part 3. Modification of N-$tert$-butylhydrazine moiety. *Pest Manag. Sci.* **59**, 49–57.

Sawada, Y., Yanai, T., Nakagawa, H., Tsukamoto, Y., Yokoi, S., Yanagi, M., Toya, T., Sugizaki, H., Kato, Y., Shirakura, H., Watanabe, T., Yajima, Y., Kodama, S., and Masui, A. (2003b). Synthesis and insecticidal activity of benzoheterocyclic analogues of N'-benzoyl-N-($tert$-butyl)benzohydrazide: Part 1. Design of benzoheterocyclic analogues. *Pest Manag. Sci.* **59**, 25–35.

Sawada, Y., Yanai, T., Nakagawa, H., Tsukamoto, Y., Yokoi, S., Yanagi, M., Toya, T., Sugizaki, H., Kato, Y., Shirakura, H., Watanabe, T., Yajima, Y., Kodama, S., and Masui, A. (2003c). Synthesis and insecticidal activity of benzoheterocyclic analogues of N'-benzoyl-N-($tert$-butyl) benzohydrazide: Part 2. Introduction of substituents on the benzene rings of the benzoheterocycle moiety. *Pest Manag. Sci.* **59**, 36–48.

Seth, R. K., Kaur, J. J., Rao, D. K., and Reynolds, S. E. (2004). Effects of larval exposure to sublethal concentrations of the ecdysteroid agonists RH-5849 and tebufenozide (RH-5992) on male reproductive physiology in *Spodoptera litura*. *J. Insect Physiol.* **50**, 505–517.

Shilhacek, D. L., Oberlander, H., and Porcheron, P. (1990). Action of RH-5849, a nonsteroidal ecdysteroid mimic, on *Plodia interpunctella* (Hübner) *in vivo* and *in vitro*. *Arch. Insect Biochem. Physiol.* **15**, 201–212.

Shimizu, B., Nakagawa, Y., Hattori, K., Nishimura, K., Kurihara, N., and Ueno, T. (1997). Molting hormonal and larvicidal activities of aliphatic acyl analogs of dibenzoylhydrazine insecticides. *Steroids* **62**, 638–642.

Shockett, P., Difilippantonio, M., Hellman, N., and Schatz, D. G. (1995). A modified tetracycline-regulated system provides autoregulatory, inducible gene expression in cultured cells and transgenic mice. *Proc. Natl. Acad. Sci. USA* **92**, 6522–6526.

Slama, K. (1995). Hormonal status of RH-5849 and RH-5992 synthetic ecdysone agonists (ecdysoids) examined on several standard bioassays for ecdysteroids. *Eur. J. Entomol.* **92**, 317–323.

Smagghe, G., and Degheele, D. (1992). Effects of RH-5849, the first nonsteroidal ecdysteroid agonist, on larvae of *Spodoptera littoralis* (Boisd.) (Lepidoptera: Noctuidae). *Arch. Insect Biochem. Physiol.* **21**, 119–128.

Smagghe, G., and Degheele, D. (1993). Metabolism, pharmacokinetics, and toxicity of the first nonsteroidal ecdysteroid agonist RH 5849 to *Spodoptera exempta* (Walker), *Spodoptera exigua* (Hubner), and *Leptinotarsa decemlineata* (Say). *Pestic. Biochem. Physiol.* **46**, 149–160.

Smagghe, G., and Degheele, D. (1994a). Action of a novel nonsteroidal ecdysteroid mimic, tebufenozide (RH-5992), on insects of different orders. *Pestic. Sci.* **42**, 85–92.

Smagghe, G., and Degheele, D. (1994b). Action of the nonsteroidal ecdysteroid mimic RH-5849 on larval development and adult reproduction of insects of different orders. *Invert. Reprod. Dev.* **25**, 227–236.

Smagghe, G., and Degheele, D. (1994c). The significance of pharmacokinetics and metabolism to the biological activity of RH-5992 (Tebufenozide) in *Spodoptera exempta*, *Spodoptera exigua* and *Leptinotarsa decemlineata*. *Pestic. Biochem. Physiol.* **49**, 224–234.

Smagghe, G., and Degheele, D. (1995a). Biological activity and receptor-binding of ecdysteroids and the ecdysteroid agonists RH-5849 and RH-5992 in imaginal wing discs of *Spodoptera exigua* (Lepidoptera: Noctuidae). *Eur. J. Entomol.* **92**, 333–340.

Smagghe, G., and Degheele, D. (1995b). Selectivity of nonsteroidal ecdysteroid agonists RH5849 and RH5992 to nymphs and adults of predatory soldier bugs, *Podisus nigrispinus* and *P. maculiventris* (Hemiptera: Pentatomidae). *J. Econ. Entomol.* **88**, 40–45.

Smagghe, G., Bohm, G.-A., Richter, K., and Degheele, D. (1995c). Effect of nonsteroidal ecdysteroid agonists on ecdysteroid titer in *Spodoptera exigua* and *Leptinotarsa decemlineata*. *J. Insect Physiol.* **41**, 971–974.

Smagghe, G., Eelen, H., Verschelde, E., Richter, K., and Degheele, D. (1996a). Differential effcts of nonsteroidal ecdysteroid agonists in Coleoptera and Lepidoptera: Analysis of evagination and receptor binding in imaginal discs. *Insect Biochem. Mol. Biol.* **26**, 687–695.

Smagghe, G., Vinuela, E., Budia, F., and Degheele, D. (1996b). *In vivo* and *in vitro* effects of the nonsteroidal ecdysteroid agonist tebufenozide on cuticle formation in *Spodoptera exigua*: An ultrastructural approach. *Arch. Insect Biochem. Physiol.* **32**, 121–134.

Smagghe, G., Viñuela, E., Limbergen, V. H., Budia, F., and Tirry, L. (1999a). Nonsteroidal moulting hormone agonists: Effects on protein synthesis and cuticle formation in Colorado potato beetle larvae. *Entomologia Experimentalis et Applicata* **93**, 1–8.

Smagghe, G., Nakagawa, Y., Carton, B., Mourad, A. K., Fujita, T., and Tirry, L. (1999b). Comparative ecdysteroid action of ring-substituted dibenzoylhydrazines in *Spodoptera exigua*. *Arch. Insect Biochem. Physiol.* **41**, 42–53.

Smagghe, G., Carton, B., Heirman, A., and Tirry, L. (2000). Toxicity of four dibenzoylhydrazine correlates with evagination-induction in the cotton leafworm. *Pestic. Biochem. Physiol.* **68**, 49–58.

Smagghe, G., Carton, B., Decombel, L., and Tirry, L. (2001). Significance of absorption, oxidation, and binding to toxicity of four ecdysone agonists in multi-resistant cotton leafworm. *Arch. Insect Biochem. Physiol.* **46**, 127–139.

Smagghe, G., Decombel, L., Carton, B., Voigt, B., Adam, G., and Tirry, L. (2002a). Action of brassinosteroids in the cotton leafworm *Spodoptera littoralis*. *Insect Biochem. Mol. Biol.* **32**, 199–204.

Smagghe, G., Dhadialla, T. S., and Lezzi, M. (2002b). Comparative toxicity and ecdysone receptor affinity of nonsteroidal ecdysone agonists and 20-hydroxyecdysone in *Chironomus tentans*. *Insect Biochem. Mol. Biol.* **32**, 187–192.

Smirle, M. J., Lowery, D. T., and Zurowski, C. L. (2002). Resistance and cross-resistance to four insecticides in populations of obliquebanded leafroller (Lepidoptera: Tortricidae). *J. Econ. Entomol.* **95**, 820–825.

Smith, H. C., Cavanaugh, C. K., Friz, J. L., Thompson, C. S., Saggers, J. A., Michelotti, E. L., Garcia, J., and Tice, C. M. (2003). Synthesis and SAR of *cis*-1-benzoyl-1,2,3,4-tetrahydroquinoline ligands for control of gene expression in ecdysone responsive systems. *Bioorg. Med. Chem. Lett.* **13**, 1943–1946.

Sobek, L., Bohm, G.-A., and Penzlin, H. (1993). Ecdysteroid receptors in last instar larvae of the wax moth *Galleria mellonella* L. *Insect Biochem. Mol. Biol.* **23**, 125–129.

Sohi, S. S., Palli, S. R., Cook, B. J., and Retnakaran, A. (1995). Forest insect cell lines responsive to 20-hydroxyecdysone and two nonsteroidal ecdysone agonists, RH-5849 and RH-5992. *J. Insect. Physiol.* **41**, 457–464.

Song, M. Y., Stark, J. D., and Brown, J. J. (1997). Comparative toxicity of four insecticides, including imidacloprid and tebufenozide, to four aquatic arthropods. *Environ. Toxicol. Chem.* **16**, 2494–2500.

Spindler-Barth, M., Turberg, A., and Spindler, K. D. (1991). On the action of RH 5849, a nonsteroidal ecdysteroid agonist, on a cell line from *Chironomus tentans*. *Arch. Insect Biochem. Physiol.* **16**, 11–18.

Spindler-Barth, M., and Spindler, K. D. (1998). Ecdysteroid resistant subclones of the epithelial cell line from *Chironomus tentans* (Insecta, Diptera). I. Selection and characterization of resistant clones. *In Vitro Cell Dev. Biol. Anim.* **34**, 116–122.

Sun, X., Song, Q., and Barrett, B. (2003a). Effect of ecdysone agonists on vitellogenesis and the expression of EcR and USP in codling moth (*Cydia pomonella*). *Arch. Insect Biochem. Physiol.* **52**, 115–129.

Sun, X., Song, Q., and Barrett, B. (2003b). Effects of ecdysone agonists on the expression of EcR, USP and other specific proteins in the ovaries of the codling moth (*Cydia pomonella* L.). *Insect Biochem. Mol. Biol.* **33**, 829–840.

Sundaram, K. M. S. (1995). Persistence and fate of tebufenozide (RH-5992) insecticide in terrestrial microcosms of a forest environment following spray application of two mimic formulations. *J. Environ. Sci. Health* **B30**, 321–358.

Sundaram, K. M. S., Nott, R., and Curry, J. (1996a). Deposition, persistence and fate of tebufenozide (RH-5992) in some terrestrial and aquatic components of a boreal forest environment after aerial application of mimic. *J. Environ. Sci. Health* **B31**, 699–750.

Sundaram, K. M. S., Sundaram, A., and Sloane, L. (1996b). Foliar persistence and residual activity of tebufenozide against spruce budworm larvae. *Pestic. Sci.* **47**, 31–40.

Sundaram, K. M. S. (1997a). Persistence and mobility of tebufenozide in forest litter and soil ecosystems under field and laboratory conditions. *Pestic. Sci.* **51**, 115–130.

Sundaram, K. M. S. (1997b). Persistence of tebufenozide in aquatic ecosystems under laboratory and field conditions. *Pestic. Sci.* **51**, 7–20.

Sundaram, M., Palli, S. R., Ishaaya, I., Krell, P. J., and Retnakaran, A. (1998a). Toxicity of ecdysone agonists correlates with the induction of CHR3 mRNA in the spruce budworm. *Pestic. Biochem. Physiol.* **62**, 201–208.

Sundaram, M., Palli, S. R., Krell, P. J., Sohi, S. S., Dhadialla, T. S., and Retnakaran, A. (1998b). Basis for selective action of a synthetic molting hormone agonist, RH-5992 on lepidopteran insects. *Insect Biochem. Mol. Biol.* **28**, 693–704.

Sundaram, M., Palli, S. R., Smagghe, G., Ishaaya, I., Feng, Q. L., Primavera, M., Tomkins, W. L., Krell, P. J., and Retnakaran, A. (2002). Effect of RH-5992 on adult development in the spruce budworm, *Choristoneura fumiferana*. *Insect Biochem. Mol. Biol.* **32**, 225–231.

Swevers, L., Drevet, J. R., Lunke, M. D., and Iatrou, K. (1995). The silkmoth homolog of the *Drosophila* ecdysone receptor (B1 isoform): Cloning and analysis of expression during follicular cell differentiation. *Insect Biochem. Mol. Biol.* **25**, 857–866.

Swevers, L., and Iatrou, K. (1999). The ecdysone agonist tebufenozide (RH-5992) blocks the progression into the ecdysteroid-induced regulatory cascade and arrests silkmoth oogenesis at mid-vitellogenesis. *Insect Biochem. Mol. Biol.* **29**, 955–963.

Swevers, L., Kravariti, L., Ciolfi, S., Xenou-Kokoletsi, M., Ragoussis, N., Smagghe, G., Nakagawa, Y., Mazomenos, B., and Iatrou, K. (2004). A cell-based high-throughput screening system for detecting ecdysteroid agonists and antagonists in plant extracts and libraries of synthetic compounds. *FASEB J.* **18**, 134–136.

Taibi, F., Smagghe, G., Amrani, L., and Soltani-Mazouni, N. (2003). Effect of ecdysone agonist RH-0345 on reproduction of mealworm, *Tenebrio molitor*. *Comp. Biochem. Physiol.* **135C**, 257–267.

Takeuchi, H., Chen, J.-H., O'Reilly, D. R., Rees, H. H., and Turner, P. C. (2000). Regulation of ecdysteroid signalling: molecular cloning, characterization and expression of 3-dehydroecdysone 3a-reductase, a novel eukaryotic member of the short-chain dehydrogenases/reductases superfamily from the cotton leafworm, *Spodoptera littoralis*. *Biochem. J.* **349**, 239–245.

Takeuchi, H., Chen, J. H., O'Reilly, D. R., Turner, P. C., and Rees, H. H. (2001). Regulation of ecdysteroid signaling: Cloning and characterization of ecdysone oxidase. A novel steroid oxidase from the cotton leafworm, *Spodoptera littoralis*. *J. Biol. Chem.* **276**, 26819–26828.

Tanaka, K., Tsukamoto, Y., Sawada, Y., Kasuya, A., Hotta, H., Ichinose, R., Watanabe, T., Toya, T., Yokoi, S., Kawagishi, A., Ando, M., Sadakane, S., Katsumi, S., and Masui, A. (2001). Chromafenozide: A novel lepidopteran insect control agent. *Annu. Rep. Sankyo Res. Lab.* **53**, 1–49.

Tice, C. M., Hormann, R. E., Thompson, C. S., Friz, J. L., Cavanaugh, C. K., Michelotti, E. L., Garcia, J., Nicolas, E., and Albericio, F. (2003a). Synthesis and SAR of a-acylaminoketone ligands for control of gene expression. *Bioorg. Med. Chem. Lett.* **13**, 475–478.

Tice, C. M., Hormann, R. E., Thompson, C. S., Friz, J. L., Cavanaugh, C. K., and Saggers, J. A. (2003b). Optimization of a-acylaminoketone ecdysone agonists for control of gene expression. *Bioorg. Med. Chem. Lett.* **13**, 1883–1886.

Toya, T., Fukasawa, H., Masui, A., and Endo, Y. (2002a). Potent and selective partial ecdysone agonist activity of chromafenozide in Sf9 cells. *Biochem. Biophys. Res. Commun.* **292**, 1087–1091.

Toya, T., Yamaguchi, K., and Endo, Y. (2002b). Cyclic dibenzoylhydrazines reproducing the conformation of ecdysone agonists, RH-5849. *Bioorg. Med. Chem.* **10**, 953–961.

Trisyono, A., and Chippendale, M. (1997). Effect of the nonsteroidal ecdysone agonists, methoxyfenozide and tebufenozide, on the European corn borer (Lepidoptera: Pyralidae). *J. Econ. Entmol.* **90**, 1486–1492.

Trisyono, A., and Chippendale, M. (1998). Effect of the ecdysone agonists, RH-2485 and tebufenozide, on the southwestern corn borer, *Diatraea grandiosella*. *Pestic. Sci.* **53**, 177–185.

Verloop, A., Hoogenstraaten, W., and Tipker, J. (1976). Development and application of new steric substituent parameters in drug design. In "Drug Design" (E. J. Ariens, Ed.), vol. 4, pp. 165–206. Academic Press, New York.

Verloop, A. (1983). The STERIMOL approach: Further development of the method and new applications. In "Pesticide Chemistry, Human Welfare and Environment" (J. Miyamoto and P. C. Kearney, Eds.), vol. 1, pp. 339–344. Pergamon Press, Oxford.

Verras, M., Mavroidis, M., Kokolakis, G., Gourzi, P., Zacharopoulou, A., and Mintzas, A. C. (1999). Cloning and characterization of CcEcR. An ecdysone receptor homolog from the mediterranean fruit fly *Ceratitis capitata*. *Eur. J. Biochem.* **265**, 798–808.

Waldstein, D. E., and Reissig, W. H. (2000). Synergism of tebufenozide in resistant and susceptible strains of obliquebanded leafroller (Lepidoptera: Tortricidae) and resistance to new insecticides. *J. Econ. Entomol.* **93**, 1768–1772.

Watanabe, B., Nakagawa, Y., and Miyagawa, H. (2003). Synthesis of a castasterone/ponasterone hybrid compound and evaluation of its molting hormone-like activity. *J. Pestic. Sci.* **28**, 188–193.

Watanabe, B., Nakagawa, Y., Ogura, T., and Miyagawa, H. (2004). Stereoselective synthesis of ($22R$)- and ($22S$)-castasterone/ponasterone A hybrid compounds and evaluation of their molting hormone activity. *Steroids* **48**, 483–493.

Wearing, C. H. (1998). Cross-resistance between azinphosmethyl and tebufenozide in the greenheaded leafroller, *Planotortrix octo*. *Pestic. Sci.* **54**, 203–211.

Wehling, M. (1997). Specific, nongenomic actions of steroid hormones. *Annu. Rev. Physiol.* **59**, 365–393.

Williams, D. R., Chen, J.-H., Fischer, M. J., and Rees, H. H. (1997). Induction of enzymes involved in molting hormone (ecdysteroid) inactivation by ecdysteroids and an agonist, 1,2-dibenzoyl-1-*tert*-butylhydrazine (RH-5849). *J. Biol. Chem.* **272**, 8427–8432.

Williams, D. R., Fisher, M. J., and Rees, H. H. (2000). Characterization of ecdysteroid 26-hydroxylase: An enzyme involved in molting hormone inactivation. *Arch. Biochem. Biophys.* **376**, 389–398.

Williams, D. R., Fischer, M. J., Smagghe, G., and Rees, H. H. (2002). Species specificity of changes in ecdysteroid metabolism in response to ecdysteroid agonists. *Pestic. Biochem. Physiol.* **72**, 91–99.

Wing, K. D. (1988). RH 5849, a nonsteroidal ecdysone agonist: Effects on a *Drosophila* cell line. *Science* **241**, 467–469.

Wing, K. D., Slawecki, R. A., and Carlson, G. R. (1988). RH-5849: a nonsteroidal ecdysone agonist: Effects on larval lepidoptera. *Science* **241**, 470–472.

Wurtz, J.-M., Guillot, B., Fagart, J., Moras, D., Tietjen, K., and Schindler, M. (2000). A new model for 20-hydroxyecdysone and dibenzoylhydrazine binding: A homology modeling and docking approach. *Protein Sci.* **9**, 1073–1084.

Yao, T.-P., Forman, B. M., Jiang, Z., Cherbas, L., Chen, J.-D., McKeown, M., Cherbas, P., and Evans, R. M. (1993). Functional ecdysone receptor is the product of *EcR* and *ultraspiracle* genes. *Nature* **366**, 476–479.

Zhang, D.-Y., Zheng, S.-C., Zheng, Y.-P., Ladd, T. R., Pang, A. S. D., Davey, K. G., Krell, P. J., Arif, B. M., Retnakaran, A., and Feng, Q.-L. (2004). An ecdysone-inducible putative "DEAD box" RNA helicase in the spruce budworm (*Choristoneura fumiferana*). *Insect Biochem. Mol. Biol.* **34**, 273–281.

Zhao, X.-F., Wang, J.-X., Xu, X.-L., Li, Z. M., and Kang, C. J. (2004). Molecular cloning and expression patterns of the molt-regulating transcription factor HHR3 from *Helicoverpa armigera*. *Insect Mol. Biol.* **13**, 407–412.

6

Juvenile Hormone Molecular Actions and Interactions During Development of Drosophila melanogaster

Edward M. Berger and Edward B. Dubrovsky

Department Of Biology, Dartmouth College, Hanover, New Hampshire 03755

    I. Introduction
   II. Hormones and Development
       A. Prothoracicotropic Hormone
       B. Ecdysone
       C. Juvenile Hormone
  III. Hormones and Reproduction
       A. Spermatogenesis
       B. Oogenesis
  IV. Molecular Biology of Hormone Action
       A. Ecdysone Receptor
       B. Juvenile Hormone Receptor
       C. Genes Regulated by JH
   V. Hormonal Cross-Talk
       A. BR-C *as a Major Target of JH "Status Quo" Action*
       B. E75A *as a Mediator of JH "Status Quo" Action*
     References

I. INTRODUCTION

This review summarizes our understanding of the molecular mechanism of juvenile hormone (JH) action in the fruit fly, *Drosophila melanogaster*. It details the growing evidence that JH acts primarily at the level of gene expression and does so in apparently two different ways—directly, as a transcriptional regulator of JH "target genes," and indirectly, as a suppressor of 20-hydroxyecdysone (20E)-dependent transcriptional regulation. Since *Drosophila* has emerged as a model genetic system for the analysis of insect hormone action, particular emphasis in this chapter is placed on the analysis of mutants that have proved useful for acquiring understanding of insect hormone action during development and reproduction.

During its life cycle, *Drosophila melanogaster* progresses through four structurally and functionally distinct stages—the embryo, larva, pupa, and adult. During embryogenesis, both maternal and zygotic gene expression contribute to the creation of the insect body plan; first establishing polarity, then segmentation, and finally, segment identity (Lawrence, 1992). After hatching, a sequence of three larval stages, or instars, occurs as the insect feeds and grows. Larvae are well adapted for feeding, being primarily composed of a tubular digestive system and an extensively anastomosing fat body whose function is analogous to the mammalian liver. Because insects, like all Arthropods, possess an external and rigid chitinous exoskeleton (the cuticle), larval growth is restricted and must be accompanied by a sequence of molts, when the existing larval cuticle detaches from the underlying epidermis (apolysis), is partially digested, and eventually shed (ecdysis). Following ecdysis, a new cuticle is synthesized and secreted by the underlying epidermis. When a larva attains full size, it emerges from the food and enters a wandering stage. Within a few hours the larva ceases movement, becomes immobile, and secretes a gluelike substance produced by the salivary glands. The glue proteins allow the larva to adhere to a dry substrate. At this time the insect initiates puparium formation. The third-instar larval (L3) cuticle becomes detached, and a pupal cuticle is secreted. In this molt, however, the L3 cuticle is not discarded but rather is retained and tanned to serve as the external pupal case, which acts to protect the pupa and later the developing adult from infection and desiccation. The pupal stage is followed by adult development, a process known as metamorphosis. The development of the adult fly involves both the histolysis of many larval tissues and organs, the growth and differentiation of adult structures derived from nests of abdominal histoblast cells, and from the growth, expansion, and differentiation of organ anlagen, the imaginal discs. Paired imaginal discs give rise to the adult wings, legs, head, and thoracic structures, and a terminal unpaired disc develops into the external genitalia. The adult that emerges from the pupal case is elegantly adapted both for dispersal (flight) and for reproduction. A detailed review of *Drosophila* development, from egg to adult, and its hormonal basis can be found in Riddiford (1993), so what follows is a brief summary.

The episodic progress of *Drosophila* development is governed largely by the action and interaction of three hormones—prothoracicotropic hormone (PTTH), a peptide; a cholesterol derived steroid, 20E also referred to in the literature as beta-ecdysone or ecdysterone; and a sesquiterpene, JH. The basic formula for development in larvae is that PTTH triggers the synthesis and release of ecdysone from the prothoracic gland, which is then converted to its active form, 20E, by a 20-monooxidase located in the mitochondria of larval fat body and in other target tissues (Mitchell and Smith, 1988). 20-Hydroxyecdysone subsequently initiates the molt cycle. The type of molt that takes place depends on the titer of JH, which fluctuates during the life cycle (Bownes and Rembold, 1987; Sliter *et al.*, 1987). In the presence of high JH titers, a larval cuticle is replaced by another larval cuticle. At the end of the L3 stage, the JH titer is low and after several minor peaks of 20E secretion, a major peak of 20E initiates puparium formation and the synthesis of the pupal cuticle. After head eversion, the pupal stage is complete. As with *Manduca*, low JH titers during the larva–pupa transition likely prevent the precocious adult differentiation of imaginal discs (Kiguchi and Riddiford, 1978). The onset of adult development requires a major peak of 20E secretion in the total absence of JH. As discussed in the later section, 20E and JH also have important functions in the adult, promoting both reproductive maturation (Bownes, 1989) and the stereotyped patterns of mating behavior (Manning, 1966). Later sections discuss features of the three hormones that are relevant for understanding of JH function.

II. HORMONES AND DEVELOPMENT

A. PROTHORACICOTROPIC HORMONE

Prothoracicotropic hormone (PTTH) is produced by a pair of neurosecretory cells located in the dorsomedial region of the larval brain (Agui *et al.*, 1979; Gilbert *et al.*, 2002). Immunolocalization studies find that PTTH is transferred from its site of synthesis to its site of action, the brain-associated ring gland, through neurons (Cha *et al.*, 1998), although a significant fraction of PTTH activity was found outside the brain in the ventral ganglion (Henrich *et al.*, 1987a, 1999). *Drosophila* PTTH isolated from late L3 animals is a protein of 45 kDa, and its complete amino acid sequence has been determined (Kim *et al.*, 1997). The native form of PTTH is larger, about 66 kDa, due to multiple N-linked glycosylations. Although many glycoproteins require the carbohydrate moieties for biological activity, fully unglycosylated 45-kDa PTTH was shown to be functionally indistinguishable from the native 66-kDa form, using a ring-gland stimulation assay. Earlier studies

(Henrich et al., 1987a; Pak et al., 1992) had indicated that *Drosophila* PTTH was of lower molecular weight and heterogeneous in composition; however, these results are likely due to extensive proteolysis that occurs during PTTH purification (Kim et al., 1997). Homology searches of the *Drosophila* genome sequence database using any of the lepidopteran PTTH amino acid sequences yielded no matches (Gilbert et al., 2002), suggesting that either PTTH sequences have diverged extensively during insect evolution or the steroidogenesis inducing PTTHs from different insect orders have arisen by convergent evolution. The convergence argument is supported by evidence that lepidopteran and *Drosophila* PTTH differ in their mechanism of action. In the moth *Manduca sexta*, PTTH has been shown to induce steroidogenesis in the prothoracic gland via a signal-transduction cascade, analogous to the regulation of the vertebrate adrenocorticotropic axis (Henrich, 1995). Using *in vitro* assays, PTTH was found to evoke a rapid (within 5 min) rise in cyclic-AMP (cAMP) levels in the larval prothoracic gland (Smith et al., 1984, 1985), as well as in the pupal prothoracic gland (Smith and Pasquarello, 1989). Subsequent studies showed that this rise in cAMP depends on the influx of calcium ions, indicating the participation of a Ca^{++}-calmodulin–dependent adenylyl cyclase (Meller et al., 1988, 1990; Smith and Gilbert, 1986; Smith et al., 1985, 1986). The mechanism by which PTTH controls the flow of calcium through ion channels and the possible presence and role of a PTTH receptor on the gland's cell surface are yet to be determined (Henrich et al., 1999). The rise in cAMP in the *Manduca* prothoracic gland leads to the rapid activation of a cAMP-dependent protein kinase A (PKA), whose activity is essential for propagating the PTTH cascade (Smith et al., 1986; Watson et al., 1993). In *Drosophila*, however, ecdysteroidogenesis, while dependent on external calcium, is not stimulated by cAMP or its analogues (Henrich, 1995). The mechanism that determines the timing of PTTH release in *Drosophila* is also poorly understood. As in other insects (Truman, 1972), photoperiod appears to be a component that dictates the onset of the wandering stage in *Drosophila* in uncrowded conditions (Roberts et al., 1987). Also the timing mechanism might involve the critical size attainment at final instar, as well as the cessation of imaginal disc growth (Edgar and Nijhout, 2004; Nijhout, 2003). A complete discussion of PTTH and its mechanisms of action may be found in several reviews (Gilbert et al., 2002, Henrich et al., 1999).

B. ECDYSONE

In *Drosophila* larvae, ecdysone synthesis takes place in the mitochondria of large polyploid cells comprising the paired prothoracic glands. The prothoracic glands in *Drosophila* are lateral projections of a larger neurohemal organ, the ring gland, which also contains the corpus allatum, the site of JH synthesis, and the corpus cardiacum, a storage gland (Dai and Gilbert, 1991;

Sedlak, 1985). Larval prothoracic glands degenerate during metamorphosis, and in adult *Drosophila*, ecdysteroid synthesis is carried out in the follicular epithelium and nurse cells of ovarian egg chambers (Bownes, 1989; Buszczak et al., 1999; Garen et al., 1977; Richard et al., 1998; Schwartz et al., 1985, 1989; Warren et al., 1996). *Drosophila* eggs contain maternally derived ecdysteroids that are deposited during oogenesis and stored as inactive conjugates in the yolk (Bownes et al., 1988b; Maroy et al., 1988). Evidence suggests that prior to the development of a functional ring gland during embryogenesis these inactive ecdysteroids are converted into active forms by enzymes located in the yolk and/or the extra embryonic amnioserosa (Bownes et al., 1988b; Kozlova and Thummel, 2003).

In vertebrates the ultimate precursor for steroid hormone biosynthesis is cholesterol. Insects, however, are unable to synthesize cholesterol and must acquire it from their diet. In *Drosophila*, a phytophagous insect, the major sterol ingested is sitosterol, which is then enzymatically dealkylated to cholesterol (Gilbert et al., 2002). The biosynthetic pathway leading to 20E formation, and the cellular components that regulate ecdysteroid synthesis and release, are well understood, in large part due to the use of bioinformatics and the isolation and characterization of mutants defective in development. START domain proteins in mammalian systems are used to transfer cholesterol to mitochondria in steroidogenic cells. In a recent study, Roth et al. (2004) reported the isolation of a *Drosophila* gene, *Start1*, whose putative protein product shows extensive homology with the vertebrate cholesterol transporter, MLN64. *In situ* hybridization studies found that *Start1* transcripts are highly abundant in prothoracic gland cells and ovarian nurse cells and that in the ecdysone deficient mutant *ecd-1*, *Start1* expression is severely reduced. This latter observation suggests the presence of a feedback mechanism, in which *Start1* synthesis itself may be dependent on the presence of ecdysone.

Using a reverse genetic approach and P-element mutagenesis, Freeman et al. (1999) identified a *Drosophila* gene *dare* that encodes adrenodoxin reductase (AR). Adrenodoxin reductase is a mitochondrial enzyme that transports electrons in vertebrates to all known P450 enzymes, including the one essential for the synthesis of all steroid hormones. Null mutations in the *dare* gene undergo developmental arrest during the second larval molt, while less severe hypomorphic mutations lead to delayed pupariation. These defects can be largely rescued by feeding mutant larvae 20E, indicating that the AR encoded by *dare* is required for ecdysteroid biosynthesis. *In situ* hybridization analysis revealed that *dare* transcripts are restricted to the larval prothoracic glands and ovarian nurse cells in the adult. The co-localization of *Start1* and *dare* transcripts in ovarian nurse cells at the time of vitellogenesis suggests that: (1) ecdysteroid biosynthesis, *in situ*, is essential for the development of mature egg chambers; (2) inactive ecdysteroid conjugates deposited in oocytes are produced by nurse cells during this period; and (3) these maternally derived transcripts support ecdysteroid biosynthesis during embryogenesis prior to the development of a functional larval prothoracic glands. These suggestions, if confirmed, would help explain why null mutants of *dare* survive to the L2 stage and why germ line clones of loss of function *dare* alleles show arrest by mid-oogenesis (Buszczak *et al.*, 1999).

Three members of the Halloween gene family encode P450 enzymes that function in the ecdysteroid biosynthetic pathway. Halloween genes were uncovered initially in a screen for embryonic lethal mutants (Jurgens *et al.*, 1984; Nusslein-Volhard *et al.*, 1984; Wieschaus *et al.*, 1984). Through reverse genetics, functional assays, and biochemistry, the products of the *disembodied* (*dib*) and *shadow* (*sad*) were shown to be two mitochondrial P450 enzymes as the C22 and C2 hydroxylases, respectively (Chavez *et al.*, 2000; Gilbert, 2004; Warren *et al.*, 2002). Analysis of the *shade* (*shd*) gene product (Gilbert, 2004; Petryk *et al.*, 2003) led to its assignment as the 20-monooxygenase responsible for the conversion of ecdysone to 20E. Localization studies found that both *dib* and *sad* expression is restricted to mitochondria of the larval prothoracic glands (Gilbert, 2004), consistent with their role in the ecdysone biosynthetic pathway, while *shd* expression was absent in the prothoracic glands but abundant in the gut, fat body, and Malpighian tubules, all sites of E to 20E conversion. It is curious that, while *Start1* and *dare* expression is localized to the nurse cells of developing egg chambers, cells in which ecdysteroid-dependent gene expression has been shown to occur (Kozlova and Thummel, 2000), the expression of *dib* and *sad* is restricted to the somatically derived follicle cells that surround the developing oocyte–nurse cell complex (Roth *et al.*, 2004). Since there is evidence that mature eggs and early embryos also contain maternally derived mRNA expressed from *Start1*, *dare*, *sad*, and *dib* (Roth *et al.*, 2004), it will be important to determine whether the transcripts localized in either nurse cells or follicle cells are actually translated in those cells or become transferred to the oocyte for use during early embryogenesis.

In addition, several other *Drosophila* mutants associated with ecdysteroid metabolism are known (Holden *et al.*, 1986; Klose *et al.*, 1980). The recessive temperature sensitive *ecdysoneless* mutation, *ecd-1*, disrupts 20E production (Garen *et al.*, 1977; Henrich *et al.*, 1987b) and causes a host of development defects (Audit-Lamour and Busson, 1981; Brennan *et al.*, 1998; Lepesant *et al.*, 1978; Li and Cooper, 2001; Redfern and Bownes, 1983). While the amino acid sequence of the Ecd protein has been determined (Gaziova *et al.*, 2004) and found to be evolutionarily conserved, the function of Ecd has not been identified, although the suspicion is that this protein is required for the intracellular transport of 7-dehydrocholesterol from its site of synthesis in the endoplasmic reticulum, to the site of subsequent oxidation to ecdysteroids in the mitochondria membranes (Warren *et al.*, 1996). The *lethal (1) giant* (*l(1)g*) mutant is also a nonpupariating conditional mutant, but in this case, mutant animals lack nerve endings in their enlarged prothoracic glands and neurosecretory cells, suggesting that this mutant may not be defective in ecdysteroid synthesis but rather in PTTH signaling (Klose et al., 1980). Molecular analysis of the wild-type *giant* gene indicates that it encodes a b-ZIP transcription factor expressed in a variety of tissues throughout development (Schwartz et al., 1984). Mutations in the *without children* (*woc*) gene result in ecdysteroid deficiencies, larval lethality, and abnormalities in a wide variety of tissues (Wismar et al., 2000). In this case as well, the wild-type gene does not encode an enzyme but rather a zinc-finger transcription factor that probably activates the expression of the cholesterol 7,8-dehydrogenase responsible for carrying out the first step of ecdysone biosynthesis, the conversion of cholesterol to 7-dehydrocholesterol (Warren et al., 2001).

The defective regulation of steroidogenesis in the prothoracic glands is also observed in mutants of *dre4*, *iptr*, and *E75A* (Bialecki et al., 2002; Sliter and Gilbert, 1992; Venkatesh and Hasan, 1997), resulting in the precocious formation of L2 prepupae. This phenotype is also observed following the ectopic expression of BR-Z3, one of several protein isoforms produced from the *Broad-Complex* (*BR-C*) gene, in the prothoracic glands (Zhou et al., 2004).

The inactivation of 20E occurs through either an irreversible hydroxylation of 20E to 20,26-dihydroecdysone or by the sequestration of 20E into polar conjugates, such as 20E phosphate. Such conjugates may be either secreted or stored in the egg and embryo for later activation (Rees, 1995).

C. JUVENILE HORMONE

The juvenoids are a family of sesquiterpenoid hormones produced by the corpus allatum. Juvenile hormones are absent in Crustacea and ticks and are likely unique to insects (Wilson, 2004). Although the existence of JH was hypothesized nearly 70 years ago (Wigglesworth, 1934, 1936), based on physiological experiments, the structure of the first naturally occurring JH was not solved until fairly recently (Judy et al., 1973; Meyer et al., 1970; Roller et al., 1967). At least four forms of JH are known. Among the Lepidoptera, JHI and JHII are the predominant forms. In most other insect orders the major form is JHIII (epoxy farnesoic acid methyl ester). In L3 larvae of *Drosophila* and perhaps in other cyclorrhaphous Dipterans, the predominant larval JH is JH bisepoxide (JHB3; methyl 6,7,10,11-bisepoxyfarnesoate; Richard et al., 1989a,b). Although the original evidence supporting the biological activity of JHB3 in *Drosophila* was indirect, based on a white puparia bioassay (Richard et al., 1989b), experiments (see Section IV.C) confirm that JHB3 and JHIII are both potent inducers of *Drosophila* *E75A*, a gene shown to be a direct target for JH and methoprene transcriptional induction (Dubrovskaya et al., 2004; Dubrovsky et al., 2004a). The regulation of JH synthesis appears to involve both inhibitory (allatostatins) and stimulatory (allatotropins) peptides that are synthesized by the brain and other tissues (Gilbert et al., 1996; Rachinsky and Tobe, 1996). The mechanism of allatostatin action is not well understood, but it is believed to involve second-messenger events, such as changes in the levels of intracellular calcium and cAMP (Rachinsky and Tobe, 1996; Richard et al., 1990). Using a bioinformatics approach, in which the sequences of various insect allatostatins were matched to the *Drosophila* database, a putative allatostatin gene has been identified (Lenz et al., 2000c). PCR amplification of this putative drosostatin resulted in a pre-prohormone that contains four amino acid sequences that, after processing, would give rise to four allatostatins designated drosostatin-1–4. These results, together with the cloning of two allatostatin receptor genes, *DAR-1* and *DAR-2* (Lenz et al., 2000a,b), indicate that a fuller understanding of allatostatin regulated JH biosynthesis is on the horizon.

The regulation of JH titer is critical for the normal progress of *Drosophila* development and is achieved by a balance between biosynthesis and degradation, although mechanisms may also exist for controlling excretion and sequestration. The major degradation pathways for JH involve the hydrolysis of the ester group to the corresponding carboxylic acid, by JH esterase (JHE), and hydrolysis of the epoxide group by JH epoxyhydrolase (JHEH) to the JH diol (Roe and Venkatesh, 1990). In larvae, the predominant enzyme is a JHEH that is associated exclusively with membranes in the mitochondrial and microsomal fractions (Casas et al., 1991; Harshman et al., 1991; Klages and Emmerlich, 1979; Wilson and Gilbert, 1978). At or soon after puparium formation JHEH levels decline precipitously, simultaneous with a large increase in the level of soluble JHE synthesis in the fat body and secretion into the hemolymph that persists through eclosion (Campbell et al., 1992). Juvenile hormone esterase is a highly aggressive enzyme that rapidly clears the animal of endogenous JH in preparation for adult development. The *Drosophila* JHE has been purified and characterized (Campbell et al., 1998), and a complete cDNA clone was obtained using a reverse genetic approach (Campbell et al., 2001). Juvenile hormone esterase cDNA sequences have been reported for a number of insect species, including the fly *D. melanogaster* (Campbell et al., 2001), the cabbage looper *Trichoplusia ni* (Jones et al., 1994), the spruce budworm *Choristoneura fumiferana* (Feng et al., 1999), the tobacco hornworm *M. sexta* (Hinton and Hammock, 2001), the yellow mealworm *Tenebrio molitor* (Hinton and Hammock, 2003; Thomas et al., 2000), the Colorado potato beetle *Leptinotarsa decemlineata* (Vermunt et al., 1997) and the tobacco budworm *Heliothis virescens* (Hanzlik et al., 1989). Each of the predicted protein sequences contains a putative 19–23 amino acid N-terminal signal peptide sequence that presumably facilitates JHE secretion from the fat body into the hemolymph. In several cases insertion of the cDNA into an expression vector was shown to result in the synthesis and secretion of active enzyme (Hinton and Hammock, 2003). Juvenile hormone esterase gene expression appears to be developmentally regulated by 20E and/or JH in several insect species (Feng et al., 1999; Jones et al., 1994; Venkataraman et al., 1994). There is great interest in identifying specific inhibitors of JHE activity that can be used effectively for insect pest control.

Unfortunately, there are few *Drosophila* genes known which, when mutated, affect JH synthesis, hydrolysis, or activity. The *apterous* (*ap*) gene is one of these. It encodes a LIM homeodomain zinc-finger transcription factor that is expressed in a number of tissues, throughout development (Cohen et al., 1992). One highly pleiotropic mutation, *ap-4*, has garnered special interest because *ap-4* females are short lived and defective in oogenesis, while *ap-4* males have fertile gametes but are behaviorally sterile (Ringo et al., 1991, 1992; Tompkins, 1990). In this mutant the adult but not larval, corpus allatum secretes only 10–15% of the JH normally found at this stage (Altaratz et al., 1991; Bownes, 1989; Richard et al., 2001), a defect based on ultrastructural abnormalities in the mitochondria of JH secreting cells (Altaratz et al., 1991; Dai and Gilbert, 1993). The ovaries of *ap-4* females are poorly developed and contain no vitellogenic oocytes (Butterworth and King, 1965; Dubrovsky et al., 2002; Stevens and Bryant, 1985). However, when *ap-4* females are topically treated with JH or the analogue, methoprene, or when their ovaries are transplanted into wild-type female hosts, or cultured in medium containing JH, yolk protein uptake by the ovaries and vitellogenesis is partially restored (King and Bodenstein, 1965; Postlethwait and Weiser, 1973).

A second gene of interest is *circklet* (*clt*). This gene encodes a carboxyl esterase that shares significant amino acid sequence homology with JHE (Campbell et al., 2001). Mutations in *clt* result in defects in several functions, which are JH dependent, including yolk protein synthesis and uptake into oocytes, and degeneration of the larval fat body in adults. These defects are similar to those seen in *ap-4* mutants, but unlike *ap-4* mutants, they are not restored by the application of ectopic methoprene (Shirras and Bownes, 1989). Moreover, wild-type ovaries transplanted into *clt* mutant females are able to effectively take up yolk proteins, indicating that JH is present. The authors argue that the *clt* gene product may encode a protein with either JH receptor activity or some other function required for JH responsiveness. However, as Riddiford (1993) points out, *clt* mutants have extended larval stages and many that pupariate fail to eclose, showing defects in abdominal cuticle. Postlethwait (1974) has reported similar anomalies associated with the application of exogenous JH to larvae or at the time of pupariation. These features would suggest that *clt* mutant larvae have either higher than normal levels of JH or an enhanced responsiveness to JH.

In addition, there are two other genes, *Methoprene tolerant* (*Met*) and *ultraspiracle* (*usp*), that have been implicated in the JH regulatory hierarchy.

Each of these encodes a protein with properties consistent with their being a JH receptor. The evidence supporting these claims is discussed in Section IV.B.

III. HORMONES AND REPRODUCTION

A. SPERMATOGENESIS

In adult *Drosophila*, 20E and JH both serve to regulate and coordinate various aspects of reproductive maturation and sexual behavior in males and females. This section focuses on the functions of JH. Juvenile hormone plays no role in spermatogenesis, which is initiated during pupal development and therefore in the absence of JH (Riddiford, 1993). However, the production of male accessory gland proteins is known to be under JH control. Yamamoto *et al.* (1988) found that either mating or low doses of JH stimulate accessory gland protein synthesis through a mechanism that requires both calcium and protein kinase C, suggesting that the stimulation involves a JH-dependent signal-transduction cascade initiated at the surface of accessory gland cells. Subsequent studies confirmed and extended these results (Herndon *et al.*, 1997; Shemshedini *et al.*, 1990). Conversely, studies in which JH levels were reduced, either pharmacologically or genetically, showed that a deficiency of JH results in lowered levels of protein accumulation in male accessory glands (Gavin and Williamson, 1976; Shemshedini *et al.*, 1990; Whalen and Wilson, 1986). Accessory gland proteins are secreted into the seminal fluid and deposited in the female reproductive tract during copulation. They depress female receptivity, stimulate oogenesis and oviposition, provide antimicrobial and nutritive properties, and promote sperm storage in the female (Wilson *et al.*, 2003). Moshitzky *et al.* (1996) found that one accessory gland protein, the sex-peptide, actually activates the corpus allatum of virgin females to produce and secrete JH, thereby stimulating oogenesis and promoting mating behavior. The importance of JH in promoting normal courtship behavior in males was shown in a study by Tompkins (1990), who found that while JH-deficient *apterous-4* mutant males are able to carry out courtship behavior, they do so at a reduced rate and with less success. However, since the *apterous-4* mutation produces a variety of pleiotropic defects, including structural defects in the appendages involved in mating behavior, the connection between reduced JH and reduced mating behavior remains ambiguous. In the analysis of a methoprene-tolerant mutant, *Met27*, that has no such observable structural defects, Wilson *et al.* (2003) found that mutant males had both reduced levels of protein accumulation in their accessory glands and were less efficient at mating. These phenotypes could be partially rescued by topical methoprene application. Cytosolic extracts from these mutants showed a

10-fold reduction in the level of JH binding, compared to wild-type (Shemshedini and Wilson, 1990), again suggesting a connection between JH and male mating behavior. These findings are consistent with the study by Pursley *et al.* (2000), who localized the normal Met protein to both male accessory glands and ejaculatory duct tissue. Collectively, these studies suggest that, in male *Drosophila*, JH is important both for accessory gland function and conferring normal mating behavior. The molecular mechanism underlying these effects is unknown.

Finally, using transgender neuron transplantation techniques, Belgacem and Martin (2002) demonstrated that the sexually dimorphic pattern of walking behavior in *Drosophila* is under the control of several neurons, which are located in the pars intercerebralis (PI) region of the brain, suggesting that those neurons may act through a humoral factor to transmit their effects. They went on to show that feeding males fluvastatin, a JH inhibitor, led to the feminization of walking behavior, and this effect could be reversed by the simultaneous topical application of methoprene. The authors conclude that JH could serve as a humoral link between the PI neurons and the locomotor center, but the mechanism underlying this link is not known.

B. OOGENESIS

The role of JH in promoting reproductive maturation and mating behavior in female *Drosophila* has also been investigated in detail. Juvenile hormone is known to be involved in both yolk protein synthesis and secretion and yolk protein uptake by the ovary, a process collectively referred to as vitellogenesis (Bownes, 1994; Gavin and Williamson, 1976; King, 1970; Postlethwait and Handler, 1979). Ablation of the corpus allatum in adult females significantly reduces egg production (Bouletreau-Merle, 1974), as does JH deficiency produced by the *apterous-4* mutation (Altaratz *et al.*, 1991). Conversely, topical application of the JH analogue, methoprene, to immature females promotes vitellogenesis and egg production (Ringo and Pratt, 1978), as does implantation of mature corpus allatum to immature females (Manning, 1966) and topical JH application to JH deficient *apterous-4* mutant females (Dubrovsky *et al.*, 2002).

At the molecular level, the analysis of JH action during oogenesis has focused largely on the regulation of yolk protein synthesis and secretion by the fat body and ovarian follicle cells and on the uptake of yolk protein by the developing oocyte, processes collectively referred to as vitellogenesis. In *D. melanogaster* there are three yolk proteins, YP1, YP2, and YP3, produced in both the adult female fat body (Jowett and Postlethwait, 1980) and ovarian follicle cells (Brennan *et al.*, 1982; Isaac and Bownes, 1982). YP1, YP2, and YP3 are each encoded by a single copy gene on the X-chromosome. Two of these genes, *YP1* and *YP2*, are tightly linked and transcribed divergently, and an intergenic region approximately 1 kb in length separates their transcription units. The *YP3* gene lies about 10 kb from the *YP1–YP2* cluster. The molecular basis of *YP* gene expression is complex and involves hormonal inputs as well as tissue-specific and sex-specific factors. Yolk protein synthesis in the female fat body is induced by 20E (and perhaps JH), although induction is not direct and requires a concurrent protein synthesis (Bownes *et al.*, 1987). YP synthesis in ovarian follicle cells is exclusively under JH control (Bownes and Blair, 1986; Bownes *et al.*, 1988a; Jowett and Postlethwait, 1980).

YP gene expression occurs normally in adult females but not in males, due to the sexually dimorphic expression of the *doublesex* gene (*dsx*). The male DSX protein (DSXM) represses *YP* gene expression in the male fat body (Bownes and Nothiger, 1981), while the female DSX protein (DSXF) enables and likely stimulates *YP* gene expression in the female fat body (Coschigano and Wensink, 1993). Burtis *et al.* (1991) have successfully localized *cis*-acting DNA-binding sites for DSXF and DSXM in the *YP1–YP2* intergenic region (Shirras and Bownes, 1987). The *YP1–YP2* intergenic region also contains a 125-nucleotide sequence referred to as the FBE that binds a number of additional *trans*-acting factors, which are believed to confer fat body and follicle cell specific expression (Abel *et al.*, 1992; Abrahamsen *et al.*, 1993; Bownes, 1994; Falb and Maniatis, 1992; Garabedian *et al.*, 1985; Logan and Wensink, 1990; Ronaldson and Bownes, 1995; Tamura *et al.*, 1985). In terms of hormonal control, Pongs (1989) initially identified a region in the *YP1–YP2* spacer, which selectively binds iodinated ponasterone, a potent analogue of 20E, suggesting that this region also contains one or more functional ecdysone response elements (EcREs) that serve to bind the EcR/USP receptor complex. Germ line transformation studies functionally mapped several of these EcREs to regions located 5′ and 3′ of *YP3*, and sites within the coding region and introns, as well as to the intergenic spacer region between *YP1* and *YP2* (Bownes *et al.*, 1996). To date, there is no evidence for the presence of a specific DNA region that confers JH-dependent *YP* gene expression in the ovarian follicle cells.

In sexually mature mated female *Drosophila*, egg production occurs continuously. However, sexually mature virgin females rarely oviposit, and in these flies egg production is terminated at stage nine of oogenesis, by a process that involves both apoptosis and egg resorption. Based on a series of studies Soller *et al.* (1999) proposed a model to account for these observations. This model involves the activities of both JH and 20E, in conjunction with the seminal fluid component, sex-peptide. According to this model, apoptosis and the subsequent resorption of immature stage nine oocytes observed in unmated females are mediated by high levels of 20E in the hemolymph. Upon mating, sex-peptide introduced in the seminal fluid stimulates JH production in the corpus allatum, thereby promoting both vitellogenesis and oocyte progression though the stage nine checkpoint. These events also result in a reduction in 20E levels, although the model does not speculate on how this reduction occurs. Because the application of 20E has been shown to slow down the progression of stage nine oocytes, even when sex-peptide and 20E are present, the authors propose that 20E could serve as an intrinsic feedback signal that reduces egg production during unfavorable conditions.

IV. MOLECULAR BIOLOGY OF HORMONE ACTION

Hormones typically function by initially binding to specific protein receptors, either on the cell surface or within the nucleus. The identification of hormone receptors in *Drosophila* has been facilitated by the application of reverse genetics and by the application of bioinformatics algorithms to the *Drosophila* genome sequence database. Although great progress has been made in understanding the structure and function of the receptors for ecdysone, far less is known about the putative receptor for JH.

A. ECDYSONE RECEPTOR

The ecdysone receptor is a heterodimer composed of EcR, a member of the nuclear hormone receptor superfamily (Koelle *et al.*, 1991), and ultraspiracle (USP), the insect orthologue of the vertebrate retinoid X receptor (RXR) (Christianson *et al.*, 1992; Oro *et al.*, 1990; Yao *et al.*, 1992, 1993). EcR acts as a ligand-dependent transcription factor (Freedman, 1997; Mangelsdorf and Evans, 1995) that together with USP binds to a palindromic ecdysone response element (EcRE) found in the promoter of many 20E inducible genes (Cherbas *et al.*, 1991; Hoffmann and Corces, 1986; Mestril *et al.*, 1986; Riddihough and Pelham, 1986, 1987; Rudolph *et al.*, 1991). Like the vertebrate thyroid hormone receptor, the EcR/USP receptor is associated with its hormone response element in the absence of ligand, where it functions to repress basal expression of the adjacent gene (Cherbas *et al.*, 1991; Dobens *et al.*, 1991). The presence of the ligand, 20E, promotes EcR/USP dimerization and converts the repressor into a transcriptional activator (Cherbas and Cherbas, 1996; Yao *et al.*, 1993).

In *Drosophila*, both EcR and USP exist as multiple isoforms (Henrich *et al.*, 1990, 1994; Oro *et al.*, 1990; Shea *et al.*, 1990). In the case of EcR, there is a single gene (Koelle *et al.*, 1991) that encodes three protein isoforms, EcR-A, EcR-B1, and EcR-B2 (Talbot *et al.*, 1993). The A and B isoforms are generated from the use of alternative promoters, while the B1 and B2 isoforms are the outcome of alternative mRNA splicing. The isoforms contain identical DNA- and ligand-binding domains and unique N-terminal domains. Analyses of spatial and temporal distribution of each isoform, using isoform specific antibodies, have revealed an interesting pattern of expression. EcR-A is found in the ring gland, and it is the predominant form in imaginal discs, the precursors of adult structures. EcR-A is also the isoform found in those larval nerve cells that persist to the adult stage (Robinow et al., 1993). EcR-B1 is expressed primarily in larval tissues that will be histolyzed during metamorphosis, such as the salivary glands, and in a distinct set of larval nerve cells (Truman et al., 1994). By analyzing mutations affecting only individual EcR isoforms, isoform-specific functions have been identified (Bender et al., 1997; Cherbas et al., 2003; Riddiford et al., 2001).

A single gene in *Drosophila* encodes the USP polypeptide, and the two isoforms, of molecular weight 54 and 56 kDa, correspond to unphosphorylated and phosphorylated variants (Song et al., 2003). The relative abundance of the two isoforms varies in wandering larvae and prepupae and might be regulated by 20E. However, the rate of *usp* transcription is rather uniform during development, and is not linked to changes in 20E titer (Andres et al., 1993; Henrich et al., 1999). Ultraspiracle phosphorylation may be a mechanism for regulating EcR/USP function and 20E responsiveness; however, the functional significance of the USP isoforms is not known.

Mutations affecting both USP and each of the individual EcR isoforms have been characterized (Henrich et al., 1999). Lesions in either the DNA- or the ligand-binding domain of EcR result in embryonic lethality, supporting the claim that 20E plays a role in the deposition of the L1 cuticle (Kozlova and Thummel, 2000, 2003). Lesions in the unique N-terminal region of EcR-B1 lead to a disruption of the puffing pattern in L3 salivary gland polytene chromosomes and death at or soon after puparium formation (Bender et al., 1997). Mutations in the DNA-binding domain of the orphan nuclear receptor, USP, result in lethality during the L1 to L2 molt, while the absence of functional maternal USP produces death during embryogenesis (Henrich et al., 1994; Oro et al., 1990; Perrimon et al., 1985). These results suggest that, in the absence of zygotic USP, maternally derived protein enables survival through the L1 stage. Why mutations in *EcR* result in an earlier lethality than mutations in *usp* is unknown. One possibility is that EcR may have binding partners other than USP during embryogenesis, and these heterodimers are able to fulfill vital functions during very early development. In support of this interpretation, DHR3, another orphan nuclear receptor in *Drosophila* (Koelle et al., 1992), is known to interact with EcR *in vitro* (White et al., 1997), and mutations in *DHR3* produce embryonic lethality (Carney et al., 1997).

Drosophila USP has also been shown to have a binding partner other than EcR. *Drosophila* DHR38 is a member of the orphan nuclear receptor superfamily and will form a heterodimer with USP (Sutherland et al., 1995). DHR38/USP complexes respond to ecdysteroids by enhancing ligand-dependent transcription, but, remarkably, this enhancement does not involve direct ligand binding (Baker et al., 2003). *DHR38* mutants, like *usp* mutants, show defects in cuticle formation that are not seen with *EcR* mutations, suggesting that DHR38/USP complexes may govern an ecdysteroid signaling pathway that is distinct from EcR (Baker *et al.*, 2003; Hall and Thummel, 1998; Kozlova *et al.*, 1998). In their model, Baker *et al.* (2003) suggest that since the DHR38/USP complex requires an activated USP for transcriptional activation, native USP could be capable of becoming activated by an as yet unknown ligand. This hypothesis is supported by X-ray crystallography data that identifies a large hydrophobic pocket in the USP ligand-binding domain that could be occupied by a lipophilic ligand (Billas *et al.*, 2001; Clayton *et al.*, 2001). DHR38, in contrast, appears to lack a functional ligand-binding domain, implying that in the EcR/USP complex it is EcR that is activated by ligand binding, while in the DHR38/USP complex transcriptional activation may occur through as yet unknown ligand binding to USP.

Not unexpectedly, EcR/USP is known to interact with other factors in the cell to either enhance or repress target gene expression. Among these is a protein co-repressor, SMRTER, an orthologue of vertebrate SMRT (Tsai *et al.*, 1999). SMRTER has been shown to bind to EcR *in vitro*, and USP and SMRTER co-localize at specific bands and interbands on the *Drosophila* polytene chromosomes in immunostaining studies. SMRTER also binds to dSin3A, a protein associated with the nucleosome/histone modifying proteins Rad3/HAAC. The model is that in the absence of 20E, the EcR/USP/SMRTER complex is loosely associated with a target gene EcRE. This leads to the recruitment of a histone deacetylase complex that then chemically modifies the chromatin structure resulting in the suppression of basal transcription. It goes without saying, then, that the transactivation function of ligand-bound EcR/USP likely involves the removal of SMRTER and the binding of as yet unidentified co-activators, which serve to further recruit a protein complex with histone acetyltransferase activity.

B. JUVENILE HORMONE RECEPTOR

Juvenile hormone is involved in regulation of an unusually extensive range of physiological and developmental processes in insects. As the "status quo" hormone, its presence during larval stages ensures that peaks of 20E release promote larva-to-larva molts rather than initiating metamorphosis. During puparium formation the low titer of JH prevents the premature differentiation of imaginal discs, and in adult females JH is required for vitellogenesis and egg production. In adult males, JH is required for normal mating behavior, and it stimulates protein synthesis in male accessory glands apparently through a cell surface initiated signal-transduction mechanism that involves calcium and protein kinase C activity (Sevala and Davey, 1993; Sevala *et al.*, 1995; Yamamoto *et al.*, 1988). In spite of what is known about the biology of these JH actions, very little is known about the molecular mode of JH action, and the identity of the JH receptor remains elusive. Because naturally occurring juvenoids are highly lipophilic (i.e., they are essentially insoluble in water) they must be transported within and between cells in association with abundant carrier proteins that lack the properties of conventional nuclear hormone receptors. As a consequence the numerous attempts to identify and isolate authentic JH receptors have invariably resulted in the purification of one or several of the JH carrier proteins or of enzymes that use JH as a substrate (Chang et al., 1980; Palli et al., 1990, 1994; Riddiford, 1996). A second problem is that in the absence of a functionally defined JH response element, there is no unambiguous assay for JH receptor activity.

In recent years, two candidates for the JH receptor in *Drosophila* have emerged. Loss of function mutations in the *Methoprene tolerant* (*Met*) gene result in a 50–100 fold resistance to the toxic and morphogenetic effects of exogenous JH or methoprene, a potent synthetic JH analogue (Riddiford and Ashburner, 1991; Wilson and Fabian, 1986). *Met* mutations are semi-dominant and extracts from *Met* flies show a 10-fold lower binding affinity for JH, compared to a wild-type control (Shemshidini and Wilson, 1990). The *Met* gene has been cloned and sequenced, and the MET protein identified as a member of the bHLH-PAS transcription factor family (Ashok et al., 1998; Wilson and Ashok, 1998), which is intriguing because PAS proteins typically function as heterodimers with nuclear receptor proteins or other PAS proteins as transcriptional activators (Wilson, 2004). The MET protein is localized in the nuclei of a number of JH target tissues, including ovarian follicle cells (Pursley et al., 2000; Wilson, 2004). Recently, Miura et al. (2005) have shown that MET proteins binds to JH ($K_d \sim 5$ nM), and activates transcription in a JH-dependent manner. However, flies homozygous for null *Met* mutations are viable, suffering only about 20% reduction in female fertility. Wilson has suggested that the viability of *Met* mutants could be attributed to the presence of gene products whose function is redundant with MET.

A second candidate for the JH receptor is the EcR binding partner, USP (Jones and Sharp, 1997). Although initially rejected as a JH receptor based on its failure to bind either of the ester forms of JH analogues, methoprene and hydropene (Harmon et al., 1995), subsequent fluorescence-based binding assays did detect specific, albeit low, affinity binding of JHIII and JHIII-acid to USP (Jones and Sharp, 1997). Furthermore, in transfected Sf9 cells, JHIII was shown to activate the expression of a luciferase reporter construct driven by the *Juvenile hormone esterase* (*Jhe*) core promoter and a putative JH response element (JHRE) to which recombinant USP bound in a gel shift assay. Co-transfection experiments, in which mutant *Drosophila* USP defective in JHIII binding is also introduced, showed suppression of this activation, leading to the conclusion that mutant USP behaves as a competitive inhibitor of wild-type USP (Jones et al., 2001; Xu et al., 2002). Based on crystallographic data, the controversy remains regarding the ability of JH to associate with the ligand-binding domain of USP (Billas *et al.*, 2001; Clayton *et al.*, 2001; Jones and Jones, 2000; Sasorith *et al.*, 2002). However, functional studies have been supportive. Maki *et al.* (2004) report that in a transient expression assay using *Drosophila* S2 cells, JHIII and a synthetic juvenoid, methoprene acid, strongly reduce the 20E-dependent transactivation of a luciferase reporter construct driven by an *hsp27* EcRE, a result consistent with previous studies (Berger *et al.*, 1992). Using a two-hybrid assay, they went on to show that the JH-dependent transrepression was neither due to the prevention of EcR/USP heterodimerization nor due to the failure of receptor binding to the EcRE. Based on additional studies with TSA, a histone transacetylase inhibitor, the authors propose that JH-bound USP attenuates 20E-dependent transactivation through the recruitment of a histone deacetylase co-repressor complex, which in principle could involve the activity of factors, such as SMRTER (Tsai *et al.*, 1999). Considering that JH functions as both a repressor of the EcR/USP mediated 20E activation, and as a transcriptional activator independent of 20E (discussed in Section IV.C), the possibility remains that there are multiple JH receptors in *Drosophila*.

C. GENES REGULATED BY JH

While the issues of JH signal reception and transduction still remain in the dark, there has been significant progress in identifying JH-responsive genes in different insect species.

There are various examples showing JH-mediated modulation of the expression of genes activated by 20E. In holometabolous insects, a metamorphic transition from larva-to-pupa is associated with 20E-mediated activation of the regulatory hierarchy that includes genes encoding transcription factors, such as EcR, USP, E74, E75, and Broad-Complex (BR-C). In *Manduca*, metamorphosis involves reprogramming of the larval epidermis, when, in response to 20E, cells acquire pupal commitment and switch the expression profile from larval-specific to pupal-specific genes encoding cuticle proteins. This switch can be prevented by exogenous JH. When added before pupal commitment, JH abrogates 20E activation of *BR-C* and strongly enhances the accumulation of *E75A* transcripts (Zhou *et al.*, 1998a,b). Also, JH modifies 20E-induced changes in the expression of two components of the 20E receptor heterodimer, EcR and USP (Hiruma *et al.*, 1999). As a result, cells of the fifth instar *Manduca* epidermis do not become pupally committed and retain their ability to synthesize a larval cuticle (Riddiford, 1976, 1978).

In another lepidopteran, the wax moth, *Galleria mellonella*, silk glands undergo degeneration initiated by the high-level 20E pulse that occurs during the pupal molt. At least six transcripts encoded by genes from the 20E regulatory hierarchy, including *EcR*, *usp*, *HR3*, *E75*, are differentially expressed in *Galleria* silk glands. The presence of JH prevents silk gland degeneration and enhances the accumulation of *E75A* RNA (Jindra and Riddiford, 1996).

When exposed to 20E, IAL-PID2 cells derived from imaginal wing discs of the Indian meal moth *Plodia interpunctella* undergo proliferative arrest, followed by a morphological differentiation and the expression of high levels of *EcR, HR3*, and *E75* RNAs (Mottier *et al.*, 2004; Siaussat *et al.*, 2004). The simultaneous exposure of *Plodia* imaginal cells to 20E and JH inhibits this 20E-induced differentiation and modifies expression of genes from the 20E regulatory hierarchy—accumulation of *E75B* and *HR3* transcripts is increased, while the second peak of *EcRB-1* expression is inhibited (Siaussat *et al.*, 2004). To clarify the mechanism of JH action, it would very useful in future research to find a connection between JH and the three transcription factors, on one hand, and downstream gene expression linked to morphological changes in the cell line, on the other.

In the yellow fever mosquito, *Aedes aegypti*, initiation of vitellogenesis requires a blood meal, which causes an immediate decline of JH titer and an elevation of 20E titer. Consequently, the 20E signal propagated through the regulatory hierarchy activates vitellogenin (*Vg*) gene expression. The initial exposure to JH is essential for the mosquito fat body to become responsive to 20E. A study has shown that an orphan nuclear receptor, βFTZ-F1, is a JH-regulated competence factor required for the vitellogenic 20E response (Zhu *et al.*, 2003). The RNAi silencing of *βFTZ-F1* after injection of dsRNA dramatically reduced expression of the major players of the 20E regulatory hierarchy, *EcR-B, E74B*, and *E75A*, and the *Vg* gene after blood feeding. Importantly, JH regulates the synthesis of βFTZ-F1 protein in *A. aegypti*, though the mechanism remains obscure. Because female mosquitoes take several blood meals during their adult life each followed by oviposition, it appears that the post-transcriptional control of βFTZ-F1 synthesis should occur cyclically at the beginning of each vitellogenic period. Further investigations will show whether JH plays the same role at each vitellogenic cycle.

In *Drosophila* salivary glands, JH interferes with the acquisition of competence of the polytene chromosomes to respond to 20E with a stereotyped program of puffing activity (Richards, 1978). Finally, in cultured *Drosophila* cells, JH inhibits 20E induction of the four small *hsp* genes and a cluster of micro-RNA encoding genes, *mir-100, mir-125*, and *let-7* (Berger *et al.*, 1992; Sempere *et al.*, 2003).

The second type of genetic response to JH involves a direct effect of the hormone on gene expression. A number of JH-responsive genes have been described in studies of both developing and mature insects, as well as in studies with cultured insect cells. In a study of *Vg* gene expression in the fat body of the cockroach *Blattella germanica*, it was established that JH stimulates *Vg* transcript accumulation within 2 hr after *in vivo* JH application to females whose corpora allata was surgically removed (Comas *et al.*,

1999). In vitro fat body incubations revealed that 1 nM of JH was sufficient to activate Vg gene expression, while 10 nM produced the maximal effect (Comas et al., 2001). It appears, though, that while the induction of the Vg gene is rapid and efficient in terms of JH concentration, it still represents a secondary hormone response because the simultaneous presence of the protein synthesis inhibitor cycloheximide completely impairs the JH effect (Comas et al., 2001). An increase of calmodulin (CaM) gene transcripts in the follicular epithelium during the vitellogenic period is another example of JH activating action in Blattella germanica. Depletion of endogenous JH by head ligation results in a steep fall in CaM transcript levels, however, expression can be rescued to levels above control within 24 hr after JH injection (Iyengar and Kunkel, 1995).

In the fat body of the African migratory locust, Locusta migratoria, JH activates several targets including genes encoding two vitellogenins, two hemolymph proteins, and two subunits of the translation elongation factor-1 (EF-1), EF-1α and EF-1γ (Glinka and Wyatt, 1996; Zhou et al., 2002). The abundance of corresponding transcripts increases upon application of the JH agonist methoprene to JH-deprived locusts. However, these genes show very slow kinetics of induction, with a lag time of about 24 h. Moreover, the induction of the Vg mRNA in locusts is delayed by cycloheximide (Edwards et al., 1993), implying that JH-activated expression of these genes does not represent a primary response to the hormone.

The Differential Display technique (Fig. 1) enabled researches to considerably extend the list of genes transcriptionally regulated by JH and to reveal much more rapid effects of JH on gene expression (Dubrovsky et al., 2000; Feng et al., 1999; Hirai et al., 1998; Kramer et al., 2002; Royer et al., 2002, 2004). In Drosophila S2 cells, JH activates a group of genes whose developmental expression profiles tightly follow reported changes in the JH titer in flies:

JhI-26 is a novel gene (Dubrovsky et al., 2000). A database search with the predicted amino acid sequence reveals homologous proteins of unknown function from other insects, such as the fly Drosophila pseudoobscura, the mosquito Anopheles gambiae, the silkworm Bombyx mori, and the honeybee Apis mellifera. Also, the Conserved Domain Search server detects a putative choline kinase domain, CHK. Interestingly, in mammalian cells phosphocholine synthesis by choline kinase potentiates cell growth (Chung et al., 2000). However, without mutant analysis the functional significance of the CHK domain of JhI-26 remains unclear.

dRNaseZ (former *JhI-1*) is a member of the ELAC1/2 gene family, which is conserved from bacteria to man (Dubrovsky et al., 2000). It encodes a zinc-dependent phosphodiesterase, which displays endonuclease activity on nuclear and mitochondrial precursor tRNAs. It was shown that Drosophila RNase Z is involved in tRNA processing by removing a 3'-trailer from pre-tRNA. In S2 cultured cells, silencing of *dRNaseZ* with RNAi disrupts tRNA

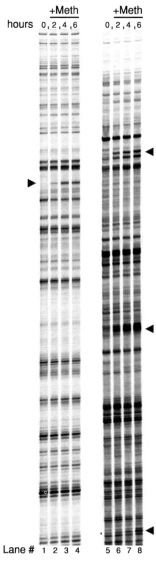

FIGURE 1. Representative mRNA differential display. S2 cells were cultured in the presence of methoprene for 0, 2, 4 and 6 h. DNA-free total RNA samples were recovered, reverse transcribed, and PCR amplified with different combinations of anchor and arbitrary primers. PCR products were separated on denaturing polyacrylamide gels and analyzed by autoradiography. Arrowheads indicate cDNA bands whose abundance increased following methoprene treatment (lanes 2–4 and 6–8).

maturation and causes accumulation of pre-tRNA molecules with 3′-extensions (Dubrovsky et al., 2004b). The human *ELAC2* orthologue is a prostate cancer susceptibility gene, whose specific role has not yet been defined (Korver et al., 2003; Tavtigian et al., 2001). By contrast, the biological role of *ELAC2* from the nematode *Caenorhabditis elegans* has been established. It appears that this gene is required for germ cell progression through mitosis. Reduction of *ELAC2* activity by RNAi results in slow-growing animals with severely under-proliferated germ lines (Smith and Levitan, 2004).

JhI-21 and **minidisks** (*mnd*) genes encode highly similar amino acid transport proteins that are homologous to transporters found in a variety of eukaryotes (Dubrovsky et al., 2002). In S2 cells, the two genes display different mechanisms of hormonal regulation—*mnd* gene expression is characteristic of a primary JH response, while *JhI-21* gene expression has features of a secondary JH response. The *mnd* gene is critical for imaginal cell proliferation. Based on mutant phenotype analysis, it has been suggested that the fat body might respond to amino acid uptake by the Mnd transporter by secreting a factor that is required for imaginal disc maturation and proliferation (Martin et al., 2000).

As discussed in Section III.B, JH plays an important role in the reproductive maturation of *Drosophila* females. Consistent with this role, transcripts of the three JH-inducible genes—*dRNaseZ*, *JhI-21*, and *mnd*—are highly abundant in ovaries. Apparently, these RNAs are synthesized in nurse cells and then transferred into maturing egg. Maternal inheritance of *dRNaseZ*, *JhI-21*, and *mnd* RNAs was confirmed by *in situ* hybridization studies (data not shown and Dubrovsky et al., 2002). Juvenile hormone–inducible transcripts were detected at high levels in nurse cells of stage 10 egg chambers, and later in nascent oocytes. In embryos, JH–inducible transcripts were most abundant during first 2 hr after fertilization during pre-cellular blastoderm stages. Once zygotic gene expression begins, these maternal RNAs are degraded. It is likely that a number of other genes whose products are needed for early embryonic development are also targets for JH-regulated expression. One opinion is that in the ovary JH might serve as a signal to stimulate both vitellogenesis and the production of maternally inherited RNAs.

mir-34 is a member of a new and quickly growing class of genes that encode small functional RNAs rather than proteins (Ruvkun, 2001). Micro RNAs (miRNAs) are transcripts of 18–25 nucleotides in length. They are believed to function as antisense translational repressors of mRNA targets. In S2 cells, *mir-34* transcript levels increased after about 6–8 h in the continuous presence of JH (Sempere et al., 2003). Conversely, the abundance of *mir-34* transcripts decreased when cells were grown in the presence of 20E. The *mir-34* gene was initially cloned in *C. elegans* and then identified in *Drosophila*; however, no mutants are available so far in either worms or flies.

E75 is a gene first described by Ashburner (1972) as an early 20E-inducible puff, 75B, in the third-instar salivary gland polytene chromosomes. After cloning, it was established that the *E75* gene is composed of several embedded transcription units (Feigl *et al.*, 1989; Segraves and Hogness, 1990). When expressed, it gives rise to several transcripts that encode four similar but distinct, proteins, designated E75A, E75B, E75C, and E75D. These proteins are orphan receptors, members of the nuclear hormone receptor superfamily. The E75B and E75D proteins are anomalous, however, as they lack a functional DNA-binding domain. In S2 cells, transcription of two *E75* encoded isoforms, *E75A* and *E75D*, is specifically activated by methoprene and JHIII at concentrations well within the physiological range found in larvae and adults (Dubrovskaya *et al.*, 2004; Dubrovsky *et al.*, 2004a). Since JHB3 is the primary secretory product of the larval *Drosophila* ring gland, we also tested this JH for its biological activity as an inducer of *E75A* transcription in cultured cells. The results of the Northern blot hybridization (Fig. 2) suggest that the efficacy of JHB3 in the *E75A* gene induction is the same as that of JHIII and methoprene. The induction is rapid and does not require concurrent protein synthesis and thus represents a primary hormone response. Consistent with JH regulation, *E75A* mRNA levels are reduced in ovaries of the *ap-4* mutants defective in JH secretion. Expression can be rescued by exogenous hormone application (Dubrovsky *et al.*, 2004a). It was found that *E75* activation by JH is conserved in Diptera and Lepidoptera. In *Manduca* CH1 cells, the *mE75D* isoform exhibits dual

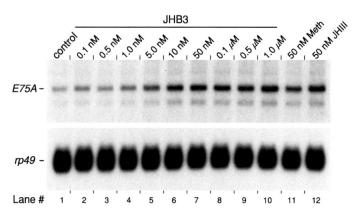

FIGURE 2. Analysis of JHB3 potency in the induction of *E75A* gene expression, as revealed by Northern blot hybridization. Each lane contains 10 μg of total RNA purified from S2 cells treated for 1 h with 5×10^{-8} M methoprene (lane 11), 5×10^{-8} M JHIII (lane 12), or with different concentrations of JHB3, from 1×10^{-10} to 1×10^{-6} M (lanes 2–10). Control is from cells treated with ethanol (lane 1). A Northern blot was successively hybridized with *E75A* and *rp49* probes, and the latter was used as a control for loading.

hormonal regulation; it can be activated by either 20E or JH (Dubrovskaya et al., 2004).

Application of the differential display technique for the RNA profiling of *Choristoneura* cultured cells allowed the cloning of the *JH esterase* (*Jhe*) gene, whose transcript abundance increased when cells were grown in the presence of JH (Feng et al., 1999). The induction was evident even in the presence of cycloheximide, indicating that this JH effect is direct. Supporting this suggestion, Kethidi et al. (2004) found that the promoter region of the *Jhe* gene could confer JH inducibility on a heterologous reporter gene transfected into the *Choristoneura* cell line. Moreover, the authors were able to identify a 30 bp JH response element (JHRE) that was necessary and sufficient for hormonal regulation (Kethidi et al., 2004). This element contains two direct repeats of AGATTA with a 4-nucleotide spacer similar to a direct repeat 4 element (DR4). Importantly, the JHRE can drive JH-dependent reporter gene expression not only in lepidopteran cultured cells but also in dipteran cells, such as fly and mosquito, as well (Kethidi and Palli, 2004). Because it takes only 3 h after hormone administration for JHRE binding activity to appear in the nuclear protein extracts, the authors suggest that JH causes protein modification rather than *de novo* protein synthesis. The authors are testing the hypothesis that JH action involves dephosphorylation of nuclear proteins that evokes their DNA-binding activity.

V. HORMONAL CROSS-TALK

As stated at the beginning of this review, the regulatory interplay between JH and 20E defines the outcome of each developmental transition in the insect life cycle. 20-hydroxyecdysone initiates each transition from larva-to-pupa-to-adult, and JH determines the nature of the transition. In *Drosophila*, at the end of the third instar, 20E acting in the absence of JH causes dramatic changes in the genetic programs leading to pupariation, and the *BR-C* gene is known to play a key role in this process.

A. *BR-C* AS A MAJOR TARGET OF JH "STATUS QUO" ACTION

Broad-Complex is one of the early response genes from the genetic regulatory hierarchy activated by 20E, and it encodes a complex set of zinc-finger DNA-binding transcription factors—Z1, Z2, Z3, and Z4 (Bayer et al., 1996). Together with proteins encoded by other early genes, such as *E74* and *E75*, they transmit the hormonal signal to a much bigger group of late genes, thus, unfolding a genetic cascade manifested at the cellular level in distinct metamorphic events including histolysis of obsolete larval tissues, remodeling of the central nervous system (CNS) and musculature, morphogenesis of imaginal discs, and pupal cuticle formation (Thummel, 2002). Immunostaining of salivary gland polytene chromosomes indicated that BR-C proteins interact directly with more than 300 loci, including 20E-inducible puffs and interbands (Gonzy et al., 2002). BR-C genetic functions are represented by three complementation groups—*rbp* (*reduced bristles on palpus*), *br* (*broad*), and *l(1)2Bc* (Belyaeva et al., 1989; Kiss et al., 1988). Amorphic mutants lacking one of these functions die during metamorphosis at prepupal and pupal stages. They exhibit severe developmental defects associated with the larva-to-pupa transition—the salivary gland and midgut cells fail to undergo programmed cell death, imaginal discs do not evaginate, and remodeling of the CNS is affected (Bayer et al., 1997; Emery et al., 1994; Jiang et al., 2000; Lee et al., 2002; Mugat et al., 2000; Restifo and White, 1991, 1992). *BR-C* null mutants defective in all three genetic functions and lacking all Z1–Z4 isoforms are unable to initiate metamorphosis. They display a prolonged third instar and eventually die without any sign of puparium formation. Given that these failures are not due to the hormone deficiency, it appears that *BR-C* is essential for the normal metamorphic response of tissues to the 20E signal (Dubrovsky et al., 1996, 2001; Gonzy et al., 2002; Renault et al., 2001).

BR-C homologues were identified in a number of nondrosophilid species. Two studies have demonstrated a conserved role of *BR-C* in the metamorphosis of lepidopteran insects. By generating transgenic *Drosophila* flies that carry the *Manduca BR-C* gene encoding the Z4 isoform, Bayer *et al.* (2003) was able to rescue *rbp* mutant lethality during *Drosophila* metamorphosis. The authors found that the *Manduca* Z4 protein has significant biological activity in *Drosophila* with respect to the rescue of *rbp*-associated lethality (Bayer et al., 2003). These data show that Z4 function has been conserved for about 250–270 million years since orders of Lepidoptera and Diptera have diverged (Whiting et al., 1997). Another report highlights the role of *BR-C* in the metamorphosis of *B. mori*. Uhlirova *et al.* (2003) employed Sindbis (SIN) virus-mediated RNA interference (RNAi) to knockdown the expression of the *Bombyx BR-C* gene. Infection of *Bombyx* larvae with a recombinant SIN virus expressing a *BR-C* antisense RNA fragment silenced the transcription of endogenous *BR-C* mRNA and disrupted the larva-to-pupa transition—the development of adult appendages from imaginal discs was abrogated, and the programmed cell death of larval silk glands was severely delayed (Uhlirova et al., 2003). Together these data clearly demonstrate that *BR-C* plays a critical role in metamorphosis of insects and, this role is evolutionarily conserved.

It was shown that, in *Drosophila*, *BR-C* is not only necessary but also sufficient to specify the pupal molt (Zhou and Riddiford, 2002; Zhou et al., 2004). At the end of larval development expression of *BR-C* is directly induced by 20E in most if not all tissues. However, it is not expressed during larva-to-larva or pupa-to-adult molts. Ectopic expression of *BR-C* under a heat-shock promoter during the pupa-to-adult transformation redirects epidermal cells from adult to pupal cuticle production, causing a supernumerary pupal molt. Similarly, ectopic expression of *BR-C* during the second larval molt precociously activates pupal-specific cuticle genes and suppresses larval-specific cuticle genes (Zhou and Riddiford, 2002). Moreover, misexpression of *BR-C* during early larval development also causes premature degeneration of the prothoracic and salivary gland cells that normally occurs during prepupal–pupal development (Zhou et al., 2004). Thus, ectopically expressed BR-C proteins are able to specify at least part of the developmental program of larva-to-pupa transition.

Studies with *Manduca* epidermis have shown that JH is an operational trigger that controls the turning on-and-off of *BR-C* expression by 20E. During the final instar, the onset of *BR-C* expression in the abdominal epidermis of *Manduca* correlates with the small rise of 20E in the absence of JH (Zhou et al., 1998b). Consequently, epidermal cells acquire pupal commitment spatially correlated with the accumulation of BR-C proteins (Zhou and Riddiford, 2001). The presence of JH prevents the 20E-dependent appearance of *BR-C* transcripts and hence the pupal commitment of epidermis. Instead, epidermal cells retain their ability to form a new larval cuticle.

By contrast, the adult cuticle is formed in the absence of *BR-C*. Apparently, during adult commitment, 20E acting in the absence of JH turns off *BR-C* so that the adult developmental program could occur. But when exogenous JH is applied, it causes *BR-C* re-expression and formation of another pupal cuticle (Zhou and Riddiford, 2002). A similar phenomenon was observed in *Drosophila*. The presence of exogenous JH before the onset of the pupa-to-adult transition leads to *BR-C* re-expression and hence reiteration of the pupal developmental program in the *Drosophila* abdomen (Zhou and Riddiford, 2002). As a result, JH-treated pupae develop into pupal-adult intermediates with an adultlike head and thorax and a pupal-like abdomen (Riddiford and Ashburner, 1991). These data suggest that JH accomplishes its "status quo" action by regulating the 20E-dependent expression of *BR-C*. So far, the mechanism whereby JH prevents *BR-C* activation at pupal commitment and *BR-C* repression at adult commitment is not clear.

B. *E75A* AS A MEDIATOR OF JH "STATUS QUO" ACTION

A study of JH response in cultured cells demonstrated that transcription of the long-known ecdysone-inducible *E75* gene could be activated by both JH and 20E in *Drosophila* and *Manduca* cells (Dubrovskaya et al., 2004; Dubrovsky et al., 2004a). In addition, in *Galleria* silk glands and in *Manduca* epidermis the 20E-inducibility of *E75A* RNA was increased in the presence of JH, although JH alone did not result in induction (Jindra and Riddiford, 1996; Zhou et al., 1998a). These studies suggest that *E75* is a common element in the JH and 20E signaling pathways and that it could act as an intermediate in the interplay between two hormones.

A functional genetic analysis showed that *E75A* is required for ecdysone biosynthesis, or release, implying one potential mechanism of hormone cross-talk at the level of hormone titer regulation (Bialecki et al., 2002). *E75* loss-of-function mutants that lack all E75 protein isoforms die as late embryos or during a prolonged first instar. Mutants lacking individual E75 isoforms display a range of viability phenotypes—while some die during second instar, pharate adult and/or adult stages, others are completely viable (Bialecki *et al.*, 2002). Most of *E75A* null mutants cease their development as second-instar larvae. However, about 20% of them skip the second larval molt and form prepupae directly from the second instar, suggesting a breakdown in the JH signaling pathway. Direct measurements of ecdysone titer showed that *E75A* homozygotes lack the peak that precedes the second to third-instar molt. Several endocrinological studies in different insect species have proposed that the ecdysone surge during the penultimate larval stage is stimulated by JH (Garcia *et al.*, 1987; Hiruma, 1986; Lonard *et al.*, 1996; Mizoguchi, 2001). It is conceivable, then, that a knockout of the E75A isoform leads to a failure in JH signaling, which eventually affects ecdysone biosynthesis or release in second-instar larvae.

Based primarily on studies in cultured *Drosophila* cells, a different role has been proposed for *E75A* as an intermediate in the 20E-JH regulatory interplay. According to the model (Fig. 3), *E75A* is a primary response gene activated either by 20E or JH. In the JH signaling pathway, E75A performs several functions—it can downregulate its own expression, it can potentiate the expression of secondary JH-response genes, and it can repress 20E activation of early genes including the *BR-C* gene (Dubrovsky *et al.*, 2004a). Based on the observation that E75A binds to the early gene puff

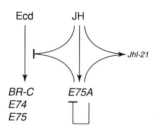

FIGURE 3. The diagram describing the hormonal regulation of *Drosophila E75* gene expression and its putative role in the JH signaling pathway. In S2 cultured cells as well as in most if not all tissues of third-instar larvae, ecdysone directly activates a small group of primary response genes, including *BR-C*, *E74*, and *E75*. In addition, in S2 cells, the *E75A* expression can be directly activated by JH. Based on experiments in cultured cells, several regulatory functions can be assigned to the E75A protein. Importantly, following induction by JH, E75A represses ecdysone responsiveness of early genes including itself. The model proposes that E75A is a part of a mechanism whereby JH prevents activation of *BR-C*, a major determinant of the pupal developmental pathway. This figure is reprinted from (Dubrovsky *et al.*, 2004) with permission of Elsevier Science.

loci (Hill *et al.*, 1993) and that ectopic E75A together with JH can abrogate early gene expression, it was suggested that E75A has a critical role during larval development (Dubrovsky, 2005). Because of additive JH and 20E action, *E75A* transcription is high during first and second instars. Apparently, *E75A* is induced before other ecdysone-responsive genes, including *EcR-B*, *E74A*, and *E75B* (Sullivan and Thummel, 2003), and in the presence of JH at larval molts E75A can repress early gene transcription thus mediating the JH "status quo" action. During the final instar, JH titer declines and the *E75A* promoter becomes less sensitive to 20E. Hence, the low titer pulse of ecdysone in the middle of the third instar activates a group of early genes including *BR-C*. Once induced, *BR-C* establishes a pupal commitment leading eventually to the larva-to-pupa transition.

ACKNOWLEDGMENTS

The authors are grateful to Dr. Veronica Dubrovsky for helpful comments and critical reading of the manuscript. We would like also to thank Dr. Laurence Gilbert for providing JHB3 used in this study. This work was supported by the grant from US Department of Agriculture (#02-35302-12356).

REFERENCES

Abel, T., Bhatt, R., and Maniatis, T. (1992). A *Drosophila* CREB/ATF transcriptional activator binds to both fat body-specific and liver-specific regulatory elements. *Genes Dev.* **6**, 466–480.

Abrahamsen, N., Martinez, A., Kjaer, T., Sondergaard, L., and Bownes, M. (1993). Cis-regulatory sequences leading to female-specific expression of the yolk protein *gene-1* and *gene-2* in the fat body of *Drosophila melanogaster*. *Mol. Gen. Genet.* **237**, 41–48.

Agui, N., Grainger, N. A., Gilbert, L. I., and Bollenbacher, W. E. (1979). Cellular localization of the insect prothoracicotropic hormone: *In vitro* assay of a single neurosecretory cell. *Proc. Natl. Acad. Sci. USA.* **76**, 5694–5698.

Altaratz, M., Applebaum, S. W., Richard, D. S., Gilbert, L. I., and Segal, D. (1991). Regulation of juvenile hormone synthesis in wild-type and *apterous* mutant *Drosophila*. *Mol. Cell. Endocrinol.* **81**, 205–216.

Andres, A. J., Fletcher, J. C., Karim, F. D., and Thummel, C. S. (1993). Molecular analysis of the initiation of insect metamorphosis: A comparative study of *Drosophila* ecdysteroid-regulated transcription. *Dev. Biol.* **160**, 388–404.

Ashburner, M. (1972). Patterns of puffing activity in the salivary glands of *Drosophila*. VI. Induction by ecdysone in salivary glands of *D. melanogaster* cultured *in vitro*. *Chromosoma* **38**, 255–281.

Ashok, M., Turner, C., and Wilson, T. G. (1998). Insect juvenile hormone resistance gene homology with the bHLH-PAS family of transcriptional regulators. *Proc. Natl. Acad. Sci. USA* **95**, 2761–2766.

Audit-Lamour, C., and Busson, D. (1981). Oogenesis defects in the *ecd-1* mutant of *Drosophila melanogaster* deficient in ecdysteroid at high temperature. *J. Insect Physiol.* **27**, 829–837.

Baker, K. D., Shewchuk, L. M., Kozlova, T., Makishima, M., Hassell, A., Wisely, B., Caravella, J. A., Lambert, M. H., Reinking, J. L., Krause, H., Thummel, C. S., Wilson,

T. M., and Mangelsdorf, D. J. (2003). The *Drosophila* orphan nuclear receptor DHR38 mediates an atypical ecdysteroid signaling pathway. *Cell* **113**, 731–742.

Bayer, C. A., Holley, B., and Fristrom, J. W. (1996). A switch in *Broad-Complex* zinc-finger isoform expression is regulated post-transcriptionally during the metamorphosis of *Drosophila* imaginal discs. *Dev. Biol.* **177**, 1–14.

Bayer, C. A., von Kalm, L., and Fristrom, J. W. (1997). Relationships between protein isoforms and genetic functions demonstrate functional redundancy at the *Broad-Complex* during *Drosophila* metamorphosis. *Dev. Biol.* **187**, 267–282.

Bayer, C., Zhou, X. F., Zhou, B. H., Riddiford, L. M., and von Kalm, L. (2003). Evolution of the *Drosophila broad* locus: The *Manduca sexta broad* Z4 isoform has biological activity in *Drosophila*. *Dev. Genes Evol.* **213**, 471–476.

Belgacem, Y. H., and Martin, J. R. (2002). Neuroendocrine control of a sexually dimorphic behavior by a few neurons of the pars intercerebralis in *Drosophila*. *Proc. Natl. Acad. Sci. USA* **99**, 15154–15158.

Belyaeva, E. S., Protopopov, M. O., Dubrovsky, E. B., and Zhimulev, I. F. (1989). Cytogenetic analysis of ecdysteroid action. *In* "Ecdysone: from Chemistry to Mode of Action" (J. Koolman, Ed.), pp. 368–376. Georg Thieme Verlag, Stuttgart.

Bender, M., Imam, F. B., Talbot, W. S., Ganetsky, B., and Hogness, D. S. (1997). *Drosophila* ecdysone receptor mutations reveal functional differences among receptor isoforms. *Cell* **91**, 777–788.

Berger, E., Goudie, K., Kliger, L., Berger, M., and De Cato, R. (1992). The juvenile hormone analogue methoprene inhibits ecdysterone induction of small heat shock protein gene expression. *Dev. Biol.* **151**, 410–418.

Bialecki, M., Shilton, A., Fichtenberg, C., Segraves, W. A., and Thummel, C. S. (2002). Loss of the ecdysteroid-inducible E75A orphan nuclear receptor uncouples molting from metamorphosis in *Drosophila*. *Dev. Cell.* **3**, 209–220.

Billas, I. M. L., Moulinier, L., Rochel, N., and Moras, D. (2001). Crystal structure of the ligand-binding domain of the ultraspiracle protein USP, the orthologue of retinoid XR receptors in insects. *J. Biol. Chem.* **276**, 7465–7474.

Bouletreau-Merle, J. (1974). Stimulation de l'ovogenèse par la copulation chez les femelles de *Drosophila melanogaster* privées de leur complexe endocrine rétrocérébral. *J. Insect Physiol.* **20**, 2035–2041.

Bownes, M. (1989). The roles of juvenile hormone, ecdysone and the ovary in the control of *Drosophila* vitellogenesis. *J. Insect Physiol.* **35**, 409–413.

Bownes, M. (1994). The regulation of the yolk protein genes, a family of sex differentiation genes in *Drosophila melanogaster*. *BioEssays* **16**, 745–752.

Bownes, M., and Blair, M. (1986). The effects of a sugar diet and hormones on the expression of the *Drosophila* yolk protein genes. *J. Insect. Physiol.* **32**, 493–501.

Bownes, M., and Nothiger, R. (1981). Sex determining genes and vitellogenin synthesis in *Drosophila melanogaster*. *Mol. Gen. Genet.* **182**, 222–228.

Bownes, M., and Rembold, H. (1987). The titer of juvenile hormone during the pupal and adult stages of the life cycle of *Drosophila melanogaster*. *Eur. J. Biochem.* **164**, 709–712.

Bownes, M., Ronaldson, E., and Mauchline, D. (1996). 20-hydroxyecdysone, but not juvenile hormone regulation of *yolk protein* gene expression can be mapped to *cis*-acting DNA sequences. *Dev. Biol.* **173**, 475–489.

Bownes, M., Scott, A., and Blair, M. (1987). The use of an inhibitor of protein synthesis to investigate the roles of ecdysteroids and sex-determination genes on the expression of the genes encoding the *Drosophila* yolk proteins. *Development* **101**, 931–941.

Bownes, M., Scott, A., and Shirras, A. (1988a). Dietary components modulate yolk protein gene transcription in *Drosophila melanogaster*. *Development* **103**, 119–128.

Bownes, M., Shirras, A., Blair, M., Collins, J., and Coulson, A. (1988b). Evidence that insect embryogenesis is regulated by ecdysteroids released from yolk proteins. *Proc. Natl. Acad. Sci. USA* **85,** 1554–1557.

Brennan, C. A., Ashburner, M., and Moses, K. (1998). Ecdysone pathway is required for furrow progression in the developing *Drosophila* eye. *Development* **125,** 2653–2664.

Brennan, M. D., Weiner, A. J., Goralski, T. J., and Mahowald, A. P. (1982). The follicle cells are a major site of vitellogenin synthesis in *Drosophila melanogaster*. *Dev. Biol.* **89,** 225–236.

Burtis, K. C., Coschigano, K. T., Baker, B. S., and Wensink, P. C. (1991). The doublesex proteins of *Drosophila melanogaster* bind directly to a sex-specific yolk protein gene enhancer. *EMBO J.* **10,** 2577–2582.

Buszczak, M., Freeman, M. R., Carlson, J. R., Bender, M., Cooley, L., and Segraves, W. A. (1999). Ecdysone response genes govern egg chamber development during mid-oogenesis in *Drosophila*. *Development* **126,** 4581–4589.

Butterworth, F. M., and King, R. C. (1965). The developmental genetics of *apterous* mutants of *Drosophila melanogaster*. *Genetics* **52,** 1153–1174.

Campbell, P. M., Healy, M. J., and Oakeshott, J. G. (1992). Characterization of juvenile hormone esterase in *Drosophila melanogaster*. *Insect Biochem. Mol. Biol.* **22,** 665–677.

Campbell, P. M., Oakeshott, J. G., and Healy, M. J. (1998). Purification and kinetic characterization of juvenile hormone esterase from *Drosophila melanogaster*. *Insect Biochem. Mol. Biol.* **28,** 501–515.

Campbell, P. M., Harcourt, R. L., Crone, E. J., Claudianos, C., Hammock, B. D., Russell, R. J., and Oakeshott, J. G. (2001). Identification of a juvenile hormone esterase gene by matching its peptide mass fingerprint with a sequence from the *Drosophila* genome project. *Insect Biochem. Mol. Biol.* **31,** 513–520.

Carney, G. E., Wade, A. A., Sapra, R., Goldstein, E. S., and Bender, M. (1997). *DHR3*, an ecdysone-inducible early-late gene encoding a *Drosophila* nuclear receptor is required for embryogenesis. *Proc. Natl. Acad. Sci, USA* **94,** 12024–12029.

Casas, J., Harshman, L. G., Messeguer, A., Kuwano, E., and Hammock, B. D. (1991). In vitro metabolism of juvenile hormone III and juvenile hormone III bisepoxide by *Drosophila melanogaster* and mammalian cytosolic epoxide hydrolase. *Arch. Biochem. Biophys.* **286,** 153–158.

Cha, G., Kim, A., Chang, K. W., Namkoong, Y., and Lee, C. C. (1998). Immunohistochemical localization of prothoracicotropic hormone-producing neurosecretory cells in the brain of *Drosophila melanogaster*. *Dros. Res. Conf.* **39,** 336A.

Chang, E. S., Coudron, T. A., Bruce, M. J., Sage, B. A., O'Connor, J. D., and Law, J. H. (1980). Juvenile hormone-binding protein from the cytosol of *Drosophila* Kc cells. *Proc. Natl. Acad. Sci. USA* **77,** 4657–4661.

Chavez, V. M., Marques, G., Delbecque, J. P., Kobayashi, K., Hollingsworth, M., Burr, J., Natzle, J. E., and O'Connor, M. B. (2000). The *Drosophila disembodied* gene controls late embryonic morphogenesis and codes for a cytochrome P450 enzyme that regulates embryonic ecdysone levels. *Development* **127,** 4115–4126.

Cherbas, L., Lee, K., and Cherbas, P. (1991). Identification of ecdysone response elements by analysis of the *Drosophila Eip 28/29* gene. *Genes Dev.* **5,** 120–131.

Cherbas, P., and Cherbas, L. (1996). Molecular aspects of ecdysteroid hormone action. *In* "Metamorphosis: Postembryonic Reprogramming of Gene Expression in Amphibian and Insect Cells" (L. I. Gilbert, J. R. Tata, and B. G. Atkinson, Eds.), pp. 175–221. Academic Press, San Diego.

Cherbas, L., Hu, X., Zhimulev, I., Belyaeva, E., and Cherbas, P. (2003). EcR isoforms in *Drosophila*: Testing tissue-specific requirements by targeted blockage and rescue. *Development* **130,** 271–284.

Christianson, A. M. K., King, D. L., Hatzivassiliou, E., Casas, J. E., Hallenbeck, P. L., Nikodem, V. M., Mitsialis, S. A., and Kafatos, F. C. (1992). DNA binding and hetermerization of the *Drosophila* transcription factor chorion factor-1/ultraspiracle. *Proc. Natl. Acad. Sci. USA* **89**, 11503–11507.

Chung, T., Huang, J. S., Mukherjee, J. J., Crilly, K. S., and Kiss, Z. (2000). Expression of human choline kinase in NIH 3T3 fibroblasts increases the mitogenic potential of insulin and insulin-like growth factor I. *Cell. Signal.* **12**, 279–288.

Clayton, G. M., Peak-Chew, S. Y., Evans, R. M., and Schwabe, J. W. (2001). The structure of the ultraspiracle ligand-binding domain reveals a nuclear receptor locked in an inactive conformation. *Proc. Natl. Acad. Sci. USA* **98**, 1549–1554.

Cohen, B., McGuffin, M. E., Pfeifle, C., Segal, D., and Cohen, S. M. (1992). *apterous*, a gene required for imaginal disc development in *Drosophila* encodes a member of the LIM family of developmental regulatory proteins. *Genes Dev.* **6**, 715–729.

Comas, D., Piulachs, M. D., and Belles, X. (1999). Fast induction of vitellogenin gene expression by juvenile hormone III in the cockroach *Blattella germanica* (L.) (Dictyoptera, Blattellidae). *Insect Biochem. Mol. Biol.* **29**, 821–827.

Comas, D., Piulachs, M. D., and Belles, X. (2001). Induction of vitellogenin gene transcription *in vitro* by juvenile hormone in *Blattella germanica*. *Mol. Cell Endocrinol.* **183**, 93–100.

Coschigano, K. T., and Wensink, P. C. (1993). Sex-specific transcriptional regulation by the male and female doublesex proteins of *Drosophila*. *Genes Dev.* **7**, 42–54.

Dai, J. D., and Gilbert, L. I. (1991). Metamorphosis of the corpus allatum and degeneration of the prothoracic glands during the larval-pupal-adult transformation in *Drosophila melanogaster*: A cytophysiological analysis of the ring gland. *Dev. Biol.* **144**, 309–326.

Dai, J.-D., and Gilbert, L. I. (1993). An ultrastructural and developmental analysis of the corpus allatum of juvenile hormone deficient mutants of *Drosophila melanogaster*. *Roux's Arch. Dev. Biol.* **202**, 85–94.

Dobens, L., Rudolph, K., and Berger, E. (1991). Ecdysterone regulatory elements function as both transcriptional activators and repressors. *Mol. Cell. Biol.* **11**, 1846–1853.

Dubrovskaya, V. A., Berger, E., and Dubrovsky, E. B. (2004). Juvenile hormone regulation of the E75 nuclear receptor is conserved in Diptera and Lepidoptera. *Gene* **340**, 171–177.

Dubrovsky, E. B. (2005). Hormonal cross talk in insect development. *Trends Endocrinol. Metab.* **16**, 6–11.

Dubrovsky, E. B., Dretzen, G., and Berger, E. M. (1996). The *Broad-Complex* gene is a tissue-specific modulator of the ecdysone response of the *Drosophila hsp23* gene. *Mol. Cell. Biol.* **16**, 6542–6552.

Dubrovsky, E. B., Dubrovskaya, V. A., Bilderback, A. L., and Berger, E. M. (2000). The isolation of two juvenile hormone-inducible genes in *Drosophila melanogaster*. *Dev. Biol.* **224**, 486–495.

Dubrovsky, E. B., Dubrovskaya, V. A., and Berger, E. M. (2001). Selective binding of *Drosophila* BR-C isoforms to a distal regulatory element in the *hsp23* promoter. *Insect Biochem. Mol. Biol.* **31**, 1231–1239.

Dubrovsky, E. B., Dubrovskaya, V. A., and Berger, E. (2002). Juvenile hormone signaling during oogenesis in *Drosophila melanogaster*. *Insect Biochem. Mol. Biol.* **32**, 1555–1565.

Dubrovsky, E. B., Dubrovskaya, V. A., and Berger, E. (2004a). Hormonal regulation and functional role of *Drosophila* E75A orphan nuclear receptor in the juvenile hormone signaling pathway. *Dev. Biol.* **268**, 258–270.

Dubrovsky, E. B., Dubrovskaya, V. A., Levinger, L., Schiffer, S., and Marchfelder, A. (2004b). *Drosophila* RNase Z processes mitochondrial and nuclear pre-tRNA 3' ends *in vivo*. *Nucleic Acids Res.* **32**, 255–262.

Edgar, B. A., and Nijhout, H. F. (2004). Growth and cell cycle control in *Drosophila*. *In* "Cell Growth: Control of Cell Size" (M. N. Hall, M. Raff, and G. Thomas, Eds.), pp. 23–83. Cold Spring Harbor Laboratory Press, Cold Spring Harbor.

Edwards, G. C., Braun, R. P., and Wyatt, G. R. (1993). Induction of vitellogenin synthesis in *Locusta migratoria* by the juvenile hormone analog, pyriproxyfen. *J. Insect Physiol.* **39**, 609–614.

Emery, I. F., Bedian, V., and Guild, G. M. (1994). Differential expression of *Broad-Complex* transcription factors may forecast tissue-specific developmental fates during *Drosophila* metamorphosis. *Development* **120**, 3275–3287.

Falb, D., and Maniatis, T. (1992). A conserved regulatory unit implicated in tissue-specific gene expression in *Drosophila* and man. *Genes Dev.* **6**, 454–465.

Feigl, G., Gram, M., and Pongs, O. (1989). A member of the steroid hormone receptor gene family is expressed in the 20-OH-ecdysone inducible puff 75B in *Drosophila melanogaster*. *Nucleic Acids Res.* **17**, 7167–7178.

Feng, Q. L., Ladd, T. R., Tomkins, B. L., Sundaram, M., Sohi, S. S., Retnakaran, A., Davey, K. G., and Palli, S. R. (1999). Spruce budworm (*Choristoneura* fumiferana) juvenile hormone esterase: Hormonal regulation, developmental expression and cDNA cloning. *Mol. Cell. Endocrinol.* **148**, 95–108.

Freedman, L. P. (1997). "Molecular Biology of Steroid and Nuclear Hormone Receptors." Birkhauser, Boston.

Freeman, M. R., Dobritsa, A., Gaines, P., Segraves, W. A., and Carlson, J. R. (1999). The *dare* gene: Steroid hormone production, olfactory behavior, and neural degeneration in *Drosophila*. *Development* **126**, 4591–4602.

Garabedian, M. J., Hung, M. C., and Wensink, P. C. (1985). Independent control elements that determine yolk protein gene expression in alternative *Drosophila* tissues. *Proc. Nat. Acad. Sci. USA* **82**, 1396–1400.

Garcia, E., Furtadu, A. F., and de Azambuja, P. (1987). Effect of allatectomy on ecdysteroid-dependent development of *Rhodnius prolixus* larvae. *J. Insect Physiol.* **33**, 729–732.

Garen, A., Kauvar, L., and Lepesant, J. A. (1977). Roles of ecdysone in *Drosophila* development. *Proc. Natl. Acad. Sci. USA* **74**, 5099–5103.

Gavin, J. A., and Williamson, J. H. (1976). Juvenile hormone induced vitellogenesis in *apterous-4*, a non-vitellogenic mutant in *Drosophila melanogaster*. *J. Insect Physiol.* **22**, 1737–1742.

Gaziova, I., Bonnette, P. C., Henrich, V. C., and Jindra, M. (2004). Cell-autonomous roles of the *ecdysoneless* gene in *Drosophila* development and oogenesis. *Development* **131**, 2715–2725.

Gilbert, L. I. (2004). Halloween genes encode P450 enzymes that mediate steroid hormone biosynthesis in *Drosophila melanogaster*. *Mol. Cell. Endocrinol.* **215**, 1–10.

Gilbert, L. I., Rybczynski, R., and Tobe, S. (1996). Regulation of endocrine function leading to insect metamorphosis. *In* "Metamorphosis: Postembryonic Reprogramming of Gene Expression in Amphibian and Insect Cells" (L. I. Gilbert, J. R. Tata, and B. G. Atkinson, Eds.), pp. 59–107. Academic Press, San Diego.

Gilbert, L. I., Rybczynski, R., and Warren, J. T. (2002). Control and biochemical nature of the ecdysteroidogenic pathway. *Annu. Rev. Entomol.* **47**, 883–916.

Glinka, A. V., and Wyatt, G. R. (1996). Juvenile hormone activation of gene transcription in locust fat body. *Insect Biochem. Mol. Biol.* **26**, 13–18.

Gonzy, G., Pokholkova, G. V., Peronnet, F., Mugat, B., Demakova, O. V., Kotlikova, I. V., Lepesant, J. A., and Zhimulev, I. F. (2002). Isolation and characterization of novel mutations of the *Broad-Complex*, a key regulatory gene of ecdysone induction in *Drosophila melanogaster*. *Insect Biochem. Mol. Biol.* **32**, 121–132.

Hall, B. L., and Thummel, C. S. (1998). The RXR homolog ultraspiracle is an essential component of the *Drosophila* ecdysone receptor. *Development* **125**, 4709–4717.

Hanzlik, T. N., Abdel-Aal, Y. A. I., Harshman, L. G., and Hammock, B. D. (1989). Isolation and sequencing of cDNA clones coding for juvenile hormone esterase from *Heliothis virescens*: Evidence for a catalytic mechanism for the serine carboxylesterases different from that of the serine proteases. *J. Biol. Chem.* **264**, 12419–12425.

Harmon, M. A., Boehm, M. F., Heyman, R.A, and Mangelsdorf, D. J. (1995). Activation of mammalian retinoid-X-receptors by the insect growth regulator methoprene. *Proc. Natl. Acad. Sci. USA* **92**, 6157–6160.

Harshman, L. G., Casas, J., Dietze, E. C., and Hammock, B.D (1991). Epoxide hydrolase activities in *Drosophila melanogaster*. *Insect Biochem.* **21**, 887–894.

Henrich, V. C. (1995). Comparison of ecdysteroid production in *Drosophila* and *Manduca*: Pharmacology and cross-species neural activity. *Arch. Insect Biochem. Physiol.* **30**, 239–254.

Henrich, V. C., Pak, M. D., and Gilbert, L. I. (1987a). Neural factors that stimulate ecdysteroid synthesis by the larval ring gland of *Drosophila melanogaster*. *J. Comp. Physiol.* **157**, 543–549.

Henrich, V. C., Tucker, R. L., Maroni, G., and Gilbert, L. I. (1987b). The *ecdysoneless* (ecd^{ts}) mutation disrupts ecdysteroid synthesis autonomously in the ring gland of *Drosophila melanogaster*. *Dev. Biol.* **120**, 50–55.

Henrich, V. C., Sliter, T. J., Lubahn, D. B., MacIntyre, A., and Gilbert, L. I. (1990). A steroid/thyroid hormone receptor superfamily member in *Drosophila melanogaster* that shares extensive sequence similarity with a mammalian homolog. *Nucleic Acids Res.* **18**, 4143–4148.

Henrich, V. C., Szekely, A. A., Kim, S. J., Brown, N. E., Antoniewski, C., Hayden, M. A., Lepesant, J. A., and Gilbert, L. I. (1994). Expression and function of the *ultraspiracle* (*usp*) gene during development of *Drosophila melanogaster*. *Dev. Biol.* **165**, 38–52.

Henrich, V. C., Rybczynski, R., and Gilbert, L. I. (1999). Peptide hormones, steroid hormones and puffs: Mechanisms and models in insect development. *Vitam. Horm.* **55**, 73–125.

Herndon, L. A., Chapman, T., Kalb, J. M., Lewin, S., Partridge, L., and Wolfner, M. F. (1997). Mating and hormonal triggers regulate accessory gland gene expression in male *Drosophila*. *J. Insect Physiol.* **43**, 1117–1123.

Hill, R. J., Segraves, W. A., Choi, D., Underwood, P. A., and Macavoy, E. (1993). The reaction with polytene chromosomes of antibodies raised against *Drosophila* E75A protein. *Insect Biochem. Mol. Biol.* **23**, 99–104.

Hinton, A. C., and Hammock, B. D. (2001). Purification of juvenile hormone esterase and molecular cloning of the cDNA from *Manduca sexta*. *Insect Biochem. Mol. Biol.* **32**, 57–66.

Hinton, A. C., and Hammock, B. D. (2003). Juvenile hormone esterase (JHE) from *Tenebrio molitor*: Full-length cDNA sequence, *in vitro* expression and characterization of the recombinant protein. *Insect Biochem. Mol. Biol.* **33**, 477–487.

Hirai, M., Yuda, M., Shinoda, T., and Chinzei, Y. (1998). Identification and cDNA cloning of novel juvenile hormone responsive genes from fat body of the bean bug, *Riptortus clavatus* by mRNA differential display. *Insect Biochem. Mol. Biol.* **28**, 181–189.

Hiruma, K. (1986). Regulation of prothoracicotropic hormone release by juvenile hormone in the penultimate and last instar larvae of *Mamestra brassicae*. *Gen. Comp. Endocrinol.* **63**, 201–211.

Hiruma, K., Shinoda, T., Malone, F., and Riddiford, L. M. (1999). Juvenile hormone modulates 20-hydroxyecdysone-inducible ecdysone receptor and ultraspiracle gene expression in the tobacco hornworm, *Manduca sexta*. *Dev. Genes Evol.* **209**, 18–30.

Hoffmann, E., and Corces, V. (1986). Sequences involved in temperature and ecdysterone-induced transcription are located in separate regions of a *Drosophila melanogaster* heat shock gene. *Mol. Cell. Biol.* **6**, 663–673.

Holden, J. J. A., Walker, V. K., Maroy, P., Watson, K. L., White, B. N., and Gausz, J. (1986). Analysis of molting and metamorphosis in the ecdysteroid deficient mutant $L(3)3^{DTS}$ of *Drosophila melanogaster*. *Dev. Genet.* **6**, 153–162.

Isaac, P. G., and Bownes, M. (1982). Ovarian and fat body vitellogenin synthesis in *Drosophila melanogaster*. *Eur. J. Biochem.* **123**, 527–534.

Iyengar, A. R., and Kunkel, J. G. (1995). Follicle cell calmodulin in *Blattella germanica*: Transcript accumulation during vitellogenesis is regulated by juvenile hormone. *Dev. Biol.* **170**, 314–320.

Jiang, C., Lamblin, A. F. J., Steller, H., and Thummel, C. S. (2000). A steroid-triggered transcriptional hierarchy controls salivary gland cell death during *Drosophila* metamorphosis. *Mol. Cell* **5**, 445–455.

Jindra, M., and Riddiford, L. M. (1996). Expression of ecdysteroid-regulated transcripts in the silk gland of wax moth, *Galleria mellonella*. *Dev. Genes Evol.* **206**, 305–314.

Jones, G., Venkataraman, V., Ridley, B., O'Mahony, P., and Turner, H. (1994). Structure, expression and gene sequence of a juvenile hormone esterase related protein from metamorphosing larvae of *Trichoplusia ni*. *Biochem. J.* **302**, 827–835.

Jones, G., and Sharp, P. A. (1997). Ultraspiracle: An invertebrate nuclear receptor for juvenile hormones. *Proc. Natl. Acad. Sci. USA.* **94**, 13499–134503.

Jones, G., and Jones, D. (2000). Considerations of the structural evidence of a ligand binding function of ultraspiracle, an insect homolog of vertebrate RXR. *Insect Biochem. Mol. Biol.* **30**, 671–679.

Jones, G., Wozniak, M., Cha, Y. X., Dhar, S., and Jones, D. (2001). Juvenile hormone III-dependent conformational changes of the nuclear receptor ultraspiracle. *Insect Biochem. Mol. Biol.* **32**, 33–49.

Jowett, T., and Postlethwait, J. H. (1980). The regulation of yolk polypeptide synthesis in *Drosophila* ovaries and fat body by 20-hydroxyecdysone and a juvenile hormone analog. *Dev. Biol.* **80**, 225–234.

Judy, K. J., Schooley, D. A., Dunham, L. L., Hall, M. S., Bergot, B. J., and Siddall, J. B. (1973). Isolation, structure and absolute configuration of a new natural insect juvenile hormone from *Manduca sexta*. *Proc. Natl. Acad. Sci. USA* **70**, 1509–1513.

Jurgens, G., Wieschaus, E., Nusslein-Volhard, C., and Kluding, H. (1984). Mutations affecting the pattern of larval cuticle in *Drosophila melanogaster*. II. Zygotic loci on the third chromosome. *Roux's Arch. Dev. Biol.* **193**, 283–295.

Kethidi, D. R., and Palli, S. R. (2004). Juvenile hormone action involves multiple signal transduction mechanisms. Abstracts from Eighth International Conference on the Juvenile hormones. *J. Insect Sci.* **4**, http://www.insectscience.org

Kethidi, D. R., Perera, S. C., Zheng, S., Feng, Q. L., Krell, P., Retnakaran, A., and Palli, S. R. (2004). Identification and characterization of a juvenile hormone (JH) response region in the JH esterase gene from the spruce budworm, *Choristoneura fumiferana*. *J. Biol. Chem.* **279**, 19634–19642.

Kiguchi, K., and Riddiford, L. M. (1978). A role of juvenile hormone in pupal development of the tobacco hornworm, *Manduca sexta*. *J. Insect Physiol.* **24**, 673–680.

Kim, A. J., Cha, G. H., Kim, K., Gilbert, L. I., and Lee, C. C. (1997). Purification and characterization of the prothoracicotropic hormone of *Drosophila melanogaster*. *Proc. Natl. Acad. Sci. USA* **94**, 1130–1135.

King, R. C. (1970). "Ovarian Development in *Drosophila melanogaster*." Academic Press, New York.

King, R. C., and Bodenstein, D. (1965). The transplantation of ovaries between genetically sterile and wild type *Drosophila melanogaster*. *Z. Naturforsch.* **20b**, 292–297.

Kiss, I., Beaton, A. H., Tardiff, J., Fristrom, D., and Fristrom, J. W. (1988). Interactions and developmental effects of mutations in the *Broad-Complex* of *Drosophila melanogaster*. *Genetics* **118**, 247–259.

Klages, G., and Emmerich, H. (1979). Juvenile hormone metabolism and juvenile hormone esterase titer in haemolymph and peripheral tissues of *Drosophila hydei*. *J. Comp. Physiol.* **132**, 319–325.

Klose, W., Gateff, E., Emmerich, H., and Beikirch, H. (1980). Developmental studies on two ecdysone deficient mutants of *Drosophila melanogaster*. *Roux's Arch. Dev. Biol.* **189**, 57–67.

Koelle, M. R., Talbot, W. S., Segraves, W. A., Bender, M. T., Cherbas, P., and Hogness, D. S. (1991). The *Drosophila EcR* gene encodes an ecdysone receptor, a new member of the steroid receptor superfamily. *Cell* **65**, 59–77.

Koelle, M. R., Segraves, W. A., and Hogness, D. S. (1992). DHR3: A *Drosophila* steroid receptor homolog. *Proc. Natl. Acad. Sci. USA* **89**, 6167–6171.
Korver, W., Guevara, C., Chen, Y., Neuteboom, S., Bookstein, R., Tavtigian, S., and Lees, E. (2003). The product of the candidate prostate cancer susceptibility gene ELAC2 interacts with the γ-tubulin complex. *Int. J. Cancer* **104**, 283–288.
Kozlova, T., Pokholkova, G. V., Tzerzinis, G., Sutherland, J. D., Zhimulev, I. F., and Kafatos, F. C. (1998). *Drosophila* hormone receptor 38 functions in metamorphosis: A role in adult cuticle formation. *Genetics* **149**, 1465–1475.
Kozlova, T., and Thummel, C. S. (2000). Steroid regulation of postembryonic development and reproduction in *Drosophila*. *Trends Endocrinol. Metab.* **11**, 276–280.
Kozlova, T., and Thummel, C. S. (2003). Essential roles for ecdysone signaling during *Drosophila* mid-embryonic development. *Science* **301**, 1911–1914.
Kramer, B., Korner, U., and Wolbert, P. (2002). Differentially expressed genes in metamorphosis and after juvenile hormone application in the pupa of *Galleria*. *Insect Biochem. Mol. Biol.* **32**, 133–140.
Lawrence, P. (1992). "The Making of a Fly: The Genetics of Animal Design." Blackwell Scientific Publications, Boston.
Lee, C. Y., Cooksey, B. A. K., and Baehrecke, E. H. (2002). Steroid regulation of midgut cell death during *Drosophila* development. *Dev. Biol.* **250**, 101–111.
Lenz, C., Sondergaard, L., and Grimmelikhuijzen, C. J. P. (2000a). Molecular cloning and genomic organization of a novel receptor from *Drosophila melanogaster* structurally related to mammalian galanin receptors. *Biochem. Biophys. Res. Commun.* **269**, 91–96.
Lenz, C., Williamson, M., and Grimmelikhuijzen, C. J. P. (2000b). Molecular cloning and genomic organization of a second probable allatostatin receptor from *Drosophila melanogaster*. *Biochem. Biophys. Res. Commun.* **273**, 571–577.
Lenz, C., Williamson, M., and Grimmelikhuijzen, C. J. P. (2000c). Molecular cloning and genomic organization of an allatostatin preprohormone from *Drosophila melanogaster*. *Biochem. Biophys. Res. Commun.* **273**, 1126–1131.
Lepesant, J. A., Kejzlarova-Lepesant, J., and Garen, A. (1978). Ecdysone inducible functions of larval fat bodies in *Drosophila*. *Proc. Natl. Acad. Sci. USA* **75**, 5570–5574.
Li, H., and Cooper, R. L. (2001). Effects of the *ecdysoneless* mutant on synaptic efficacy and structure at the neuromuscular junctions in *Drosophila* larvae during normal and prolonged development. *Neuroscience* **106**, 193–200.
Logan, S. K., and Wensink, P. C. (1990). Ovarian follicle cell enhancers from the *Drosophila* yolk protein genes: Different segments of one enhancer have different cell-type specificities that interact to give normal expression. *Genes Dev.* **4**, 613–623.
Lonard, D. M., Bhaskaran, G., and Dahm, K. H. (1996). Control of prothoracic gland activity by juvenile hormone in fourth instar *Manduca sexta* larvae. *J. Insect Physiol.* **42**, 205–213.
Maki, A., Sawatsubashi, S., Ito, S., Shirode, Y., Suzuki, E., Zhao, Y., Yamagata, K., Kouzmenko, A., Takeyama, K., and Kato, S. (2004). Juvenile hormones antagonize ecdysone actions through co-repressor recruitment to EcR/USP heterodimers. *Biochem. Biophys. Res. Commun.* **320**, 262–267.
Mangelsdorf, D. J., and Evans, R. M. (1995). The RXR heterodimers and orphan receptors. *Cell* **83**, 841–850.
Manning, A. (1966). Corpus allatum and sexual receptivity in female *Drosophila melanogaster*. *Nature* **211**, 1321–1327.
Maroy, P., Kaufmann, G., and Dubendorfer, A. (1988). Embryonic ecdysteroids of *Drosophila melanogaster*. *J. Insect Physiol.* **34**, 633–637.
Martin, J. F., Hersperger, E., Simcox, A., and Shearn, A. (2000). *minidisks* encodes a putative amino acid transporter subunit required nonautonomously for imaginal cell proliferation. *Mech. Dev.* **92**, 155–167.

Meller, V. H., Combest, W. L., Smith, W. A., and Gilbert, L. I. (1988). A calmodulin-sensitive adenylate cyclase in the prothoracic glands of the tobacco hornworm, *Manduca sexta. Mol. Cell. Endocrinol.* **59,** 67–76.

Meller, V., Sakurai, S., and Gilbert, L. I. (1990). Developmental regulation of calmodulin-dependent adenylate cyclase in an insect endocrine gland. *Cell. Reg.* **1,** 771–780.

Mestril, R., Schiller, P., Amin, J., Klapper, H., Ananthan, J., and Voellmy, R. (1986). Heat shock and ecdysterone activation of the *Drosophila melanogaster hsp23* gene: A sequence element implied in developmental regulation. *EMBO J.* **5,** 1667–1673.

Meyer, A. S., Hanzmann, E., Schneiderman, H. A., Gilbert, L. I., and Boyette, M. (1970). The isolation and identification of two juvenile hormones from the *Cecropia* silk moth. *Arch. Insect Biochem. Physiol.* **137,** 190–213.

Mitchell, M. J., and Smith, S. L. (1988). Ecdysone 20-monooxygenase activity throughout the life cycle of *Drosophila melanogaster. Gen. Comp. Endocrinol.* **72,** 467–470.

Miura, K. M., Oda, M., Makita, S., and Chinzei, Y. (2005). Characterization of the *Drosophila Methoprene-tolerant* gene product. *FEBS J.* **272,** 1169–1178.

Mizoguchi, A. (2001). Effects of juvenile hormone on the secretion of prothoracicotropic hormone in the last- and penultimate-instar larvae of the silkworm *Bombyx mori. J. Insect Physiol.* **47,** 767–775.

Mottier, V., Siaussat, D., Bozzolan, F., Auzoux-Bordenave, S., Porcheron, P., and Debernard, S. (2004). The 20-hydroxyecdysone-induced cellular arrest in G2 phase is preceded by an inhibition of cyclin expression. *Insect Biochem. Mol. Biol.* **34,** 51–60.

Moshitzky, P., Fleischmann, I., Chaimov, N., Saudan, P., Klauser, S., Kubli, E., and Applebaum, S. W. (1996). Sex-peptide activates juvenile hormone biosynthesis in *Drosophila melanogaster* corpus allatum. *Arch. Insect Biochem. Physiol.* **32,** 363–374.

Mugat, B., Brodu, V., Kejzlarova-Lepesant, J., Antoniewski, C., Bayer, C. A., Fristrom, J. W., and Lepesant, J. A. (2000). Dynamic expression of Broad-Complex isoforms mediates temporal control of an ecdysteroid target gene at the onset of *Drosophila* metamorphosis. *Dev. Biol.* **227,** 104–117.

Nijhout, H. F. (2003). The control of body size in insects. *Dev. Biol.* **261,** 1–9.

Nusslein-Volhard, C., Wieschaus, E., and Kluding, H. (1984). Mutations affecting the pattern of the larval cuticle in *Drosophila melanogaster*. I. Zygotic loci on the second chromosome. *Roux's Arch. Dev. Biol.* **193,** 267–282.

Oro, A. E., McKeown, M., and Evans, R. M. (1990). Relationship between the product of the *Drosophila ultraspiracle* locus and the vertebrate retinoid X receptor. *Nature* **347,** 298–301.

Pak, J. W., Chung, K. W., Lee, C. C., Kim, K. J., Yong, Y., and Koolman, J. (1992). Evidence for multiple forms of prothoracicotropic hormone in *Drosophila melanogaster* and indication of a new function. *J. Insect Physiol.* **38,** 167–176.

Palli, S. R., Osir, S. O., Eng, W., Boehem, M. F., Edwards, M., Kulcsar, P., Ujvary, I., Hiruma, K., Prestwich, G. D., and Riddiford, L. M. (1990). Juvenile hormone receptors in insect larval epidermis: Identification by photoaffinity labeling. *Proc. Natl. Acad. Sci. USA* **87,** 796–800.

Palli, S. R., Touhara, K., Charles, J. P., Bonning, B. C., Atkinson, J. K., Trowell, S. C., Hiruma, K., Goodman, W. G., Kyriakides, T., Prestwich, G. D., Hammock, B. D., and Riddiford, L. M. (1994). A nuclear juvenile hormone binding protein from larvae of *Manduca sexta*: A putative receptor for the metamorphic action of juvenile hormone. *Proc. Natl. Acad. Sci. USA* **91,** 6191–6195.

Perrimon, N., Engstrom, L., and Mahowald, A. P. (1985). Developmental genetics of the 2C-D region of the *Drosophila* X chromosome. *Genetics* **111,** 23–41.

Petryk, A., Warren, J. T., Marques, G., Jarcho, M. P., Gilbert, L. I., Kahler, J., Parvy, J. P., Li, Y. T., Dauphin-Villemant, C., and O'Connor, M. B. (2003). Shade is the *Drosophila* P450 enzyme that mediates the hydroxylation of ecdysone to the steroid insect molting hormone 20-hydroxyecdysone. *Proc. Natl. Acad. Sci. USA* **100,** 13773–13778.

Pongs, O. (1989). Biochemical properties of ecdysteroid receptors. *In* "Ecdysone: From Chemistry to Mode of Action" (J. Koolman, Ed.), pp. 338–344. Georg Thieme Verlag, Stuttgart.

Postlethwait, J. H. (1974). Juvenile hormone and the adult development of *Drosophila*. *Biol. Bull.* **147,** 119–135.

Postlethwait, J. H., and Handler, A. M. (1979). The roles of juvenile hormone and 20-hydroxyecdysone during vitellogenesis in isolated abdomens of *Drosophila melanogaster*. *J. Insect Physiol.* **25,** 455–460.

Postlethwait, J. H., and Weiser, K. (1973). Vitellogenesis induced by juvenile hormone in the female sterile mutant *apterous-4* in *Drosophila melanogaster*. *Nature New Biol.* **244,** 284–285.

Pursley, S., Ashok, M., and Wilson, T. G. (2000). Intracellular localization and tissue specificity of the *Methoprene-tolerant (Met)* gene product in *Drosophila melanogaster*. *Insect Biochem. Mol. Biol.* **30,** 839–845.

Rachinsky, A., and Tobe, S. S. (1996). Role of second messengers in the regulation of juvenile hormone production in insects, with particular emphasis on calcium and phosphoinositide signaling. *Arch. Insect Biochem. Physiol.* **33,** 259–282.

Redfern, C. P. F., and Bownes, M. (1983). Pleiotropic effects of the *ecdysoneless^1* mutation of *Drosophila melanogaster*. *Mol. Gen. Genet.* **189,** 432–440.

Rees, H. H. (1995). Ecdysteroid biosynthesis and inactivation in relation to function. *Eur. J. Entomol.* **92,** 9–39.

Renault, N., King-Jones, K., and Lehmann, M. (2001). Downregulation of the tissue-specific transcription factor Fork head by *Broad-Complex* mediates a stage-specific hormone response. *Development* **128,** 3729–3737.

Restifo, L. L., and White, K. (1991). Mutations in a steroid hormone-regulated gene disrupt the metamorphosis of the central nervous system in *Drosophila*. *Dev. Biol.* **148,** 174–194.

Restifo, L. L., and White, K. (1992). Mutations in a steroid hormone-regulated gene disrupt the metamorphosis of internal tissues in *Drosophila*: Salivary glands, muscle, and gut. *Roux's Arch. Dev. Biol.* **201,** 221–234.

Richard, D. S., Applebaum, S. W., and Gilbert, L. I. (1989a). Developmental regulation of juvenile hormone biosynthesis by the ring gland of *Drosophila melanogaster*. *J. Comp. Physiol.* **159,** 383–387.

Richard, D. S., Applebaum, S. W., Sliter, T. J., Baker, F. C., Schooley, D. A., Reuter, C. C., Henrich, V. C., and Gilbert, L. I. (1989b). Juvenile hormone bisepoxide biosynthesis *in vitro* by the ring gland of *Drosophila melanogaster*: A putative juvenile hormone in the higher Diptera. *Proc. Natl. Acad. Sci. USA* **86,** 1421–1425.

Richard, D. S., Applebaum, S. W., and Gilbert, L. I. (1990). Allatostatic regulation of juvenile hormone production *in vitro* by the ring gland of *Drosophila melanogaster*. *Mol. Cell. Endocrinol.* **68,** 153–161.

Richard, D. S., Watkins, N. L., Serafin, R. B., and Gilbert, L. I. (1998). Ecdysteroids regulate yolk protein uptake by *Drosophila* oocytes. *J. Insect Physiol.* **44,** 637–644.

Richard, D.S, Jones, J. M., Barbarito, M. R., Cerula, S., Detweiler, J. P., Fisher, S. J., Brannigan, D. M., and Scheswohl, D. M. (2001). Vitellogenesis in diapausing and mutant *Drosophila melanogaster*: Further evidence for the relative roles of ecdysteroids and juvenile hormones. *J. Insect. Physiol.* **47,** 905–913.

Richards, G. (1978). Sequential gene activation by ecdysone in polytene chromosomes of *Drosophila melanogaster*. VI. Inhibition by juvenile hormones. *Dev. Biol.* **66,** 32–42.

Riddiford, L. M. (1976). Hormonal control of insect epidermis cell commitment *in vitro*. *Nature* **259,** 115–117.

Riddiford, L. M. (1978). Ecdysone-induced change in cellular commitment of the epidermis of the tobacco hornworm, *Manduca sexta*, at the initiation of metamorphosis. *Gen. Comp. Endocrinol.* **34,** 438–446.

Riddiford, L. M. (1993). Hormones and *Drosophila* Development. In "The Development of *Drosophila melanogaster*" (N. Bate and A. Martinez-Arias, Eds.), pp. 899–939. Cold Spring Harbor Laboratory Press, Cold Spring Harbor.

Riddiford, L. M. (1996). Juvenile hormone: The status of its "status quo" action. *Arch. Insect Biochem. Physiol.* **32,** 271–286.

Riddiford, L. M., and Ashburner, M. (1991). Effects of juvenile hormone mimics on larval development and metamorphosis of *Drosophila melanogaster*. *Gen. Comp. Endocrinol.* **82,** 172–183.

Riddiford, L.M, Cherbas, P., and Truman, J. W. (2001). Ecdysone receptors and their biological actions. *Vitam. Horm.* **60,** 1–73.

Riddihough, G., and Pelham, H. R. B. (1986). Activation of the *Drosophila hsp27* promoter by heat-shock and by ecdysone involves independent and remote regulatory elements. *EMBO J.* **5,** 1653–1658.

Riddihough, G., and Pelham, H. R. B. (1987). An ecdysone response element in the *Drosophila hsp27* promoter. *EMBO J.* **6,** 3729–3734.

Ringo, J. M., and Pratt, N. R. (1978). A juvenile hormone analog induces precocious sexual behavior in *Drosophila grimshawi*. *Ann. Entomol. Soc. Am.* **71,** 264–266.

Ringo, J., Werczberger, R., Altaratz, M., and Segal, D. (1991). Female sexual receptivity is defective in juvenile hormone-deficient mutants of the *apterous* gene of *Drosophila melanogaster*. *Behav. Genet.* **21,** 453–468.

Ringo, J., Werczberger, R., and Segal, D. (1992). Males sexual signaling is defective in mutants of the *apterous* gene of *Drosophila melanogaster*. *Behav. Genet.* **22,** 469–487.

Roberts, B., Henrich, V., and Gilbert, L. I. (1987). Effects of photoperiod on the timing of larval wandering in *Drosophila melanogaster*. *Physiol. Entomol.* **12,** 175–180.

Robinow, S., Talbot, W. S., Hogness, D. S., and Truman, J. W. (1993). Programmed cell death in the *Drosophila* CNS is ecdysone-regulated and coupled with a specific ecdysone receptor isoform. *Development* **119,** 1251–1259.

Roe, R. M., and Venkatesh, K. (1990). Metabolism of juvenile hormones: Degradation and titer regulation. In "Morphogenetic Hormones of Arthropods" (A. P. Gupta, Ed.), Vol. 1, pp. 125–180. Rutgers University Press, New Brunswick.

Roller, H., Dahm, K. H., Sweeley, C. C., and Trost, B. M. (1967). The structure of the juvenile hormone. *Angew. Chem. Int. Ed.* **6,** 179–180.

Ronaldson, E., and Bownes, M. (1995). Two independent *cis*-acting elements regulate the sex- and tissue-specific expression of *YP3* in *Drosophila melanogaster*. *Genet. Res.* **66,** 9–17.

Roth, G. E., Gierl, M. S., Vollborn, L., Meise, M., Lintermann, R., and Korge, G. (2004). The *Drosophila* Start1: A putative cholesterol transporter and key regulator of ecdysteroid synthesis. *Proc. Natl. Acad. Sci. USA* **101,** 1601–1606.

Royer, V., Fraichard, S., and Bouhin, H. (2002). A novel putative insect chitinase with multiple catalytic domains: Hormonal regulation during metamorphosis. *Biochem. J.* **366,** 921–928.

Royer, V., Hourdry, A., Fraichard, S., and Bouhin, H. (2004). Characterization of a putative extracellular matrix protein from the beetle *Tenebrio molitor*: Hormonal regulation during metamorphosis. *Dev. Genes Evol.* **214,** 115–121.

Rudolph, K., Morganelli, C., and Berger, E. M. (1991). Regulatory elements near the *Drosophila hsp22* gene required for ecdysterone and heat shock induction. *Dev. Genet.* **12,** 212–218.

Ruvkun, G. (2001). Molecular biology: Glimpses of a tiny RNA world. *Science* **294,** 797–799.

Sasorith, S., Billas, I. M. L., Iwema, T., Moras, D., and Wurtz, J. M. (2002). Structure-based analysis of the ultraspiracle protein and docking studies of putative ligands. *J. Insect Sci.* **2,** http://www.insectscience.org

Schwartz, M. B, Imberski, R. B., and Kelly, T. J. (1984). Analysis of metamorphosis in *Drosophila melanogaster*: Characterization of *giant*, an ecdysteroid-deficient mutant. *Dev. Biol.* **103,** 85–95.

Schwartz, M. B., Kelly, T. J., Imberski, R. B., and Rubenstein, E. C. (1985). The effects of nutrition and methoprene treatment on ovarian ecdysteroid synthesis in *Drosophila melanogaster*. *J. Insect Physiol.* **31,** 947–957.

Schwartz, M. B., Kelly, T. J., Woods, C. W., and Imberski, R. B. (1989). Ecdysteroid fluctuations in adult *Drosophila melanogaster* caused by elimination of pupal reserves and synthesis by early vitellogenic ovarian follicles. *Insect Biochem.* **19,** 243–249.

Sedlak, B. J. (1985). Structure of endocrine glands. In "Comprehensive Insect Physiology, Biochemistry, and Pharmacology" (G. A. Kerkut and L. I. Gilbert, Eds.), Vol. 7, pp. 109–151. Pergamon Press, Oxford.

Segraves, W. A., and Hogness, D. S. (1990). The *E75* ecdysone-inducible gene responsible for the 75B early puff in *Drosophila* encodes two new members of the steroid receptor superfamily. *Genes Dev.* **4,** 204–219.

Sempere, L. F., Sokol, N. S., Dubrovsky, E. B., Berger, E. M., and Ambros, V. (2003). Temporal regulation of microRNA expression in *Drosophila melanogaster* mediated by hormonal signals and *Broad-Complex* gene activity. *Dev. Biol.* **259,** 9–18.

Sevala, V. L., and Davey, K. G. (1993). Juvenile hormone dependent phosphorylation of a 100 kDa polypeptide is mediated by protein kinase C in the follicle cells of *Rhodnius prolixus*. *Invert. Reprod. Dev.* **23,** 189–193.

Sevala, V. L., Davey, K. G., and Prestwich, G. D. (1995). Photoaffinity labeling and characterization of a juvenile hormone binding protein in the membranes of follicle cells of *Locusta migratoria*. *Insect Biochem. Mol. Biol.* **25,** 267–273.

Shea, M. J., King, D. L., Conboy, M. J., Mariani, B. D., and Kafatos, F. C. (1990). Proteins that bind to *Drosophila* chorion *cis*-regulatory elements: A new C2H2 zinc finger protein and a C2C2 steroid receptor-like component. *Genes Dev.* **4,** 1128–1140.

Shemshedini, L., and Wilson, T. G. (1990). Resistance to juvenile hormone and an insect growth regulator in *Drosophila* is associated with an altered cytosolic juvenile hormone binding protein. *Proc. Natl. Acad. Sci. USA* **87,** 2072–2076.

Shemshedini, L., Lanoue, M., and Wilson, T. G. (1990). Evidence for a juvenile hormone receptor involved in protein synthesis in *Drosophila melanogaster*. *J. Biol. Chem.* **265,** 1913–1918.

Shirras, A. D., and Bownes, M. (1987). Separate DNA sequences are required for normal female and ecdysone induced male expression of *Drosophila melanogaster* yolk protein 1. *Mol. Gen. Genet.* **210,** 153–155.

Shirras, A. D., and Bownes, M. (1989). *cricklet*: A locus regulating a number of adult functions of *Drosophila melanogaster*. *Proc. Natl. Acad. Sci. USA* **86,** 4559–4563.

Siaussat, D., Bozzolan, F., Queguiner, I., Porcheron, P., and Debernard, S. (2004). Effects of juvenile hormone on 20-hydroxyecdysone-inducible *EcR, HR3, E75* gene expression in imaginal wing cells of *Plodia interpunctella* lepidoptera. *Eur. J. Biochem.* **271,** 3017–3027.

Sliter, T. J., and Gilbert, L. I. (1992). Developmental arrest and ecdysteroid deficiency resulting from mutations at the *dre4* locus of *Drosophila*. *Genetics* **130,** 555–568.

Sliter, T. J., Sedlak, B. J., Baker, F. C., and Schooley, D. A. (1987). Juvenile hormone in *Drosophila melanogaster*: Identification and titer determination during development. *Insect Biochem.* **17,** 161–165.

Smith, W. A., Gilbert, L. I., and Bollenbacher, W. E. (1984). The role of cyclic AMP in the regulation of ecdysone synthesis. *Mol. Cell Endocrinol.* **37,** 285–294.

Smith, W. A., Gilbert, L. I., and Bollenbacher, W. E. (1985). Calcium cyclic AMP interactions in prothoracicotropic hormone stimulation of ecdysone synthesis. *Mol. Cell. Endocrinol.* **39,** 71–78.

Smith, W. A., Combest, W. L., and Gilbert, L. I. (1986). Involvement of cyclic AMP-dependent protein kinase in prothoracicotrophic hormone-stimulated ecdysone synthesis. *Mol. Cell Endocrinol.* **47,** 25–33.

Smith, W. A., and Gilbert, L. I. (1986). Cellular regulation of ecdysone synthesis by the prothoracic glands of *Manduca sexta*. *Insect Biochem.* **16**, 143–147.

Smith, M. M., and Levitan, D. J. (2004). The *Caenorhabditis elegans* homolog of the putative prostate cancer susceptibility gene *ELAC2*, *hoe-1*, plays a role in germline proliferation. *Dev. Biol.* **266**, 151–160.

Smith, W. A., and Pasquarello, T. J. (1989). Developmental changes in phosphodiesterase activity and hormonal response in the prothoracic glands of *Manduca sexta*. *Mol. Cell. Endocrinol.* **63**, 239–246.

Soller, M, Bownes, M., and Kubli, E. (1999). Control of oocyte maturation in sexually mature *Drosophila* females. *Dev. Biol.* **208**, 337–351.

Song, Q., Sun, X., and Jin, X. Y. (2003). 20E-regulated USP expression and phosphorylation in *Drosophila melanogaster*. *Insect Biochem. Mol. Biol.* **33**, 1211–1218.

Stevens, M. E., and Bryant, P. J. (1985). Apparent genetic complexity generated by developmental thresholds: The *apterous* locus in *Drosophila melanogaster*. *Genetics* **110**, 281–297.

Sullivan, A. A., and Thummel, C. S. (2003). Temporal profiles of nuclear receptor gene expression reveal coordinate transcriptional responses during *Drosophila* development. *Mol. Endocrinol.* **17**, 2125–2137.

Sutherland, J. D., Kozlova, T., Tzertzinis, G., and Kafatos, F. C. (1995). *Drosophila* hormone receptor 38: A second partner for *Drosophila* USP suggests an unexpected role for nuclear receptors of the nerve growth factor-induced protein B type. *Proc. Natl. Acad. Sci. USA* **92**, 7966–7970.

Talbot, W. S., Swyryd, E. A., and Hogness, D. S. (1993). *Drosophila* tissues with different metamorphic responses to ecdysone express different ecdysone receptor isoforms. *Cell* **73**, 1323–1337.

Tamura, T., Kunert, C., and Postlethwait, J. H. (1985). Sex- and cell-specific regulation of yolk polypeptide genes introduced into *Drosophila* by P-element mediated gene transfer. *Proc. Natl. Acad. Sci. USA* **82**, 7000–7004.

Tavtigian, S. V., Simard, J., Teng, D. H., Abtin, V., Baumgard, M., Beck, A., Camp, N. J., Carillo, A. R., Chen, Y., Dayananth, P., Desrochers, M., Dumont, M., Farnham, J. M., Frank, D., Frye, C., Ghaffari, S., Gupte, J. S., Hu, R., Iliev, D., Janecki, T., Kort, E. N., Laity, K. E., Leavitt, A., Leblanc, G., McArthur-Morrison, J., Pederson, A., Penn, B., Peterson, K. T., Reid, J. E., Richards, S., Schroeder, M., Smith, R., Snyder, S. C., Swedlund, B., Swensen, J., Thomas, A., Tranchant, M., Woodland, A. M., Labrie, F., Skolnick, M. H., Neuhausen, S., Rommens, J., and Cannon-Albright, L. A. (2001). A candidate prostate cancer susceptibility gene at chromosome 17p. *Nature Genet.* **27**, 172–180.

Thomas, B. A., Hinton, A. C., Moskowitz, H., Severson, T. F., and Hammock, B. D. (2000). Isolation of juvenile hormone esterase and its partial cDNA clone from the beetle, *Tenebrio molitor*. *Insect Biochem. Mol. Biol.* **30**, 529–540.

Thummel, C. S. (2002). Ecdysone-regulated puff genes 2000. *Insect Biochem. Mol. Biol.* **32**, 113–120.

Tompkins, L. (1990). Effects of the *apterous-4* mutation on *Drosophila melanogaster* males courtship. *J. Neurogenet.* **6**, 221–227.

Truman, J. W. (1972). Physiology of insect rhythms. I. Circadian organization of the endocrine events underlying the moulting cycle of larval tobacco hornworms. *J. Exp. Biol.* **57**, 805–820.

Truman, J. W., Talbot, W. S., Fahrbach, S. E., and Hogness, D. S. (1994). Ecdysone receptor expression in the CNS correlates with stage-specific responses to ecdysteroids during *Drosophila* and *Manduca* development. *Development* **120**, 219–234.

Tsai, C. C., Kao, H. Y., Yao, T. P., McKeown, M., and Evans, R. M. (1999). SMRTER, a *Drosophila* nuclear receptor coregulator, reveals that EcR-mediated repression is critical for development. *Mol. Cell* **4**, 175–186.

Uhlirova, M., Foy, B. D., Beaty, B. J., Olson, K. E., Riddiford, L. M., and Jindra, M. (2003). Use of Sindbis virus-mediated RNA interference to demonstrate a conserved role of *Broad-Complex* in insect metamorphosis. *Proc. Natl. Acad. Sci. USA* **100,** 15607–15612.

Venkataraman, V., O'Mahony, P. J., Manzcak, M., and Jones, G. (1994). Regulation of juvenile hormone esterase gene transcription by juvenile hormone. *Dev. Genet.* **15,** 391–400.

Venkatesh, K., and Hasan, G. (1997). Disruption of the IP3 receptor gene of *Drosophila* affects larval metamorphosis and ecdysone release. *Curr. Biol.* **7,** 500–509.

Vermunt, A. M. W., Koopmanschap, A. B., Vlak, J. M., and de Kort, C. A. D. (1997). Cloning and sequence analysis of cDNA encoding a putative juvenile hormone esterase from the Colorado potato beetle. *Insect Biochem. Mol. Biol.* **27,** 919–928.

Warren, J. T., Bachmann, J. S., Dai, J. D., and Gilbert, L. I. (1996). Differential incorporation of cholesterol and cholesterol derivatives into ecdysteroids by the larval ring glands and adult ovaries of *Drosophila melanogaster*: A putative explanation for the $l(3)ecd^1$ mutation. *Insect Biochem. Mol. Biol.* **26,** 931–943.

Warren, J. T., Wismar, J., Subrahmanyam, B., and Gilbert, L. I. (2001). WOC (*without children*) gene control of ecdysone biosynthesis in *Drosophila melanogaster*. *Mol. Cell. Endocrinol.* **181,** 1–14.

Warren, J. T., Petryk, A., Marques, G., Jarcho, M., Parvy, J. P., Dauphin-Villemant, C., O'Connor, M. B., and Gilbert, L. I. (2002). Molecular and biochemical characterization of two P450 enzymes in the ecdysteroidogenic pathway of *Drosophila melanogaster*. *Proc. Natl. Acad. Sci. USA* **99,** 11043–11048.

Watson, R. D., Yeh, W. E., Muehleisen, D. P., Watson, C. J., and Bollenbacher, W. E. (1993). Stimulation of ecdysteroidogenesis by small prothoracicotropic hormone: Role of cyclic AMP. *Mol. Cell. Endocrinol.* **92,** 221–228.

Whalen, M., and Wilson, T. G. (1986). Variation and genomic localization of genes encoding *Drosophila melanogaster* male accessory gland proteins separated by sodium dodecyl sulfate-polyacrylamide gel electrophoresis. *Genetics* **114,** 77–92.

White, K. P., Hurban, P., Watanabe, T., and Hogness, D. S. (1997). Coordination of *Drosophila* metamorphosis by two ecdysone-induced nuclear receptors. *Science* **276,** 114–117.

Whiting, M. F., Carpenter, J. C., Wheeler, Q. D., and Wheeler, W. C. (1997). The strepsiptera problem: Phylogeny of the holometabolous insect orders inferred from 18S and 28S ribosomal DNA sequences and morphology. *Sys. Biol.* **46,** 1–68.

Wieschaus, E., Nusslein-Volhard, C., and Jurgens, G. (1984). Mutations affecting the pattern of larval cuticle in *Drosophila melanogaster*. III. Zygotic loci on the X chromosome and fourth chromosome. *Roux's Arch. Dev. Biol.* **193,** 296–307.

Wigglesworth, V. B. (1934). The physiology of ecdysis in *Rhodnius prolixus* (Hemiptera). II. Factors controlling moulting and "metamorphosis." *Q. J. Microsc. Sci.* **77,** 191–222.

Wigglesworth, V. B. (1936). The function of the corpus allatum in the growth and reproduction of *Rhodnius prolixus* (Hemiptera). *Quart. J. Microsc. Sci.* **79,** 91–121.

Wilson, T. G. (2004). The molecular site of action of juvenile hormone and juvenile hormone insecticides during metamorphosis: How these compounds kill insects. *J. Insect Physiol.* **50,** 111–121.

Wilson, T. G., and Ashok, M. (1998). Insecticide resistance resulting from an absence of target-site gene product. *Proc. Natl. Acad. Sci. USA* **95,** 14040–14044.

Wilson, T. G., and Fabian, J. (1986). A *Drosophila melanogaster* mutant resistant to a chemical analog of juvenile hormone. *Dev. Biol.* **118,** 190–201.

Wilson, T. G., and Gilbert, L. I. (1978). Metabolism of juvenile hormone I in *Drosophila melanogaster*. *Comp. Biochem. Physiol.* **60,** 85–89.

Wilson, T. G., DeMoor, S., and Lei, J. (2003). Juvenile hormone involvement in *Drosophila melanogaster* male reproduction as suggested by the *Methoprene-tolerant*[27] mutant phenotype. *Insect Biochem. Mol. Biol.* **33,** 1167–1175.

Wismar, J., Habtemichael, N., Dai, J., Warren, J. T., Gilbert, L. I., and Gateff, E. (2000). The mutation *without children*[rg1] causes ecdysteroid deficiency in third instar larvae of *Drosophila melanogaster*. *Dev. Biol.* **226**, 1–17.

Xu, Y., Fang, F., Chu, Y. X., Jones, D., and Jones, G. (2002). Activation of transcription through the ligand-binding pocket of the orphan nuclear receptor ultraspiracle. *Eur. J. Biochem.* **269**, 6026–6036.

Yamamoto, K., Chadarevian, A., and Pellegrini, M. (1988). Juvenile hormone action mediated in male accessory glands of *Drosophila* by calcium and kinase C. *Science* **239**, 916–919.

Yao, T. P., Segraves, W. A., Oro, A. E., McKeown, M., and Evans, R. M. (1992). *Drosophila* ultraspiracle modulates ecdysone receptor function via heterodimer formation. *Cell* **71**, 63–72.

Yao, T. P., Forman, B. M., Jiang, Z. Y., Cherbas, L., Chen, J. D., McKeown, M., Cherbas, P., and Evans, R. M. (1993). Functional ecdysone receptor is the product of *EcR* and *ultraspiracle* genes. *Nature* **366**, 476–479.

Zhou, B., Hiruma, K., Jindra, M., Shinoda, T., Segraves, W. A., Malone, F., and Riddiford, L. M. (1998a). Regulation of the transcription factor E75 by 20-hydroxyecdysone and juvenile hormone in the epidermis of the tobacco hornworm, *Manduca sexta*, during larval molting and metamorphosis. *Dev. Biol.* **193**, 127–138.

Zhou, B., Hiruma, K., Shinoda, T., and Riddiford, L. M. (1998b). Juvenile hormone prevents ecdysteroid-induced expression of *Broad-Complex* RNAs in the epidermis of the tobacco hornworm, *Manduca sexta*. *Dev. Biol.* **203**, 233–244.

Zhou, B., and Riddiford, L. M. (2001). Hormonal regulation and patterning of the *Broad-Complex* in the epidermis and wing discs of the tobacco hornworm, *Manduca sexta*. *Dev. Biol.* **231**, 125–137.

Zhou, X., and Riddiford, L. M. (2002). Broad specifies pupal development and mediates the "status quo" action of juvenile hormone on the pupal-adult transformation in *Drosophila* and *Manduca*. *Development* **129**, 2259–2269.

Zhou, S., Zhang, J., Fam, M. D., Wyatt, G. R., and Walker, V. K. (2002). Sequences of elongation factors-1α and -1γ and stimulation by juvenile hormone in *Locusta migratoria*. *Insect Biochem. Mol. Biol.* **32**, 1567–1576.

Zhou, X., Zhou, B., Truman, J. W., and Riddiford, L. M. (2004). Overexpression of *broad*: A new insight into its role in the *Drosophila* prothoracic gland cells. *J. Exp. Biol.* **207**, 1151–1161.

Zhu, J. S., Chen, L., and Raikhel, A. S. (2003). Posttranscriptional control of the competence factor beta FTZ-F1 by juvenile hormone in the mosquito *A. aegypti*. *Proc. Natl. Acad. Sci. USA* **100**, 13338–13343.

7

INSECT NEUROPEPTIDE AND PEPTIDE HORMONE RECEPTORS: CURRENT KNOWLEDGE AND FUTURE DIRECTIONS

ILSE CLAEYS,* JEROEN POELS,* GERT SIMONET,
VANESSA FRANSSENS, TOM VAN LOY,
MATTHIAS B. VAN HIEL, BERT BREUGELMANS, AND
JOZEF VANDEN BROECK

*Laboratory for Developmental Physiology, Genomics and Proteomics
Department of Animal Physiology and Neurobiology, Zoological Institute
K.U.Leuven, Naamsestraat 59, B-3000 Leuven, Belgium*

I. Introduction
II. Members of the Rhodopsin (GPCR Class A) Family
   A. Tachykinin (or Neurokinin) Receptors
   B. Myokinin Receptors
   C. Sulfakinin Receptors
   D. Receptors for NPF-Like and NPY-Like Peptides
   E. Allatostatin Receptors
   F. Receptors for PRXa Peptides: Pyrokinin/PBAN, CAP2b, and ETH
   G. Proctolin Receptor
   H. Receptors for FMRFa/FLRFa-Like Peptides
   I. AKH, CRZ, and CCAP Receptors
   J. "Leu-rich Repeat-containing GPCRs" and Bursicon

*Ilse Claeys and Jeroen Poels contributed equally to this manuscript.

III. Members of Other 7TM Families
   A. Secretin Receptor (GPCR Class B) Family
   B. The Frizzled 7TM Receptor Family
IV. Single Transmembrane Receptors
   A. Receptor Tyrosine Kinases
   B. TGF-β Receptor Family
   C. Guanylyl Cyclase Receptor Family
V. Emerging Concepts and Future Developments
   A. Elements of Discussion
   B. Emerging Concepts on Receptor–ligand Co-evolution
   C. Future Prospects
   References

Peptides form a very versatile class of extracellular messenger molecules that function as chemical communication signals between the cells of an organism. Molecular diversity is created at different levels of the peptide synthesis scheme. Peptide messengers exert their biological functions via specific signal-transducing membrane receptors. The evolutionary origin of several peptide precursor and receptor gene families precedes the divergence of the important animal Phyla. In this chapter, current knowledge is reviewed with respect to the analysis of peptide receptors from insects, incorporating many recent data that result from the sequencing of different insect genomes. Therefore, detailed information is provided on six different peptide receptor families belonging to two distinct receptor categories (i.e., the heptahelical and the single transmembrane receptors). In addition, the remaining problems, the emerging concepts, and the future prospects in this area of research are discussed. © 2005 Elsevier Inc.

I. INTRODUCTION

In many biological systems (from yeast to mammals), peptides functioning as extracellular messenger substances constitute an important means of intercellular communication. They can therefore be considered as an important part of the "cellular language" (Vanden Broeck, 2001a). Well-known examples of peptide messengers are the mating pheromones in yeast, as well as the majority of paracrines and endocrines in animals. Metazoan peptides are involved in a very wide range of regulatory functions, since they play a prominent role in the physiology of the nervous system, as well as in developmental (growth factors), metabolic, reproductive, and behavioral processes.

The use of peptides as messengers provides a plethora of combinatorial possibilities, each peptide being characterized by a unique amino acid sequence. Bioactive peptide molecules are usually generated from biosynthetic precursors via enzymatic cleavage and modification pathways. These precursors are gene products that sometimes contain several distinct peptides. Molecular diversity of peptide messengers is created at different levels: (1) at the genetic level, evolution of peptide precursor genes generates orthologues and paralogues via mutations and gene duplications and (2) at the biosynthetic level, alternative RNA splicing, precursor processing, and peptide modification or degradation can generate multiple products derived from a single gene. In addition, intramolecular (3-D structure) and intermolecular (e.g., heterodimerization or homodimerization) interactions can have an important impact on the bioactivity of peptide or protein hormones. Many peptides can be grouped into families, some of which are well conserved and have an ancient origin that precedes the divergence of important animal Phyla. Since peptides exert their biological function(s) via specific interaction with receptors (i.e., plasma membrane localized signal-transducing proteins), their genetic evolution is believed to be "tuned" by the evolution of the corresponding receptor genes.

The availability of novel gene and transcript sequence information is growing very rapidly as a result of molecular cloning experiments and of metazoan genome (e.g., *Homo sapiens, Ciona intestinalis, Caenorhabditis elegans*, and the insect species *Drosophila melanogaster, Anopheles gambiae, Apis mellifera*, and *Bombyx mori*) and "expressed sequence tags" (EST) sequencing projects. These projects already have an important impact on neurobiological and endocrinological research and provide novel insights in the evolution of genes coding for important components of neuronal and endocrine signaling processes, such as peptide precursor and receptor genes (Brody and Cravchick, 2000; Hewes and Taghert, 2001; Hill *et al.*, 2002; Riehle *et al.*, 2002; Vanden Broeck, 2001a,b). For instance, an initial comparative computational analysis of the fruit fly genome (Adams *et al.*, 2000) has revealed approximately 30 precursor genes encoding more than 50 peptides as orthologues of the majority of functionally characterized insect neuropeptides (Vanden Broeck, 2001a).

The signal-transducing receptors for metazoan neuropeptides and peptide hormones can be classified into three major categories of membrane proteins: (1) ligand-gated ion channels, (2) heptahelical receptors (seven transmembrane [7TM] or G protein–coupled receptors [GPCR]), and (3) single transmembrane (TM) receptors. Each of these categories is further subdivided into different gene families, some of which appear to be well conserved in evolution and have their origin before the great divergence of protostomian and deuterostomian lineages (Vanden Broeck, 1996). With respect to insect peptide receptors, the current knowledge is limited to 7TM and single TM receptors. Although it is not unlikely that insects also could make use of peptide operated ion channels, the presence of such receptors has only been well substantiated in molecular terms in vertebrates and mollusks (Green et al., 1994; Jeziorski et al., 2000; Lingueglia et al., 1995; Nishimura et al., 2000). Therefore, this review will focus on recent, mainly postgenomic, discoveries made in the other two functional receptor categories. In general, but there are exceptions to the rule, most of the small peptides, such as many neuropeptides, act through 7TM/GPCRs, whereas larger peptides, protein hormones, and growth factors exert their functions via single TM receptor proteins. The latter are also termed "catalytic" receptors, since their cytoplasmic domain usually possesses an intrinsic enzyme activity (or can associate with a partner having such activity), which is essential for further signal transduction. This enzyme activity, such as protein tyrosine kinase, Ser/Thr kinase, phosphoprotein phosphatase, or guanylyl cyclase, will regulate the activity of downstream effectors that are components of cellular signaling pathways or networks. The former group (7TM) is also known as "G protein–coupled" or "metabotropic" receptors, since their interaction with an agonist will often lead to the activation of heterotrimeric guanine nucleotide–binding proteins (G proteins) that consequently regulate intracellular second messenger (metabolite) systems (Bockaert, 1991; Van Roey et al., 2004). G protein–coupled receptors form a very large category (superfamily) of signal-transducing membrane proteins, mediating the effects of a broad range of extracellular signals. Their mode of action is mainly based on a conformational activation mechanism that is not yet entirely understood. Based on the current genome sequence information, 7TM/GPCRs represent one of the largest functional protein categories in metazoan proteomes. Their main structural characteristic is the presence of seven putative α-helical transmembrane segments. Despite this remarkably universal structure, multiple sequence comparisons of heptahelical receptors reveal the existence of a number of distinct protein families, for which it is uncertain whether these originated separately (by convergence) or from a very ancient, common ancestor. Given the low degree of sequence similarity between these families, it is perhaps not excluded that new 7TM-families still may be discovered. This review, discusses only those 7TM/GPCR-families with members that are already known to interact with peptide ligands: (1) the largest one is the "rhodopsin family" (also termed "family 1" or "A-family" of GPCRs; Fig. 1), (2) the "secretin receptor family" (also termed "family 2" or "B-family"), and (3) the Fz/Smo family containing a number of 7TMs playing a key role in development. Previous studies have already suggested the existence of approximately 45 Drosophila genome–encoded GPCRs that show clear sequence similarities with mammalian peptide receptors belonging to the families A and B (Hewes and Taghert, 2001; Vanden Broeck, 2001b). However, at that time, most of these receptors were only predicted from the genome data and therefore were to be considered as "orphans," (i.e., receptors for which the endogenous ligand as well as the signal-transducing

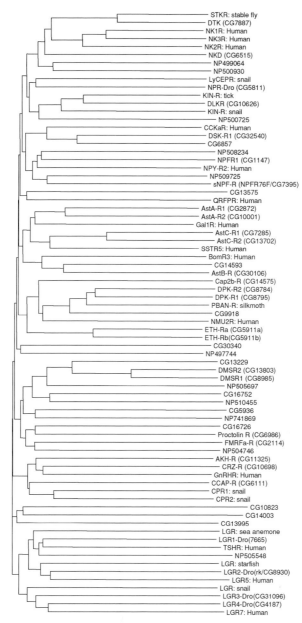

FIGURE 1. Dendrogram of a comparison of *Drosophila* GPCRs related to rhodopsin family peptide receptors. Characterized and orphan peptide receptors belonging to the rhodopsin family (GPCR class A) were compared by CLUSTALW analysis. *Drosophila* genome encoded receptors are shown in black, while the names of a number of related GPCRs from human (green), nematode *C. elegans* (violet), other arthropod (blue) and other invertebrate (dark red) origin are shown in color. One of the previously predicted fruit fly GPCRs

properties still have to be discovered). Today, naturally occurring ligands have been identified for more than half of this set of previously predicted receptors (see Tables I and II), and more detailed functional data have been reported. Therefore, these new data represent an important breakthrough in insect physiology and undoubtedly shed a new light on the (co)evolution of peptide–receptor couples. The list of putative and characterized regulatory peptide receptors with members of other 7TM and single TM receptor (sub) families in insects is extended. Finally, the remaining problems are analyzed and future prospects in this area of research are discussed.

II. MEMBERS OF THE RHODOPSIN (GPCR CLASS A) FAMILY

A. TACHYKININ (OR NEUROKININ) RECEPTORS

Tachykinins or neurokinins form a family of multifunctional brain–gut peptides that are present in both vertebrates and invertebrates. The term "tachykinin" refers to the ability of these peptides to cause tachycardia by lowering peripheral blood pressure in vertebrates. However, it is now apparent that tachykinins exert various physiological and even pathological effects. The best-known vertebrate member is substance P (SP), which was discovered in 1931 (Von Euler and Gaddum, 1931). Other members are neurokinin B (NKB; or neuromedin K) (Kangawa et al., 1983) and neurokinin A (NKA; or substance K) (Nawa et al., 1984) with its two N-terminally extended forms, neuropeptide K (NPK) (Tatemoto et al., 1985) and neuropeptide γ (NPγ) (Kage et al., 1988). The last discovered mammalian tachykinins are endokinins (EK) C and D, hemokinin-1 (HK-1), and two N-terminally extended forms of HK-1, termed EKA and EKB (Kurtz et al., 2002; Page et al., 2003; Zhang et al., 2000). All vertebrate tachykinins share a common C-terminal motif, Phe-X-Gly-Leu-Met-NH$_2$ (FXGLMa, with X being a variable amino acid residue), characteristically with an amidated methionine. Two exceptions to this rule are EKC and EKD, which possess a C-terminal FXGLLa-motif (Page et al., 2003).

(CG12610; Hewes and Taghert, 2001) was not included in this comparison, since it is only represented in the genome and protein databases as a partial amino acid sequence. NK1R, NK2R, and NK3R: neurokinin receptors; CCKaR: cholecystokinin A receptor; NPY-R2: Y2 type NPY receptor; QRFPR: receptor for QRFP, an RF-amide peptide; Gal1R: galanin 1 receptor; SSTR5: somatostatin receptor 5; BomR3: bombesin receptor 3; NMU2R: neuromedin 2 receptor; GnRHR: gonadotropin releasing hormone receptor; TSHR: thyroid stimulating hormone receptor; LGR: Leu-rich repeats containing GPCR; CPR1, CPR2: conopressin (vasopressin-like peptide) receptors. The abbreviations of the insect receptors are explained in Table I, as well as in the text. For a list of insect peptides and their sequences, recent review papers can be referred (Gäde et al., 1997; Hewes and Taghert, 2001; Nässel, 2002; Riehle et al., 2002; Vanden Broeck, 2001a). (See Color Insert.)

So far, only three vertebrate receptors for tachykinins have been cloned from multiple species and termed NK_1, NK_2, and NK_3 (Masu et al., 1987; Shigemoto et al., 1990; Yokota et al., 1989). They all belong to the rhodopsin family of GPCRs and differ in their pharmacological profiles as well as in their main distribution patterns. Although the three tachykinins that were initially discovered (SP, NKA, and NKB) can act on all three receptors, each receptor possesses a preferential binding profile (i.e., SP = NKA > NKB for NK_1, NKA > NKB > SP for NK_2 and NKB > NKA > SP for NK_3). Whether more tachykinin receptor types exist in vertebrates is currently under debate, but it is worthwhile mentioning that EKC and EKD have very weak activities at the three known NK receptors (Page et al., 2003; Pennefather et al., 2004). Other possibilities to explain some of the discrepancies that exist between tachykinin receptors and some of their alleged ligands are the existence of multiple splicing variants, the occurrence of different active states (see later section) or tissue-specific expression of the receptors and/or their ligands.

In insects, immunoreactivity to vertebrate tachykinins was demonstrated multiple times, but the identification of the first tachykinin-related peptides (TRPs, also designated as "insectatachykinins") occurred in 1990 (Schoofs et al., 1990). After this initial discovery of the locustatachykinins (derived from *Locusta migratoria*), the number of known TRPs rapidly increased, with family members found in many other arthropod species and even in other invertebrate phyla, such as mollusks (Anastasi and Erspamer, 1962; Fujisawa et al., 1994; Kanda et al., 2003; Satake et al., 2003) and echiuroid worms (Ikeda et al., 1993; Kawada et al., 2000). With the exception of tachykinin-like peptides from *Aedes* and *Stomoxys*, all known insect TRPs possess the C-terminal consensus sequence FX_1GX_2Ra (Nässel, 1999; Vanden Broeck et al., 1999). Interestingly, the tachykinin that is present in the salivary glands of the mosquito, *Aedes aegypti*, holds the vertebrate consensus sequence (Champagne and Ribeiro, 1994). Since this mosquito feeds on vertebrate blood, it cannot be excluded that this TRP mediates its effect via vertebrate NK receptors, thereby serving as a vasodilatory agent. The TRPs purified from the stable fly, *Stomoxys calcitrans*, as well as from the bivalve mollusk, *Anodonta cygnea*, contain an Ala-residue instead of the highly conserved Gly-residue (FX_1AX_2Ra) that is present in most other members of this peptide family (Nachman et al., 1999). This apparently small change of Gly to Ala in the core sequence has some important consequences that will be discussed further. Tachykinin-related peptides are processing products of larger precursor polypeptides, which have recently been identified in the fruit fly, *D. melanogaster* (Siviter et al., 2000; Vanden Broeck, 2001a; Winther et al., 2003), as well as in the honey bee, *Apis mellifera* (Takeuchi et al., 2003).

Despite the large intraspecies and interspecies diversity of TRPs in protostomian invertebrates, only four corresponding receptors have been

TABLE I. Functionally Characterized Peptide Receptors of the GPCR A-Family (Rhodopsin Family)

Species	Name	Synonyms	Accession	Agonist	Assay	EC_{50}	Reference
D. melanogaster	NKD	CG6515; Takr86C	M77168	Lom-TK II (Locustatachykinin)	IP3	2.5 nM	Monnier et al., 1992
D. melanogaster	NKD	CG6515; Takr86C	NM_079580	Drm-TK 1	Arrestin	N.D.	Johnson et al., 2003a
D. melanogaster	DTKR	CG7887; Takr99D	X62711	Vertebrate TK (substance P)	Xenopus current	N.D.	Li et al., 1991
D. melanogaster	DTKR	CG7887; Takr99D	NM_079832	Drm-TK 1	Arrestin	N.D.	Johnson et al., 2003a
S. calcitrans	STKR	Stomoxytachykinin receptor	U52347	Stomoxytachykinin (Stc-TK)	$IP3/Ca^{2+}$ cAMP	8.2 nM 74 nM	Torfs et al., 2000; Poels et al., 2004b
D. melanogaster	DLKR	CG10626	NM_139711	Drm-Kin (drosokinin)	Ca^{2+}	40 pM	Radford et al., 2002
D. melanogaster	DSK-R1	CG32540; CCKLR-17D3	NM_078681	Sulphated DSK-1	Ca^{2+}	5–10 nM range	Kubiak et al., 2002
D. melanogaster	NPFR1	CG1147; NPFR1	AF364400; NM_079521	Drm-NPF	Binding cAMP (-)	65 nM (IC50)	Garczynski et al., 2002
D. melanogaster	NPFR1	CG1147; NPFR1	AF364400; NM_079521	Drm-NPF	Arrestin	N.D.	Johnson et al., 2003a
D. melanogaster	Drm-sNPF-R	CG7395, CG18639; NPFR76F	NM_079452	Drm-sNPF-1 Drm-sNPF-2	$G\alpha16$-Xenopus current	1.4 nM 24 nM	Feng et al., 2003 Reale et al., 2004
D. melanogaster	Drm-sNPF-R	CG7395, CG18639; NPFR76F	NM_079452	Drm-sNPF-1 Drm-sNPF-2 Drm-sNPF-3 Drm-sNPF-4	$G\alpha16$-Ca^{2+}	51 nM 42 nM 31 nM 75 nM	Mertens et al., 2002
D. melanogaster	Nep Yr	CG5811; NPR	NM_079801	Vertebrate PYY; NPY	Xenopus current	N.D.	Li et al., 1992
D. melanogaster	DAR-1	CG2872; DalstR1	AF163775	Allatostatin-A (Drm-AstA 3)	Xenopus current	55 pM	Birgül et al., 1999
D. melanogaster	DAR-1	CG2872; D.AlstR1	AF163775	Allatostatin-A (Drm-AstA)	Ca^{2+} $GTP\gamma S$	10–100 nM range 1–10 nM range	Larsen et al., 2001
D. melanogaster	DAR-2	CG10001, D.AlstR2	NM_079820	Allatostatin-A (Drm-AstA)	Ca^{2+} $GTP\gamma S$	10–100 nM range 1–10 nM range	Larsen et al., 2001

Species	Receptor	CG/Gene	Accession	Ligand	Assay	Potency	Reference
D. melanogaster	DAR-2	CG10001, DAlstR2	NM_079820	Allatostatin-A (drostatins)	Ca^{2+}	10–100 nM range	Lenz et al., 2001
P. americana	AlstR		AF336364	Drm-AstA-4 Allatostatin-A (Dip-AstA 8)	Xenopus current	10 nM 1.9 nM	Auerswald et al., 2001
B. mori	BAR		AF254742	Bom-AstA (bostatins)	Ca^{2+}	10 nM	Secher et al., 2001
D. melanogaster	AstB-R	CG30106; CG14484; MIPR	NM_137397	Bom-AstA-2,3 Drm-AstB 1 (Drm-MIP 1)	Arrestin	N.D.	Johnson et al., 2003a
D. melanogaster	Drostar1	CG7285	AY017416; NM_140783	Gln-AstC; pyroGlu-AstC	Xenopus current	9.5 nM; 25.4 nM	Kreienkamp et al., 2002
D. melanogaster	Drostar1	CG7285	AY017416; NM_140783	AstC	Arrestin	N.D.	Johnson et al., 2003a
D. melanogaster	Drostar2	CG13702; ALCR2	AY046072; NM_140782	Gln-AstC; pyroGlu-AstC	Xenopus current	7.0 nM; 8.7 nM	Kreienkamp et al., 2002
D. melanogaster	Drostar2	CG13702; ALCR2	AY046072; NM_140782	AstC	Arrestin	N.D.	Johnson et al., 2003a
D. melanogaster	DAKHR	CG11325; DGRHR	AF522194; NM_058039	Drm-AKH	$G\alpha 16$-Ca^{2+}	0.8 nM	Staubli et al., 2002
D. melanogaster	DAKHR	CG11325; DGRHR	AF522194; NM_058039	Drm-AKH	Xenopus current	0.3 nM	Park et al., 2002
B. mori	BAKHR		AF403542	Hez-HrTH Mas-AKH	$G\alpha 16$-Ca^{2+}	0.3 nM 8 nM	Staubli et al., 2002
D. melanogaster	DCRZR	CG10698, DGHRH II	AF522192; NM_140314	Corazonin (Drm-CRZ)	$G\alpha 16$-Ca^{2+}	18 nM	Cazzamali et al., 2002
D. melanogaster	DCRZR	CG10698, DGHRH II	AF522192; NM_140314	Corazonin (Drm-CRZ)	Xenopus current	1.1 nM	Park et al., 2002
D. melanogaster	DCRZR	CG10698, DGHRH II	AF522192; NM_140314	Corazonin (Drm-CRZ)	Arrestin	N.D.	Johnson et al., 2003a
M. sexta	MasCRZR		AY369029	Corazonin	Xenopus current $G\alpha 16$-Ca^{2+}	200 pM 75 pM	Kim et al., 2004
D. melanogaster	Putative CCAP-R	CG6111	AF52218	CCAP AKH	Xenopus current	130 nM 240 nM	Park et al., 2002
D. melanogaster	CCAP-R	CG6111 (corrected sequence)	AF52218	CCAP	$G\alpha 16$-Ca^{2+}	0.54 nM	Cazzamali et al., 2003

(Continues)

TABLE I. (Continued)

Species	Name	Synonyms	Accession	Agonist	Assay	EC_{50}	Reference
D. melanogaster	ETH-R	CG5911a and CG5911b	NM_142703; NM_206533	ETH-1 ETH-2	$G\alpha 16$- Ca^{2+}	200 nM (a) 37 nM (b) 1.8 μM (a) 160 nM (b)	Iversen et al., 2002b
D. melanogaster	ETH-R	CG5911a and CG5911b	NM_142703; NM_206533	ETH-1 ETH-2	$G\alpha 16$- Ca^{2+}	414 nM (a) 0.9 nM (b) >2 μM (a) 2 nM (b)	Park et al., 2003
D. melanogaster	Putative Cap2b-R	CG14575; capaR	AF522193; NM_141074	Cap2b-1 Cap2b-2	$G\alpha 16$- Ca^{2+}	69 nM 110 nM	Iversen et al., 2002a
D. melanogaster	Putative Cap2b-R	CG14575; capaR	AF522193; NM_141074	Cap2b-1 Cap2b-2	Xenopus current	150 nM 230 nM	Park et al., 2002
D. melanogaster	Putative PRXa R	CG8795	AF522190; AY277899; NM_169505	hugγ	Xenopus current	N.D.	Park et al., 2002
D. melanogaster	Putative PRXa R	CG8795	AF522190; AY277899; NM_169505	Drm-PK-2 Drm-PK-1/Cap2b-3 ETH-1		N.D. 420 nM 480 nM	
D. melanogaster	DPK-R1	CG8795	AF522190; AY277899; NM_169505	Drm-PK-2 hugγ Lem-PK	$G\alpha 16$- Ca^{2+}	0.5 nM 7 nM 6 nM	Rosenkilde et al., 2003
D. melanogaster	DPK-R2	CG8784	AF522189; NM_169507	Drm-PK-2 hugγ Lem-PK	$G\alpha 16$- Ca^{2+}	1 nM 30 nM 60 nM	Rosenkilde et al., 2003
D. melanogaster	Putative PRXa R	CG9918	AF522191	Drm-PK-1/Cap2b-3	Xenopus current	> 10 μM	Park et al., 2002
B. mori	Bom-PBANR		AB181298	PBAN	Ca^{2+}	nM range	Hull et al., 2004
H. zea	Hez-PBANR		AY319852	PBAN	Ca^{2+}	25 nM	Choi et al., 2003
D. melanogaster	Proctolin receptor	CG6986	NM_131955; NM_167020	Proctolin	Ca^{2+} binding	0.3–4 nM (IC50)	Johnson et al., 2003b

Species	Receptor	Gene	Ligand	Assay	Affinity	Reference
D. melanogaster	Proctolin receptor	CG6986	Proctolin	Arrestin	N.D.	Johnson et al., 2003a
D. melanogaster	Proctolin receptor	CG6986	Proctolin	Gα16- Ca^{2+}	0.6 nM	Egerod et al., 2003a
D. melanogaster	DMS-R1	CG8985	Dromyosuppressin (DMS or Drm-MS or Drm-FLRFa) >Drm-FMRFa 2	Arrestin cAMP (-)	N.D. 1.8 nM	Johnson et al., 2003a
D. melanogaster	DMS-R1	CG8985	Dromyosuppressin (DMS or Drm-MS or Drm-FLRFa)	Gα16- Ca^{2+}	40 nM	Egerod et al., 2003b
D. melanogaster	DMS-R2	CG13803	Dromyosuppressin (DMS or Drm-MS or Drm-FLRFa) >Drm-FMRFa 2	Arrestin cAMP (-)	N.D. 0.17 nM	Johnson et al., 2003a
D. melanogaster	DMS-R2	CG13803	Dromyosuppressin (DMS or Drm-MS or Drm-FLRFa)	Gα16- Ca^{2+}	40 nM	Egerod et al., 2003b
D. melanogaster	FMRFaR	CG2114; DrmFMRFa-R	Drm-FMRFa 6 >other Drm-FMRFa, Drm-sNPF1, Drm-MS	Gα16- Ca^{2+}	0,9 nM	Cazzamali and Grimmelikhuijzen, 2002
D. melanogaster	FMRFaR	CG2114; DrmFMRFa-R	Drm-FMRFa 1–5 >Drm-SK, Drm-MS, Drm-sNPF1	Gα16- Ca^{2+}	ca. 2 nM	Meeusen et al., 2002
D. melanogaster	FMRFaR	CG2114; DrmFMRFa-R	Drm-FMRFa 2 >Drm-MS	Arrestin	N.D.	Johnson et al., 2003a
D. melanogaster	DLGR2	Rickets (rk); CG8930	Bursicon (indirect evidence)	Mutant analysis	N.D.	Baker and Truman, 2002; Dewey et al., 2004

TABLE II. Functionally Characterized Members of the GPCR B-Family (Secretin Receptor Family)

Species	Name	Synonyms	Accession	Agonist	Assay	IC/EC$_{50}$	Reference
M. sexta	*Mas*-DH-R		U03489	Mas-DH	Binding of ^{3}H-labeled	1 nM	Reagan, 1994
				Pea-DH		1 nM	
				Acd-DH		1 nM	
				Lom-DH		8 nM	
				Mas-DPII		12 nM	
				Mas-DH	cAMP	0.5 nM	
A. domesticus	*Acd*-DH-R		U15959	Acd-DH	Binding of ^{3}H-labeled	0.5 nM	Reagan, 1996
				Mas-DH		1 nM	
				Pea-DH		1 nM	
				Acd-DH	cAMP	ca. 0.5–1 nM	
D. melanogaster	*Drm*-DH-R	CG8422	NM_137116	DH44	cAMP	ca. 1 nM	Johnson et al., 2004
					Ca^{2+}	300 nM	
D. melanogaster	Mth	LD08316	AF109308	Sun A	Ca^{2+}	4 μM	Cvejic et al., 2004
				Sun B		3.8 μM	

functionally characterized. In addition, a few other, putative NK receptor orthologues have been sequenced—LTKR, a GPCR found in the cockroach brain (Johard et al., 2001); CAD27763 and EAA00906 from the malaria mosquito, *A. gambiae* (Hill et al., 2002); and the *C. elegans* receptors, NP499064 and NP500930. All these receptors belong to the rhodopsin family of GPCRs and share 35–48% sequence identity with their mammalian counterparts (Satake et al., 2003).

The first cloned insect tachykinin receptor originated from *Drosophila* and was termed "*Drosophila* tachykinin receptor" (DTKR) (Li et al., 1991). When expressed in *Xenopus* oocytes, DTKR showed pertussis toxin-sensitive responses to micromolar concentrations of SP and physalaemin (an amphibian tachykinin) but not to other vertebrate tachykinins. Utilizing a β-arrestin2-green fluorescent protein (GFP) translocation assay, Johnson et al. (2003a) recently confirmed responsiveness of DTKR to an endogenous *Drosophila* tachykinin (*Drm*-TK I). The DTKR transcripts appear to be regulated during *Drosophila* development, with the expression in adults found mainly in the CNS.

A second fruit fly tachykinin receptor, termed NKD, was cloned and expressed in a stable mammalian cell line (Monnier et al., 1992). Although SP or other tachykinins were not effective on this receptor, these investigators also tested the locustatachykinins I and II (*Lom*-TK I and II). *Lom*-TK II, in particular, was able to generate dose-dependent responses in NKD-expressing cells. In addition, *Drm*-TK I displayed activity in the β-arrestin2-GFP translocation assay (Johnson et al., 2003a). NKD transcript levels appear to be regulated during development, and expression in the adult is mainly found in the heads. This receptor protein has also been immunolocalized in the CNS of different insect species by means of antisera raised against distinct parts of the molecule (Veelaert et al., 1999). As is the case for vertebrate tachykinin receptors, both NKD and DTKR seem to be primarily coupled to the phospholipase C (PLC) pathway, which leads to an IP_3-induced release of Ca^{2+} ions from vesicular stores.

A third insect TRP receptor, STKR, was cloned from the stable fly, *Stomoxys calcitrans* (Guerrero, 1997). STKR is an orthologue of DTKR, with which it shares 79% amino acid sequence identity, compared to 40% with NKD. More extensive analyses have been performed on this GPCR, further substantiating the coupling to the PLC-IP_3-Ca^{2+} pathway (Torfs et al., 2000, 2001). Additional coupling of this receptor to the adenylyl cyclase (AC) pathway has also been detected (Torfs et al., 2000). The agonist concentration necessary for the cAMP-rise is generally 10-fold higher than the one needed to elicit a Ca^{2+} increase, a characteristic that has also been described for mammalian NK receptors (Nakajima et al., 1992). In submicromolar concentrations, STKR is not responsive to peptides belonging to the FXGLMa subfamily of tachykinins. The activity of *Lom*-TK I (belonging to the FX_1GX_2Ra family) was correspondingly reduced by changing the

C-terminal Arg to a Met residue. *Vice versa*, this chimeric insect–mammalian peptide showed increased activity on mammalian NK_1 and NK_2 receptors. In addition, Met to Arg substituted vertebrate tachykinins lose activity on NK receptors but gain responsiveness on STKR (Torfs *et al.*, 2002a,b). Similar results were obtained with UTKR, a fourth protostomian invertebrate TRP receptor that was cloned from the echiuroid worm, *Urechis unicinctus* (Kawada *et al.*, 2002). This Met/Arg preference is a major pharmacological and evolutionary difference between the receptors of insect and mammalian tachykinins.

A few tachykinin-like peptides from invertebrates, such as *S. calcitrans* (*Stc*-TK) and *Anodonta cygnea* (*Anc*-TK), contain an Ala-residue instead of the highly conserved Gly-residue in their C-terminal core sequence (FX_1-AX_2Ra). These peptides were recently shown to behave as partial agonists for the STKR-mediated Ca^{2+} response (Poels *et al.*, 2004a,b). The reverse change (i.e., replacement of Gly by Ala) turned a full agonist, such as *Lom*-TK III, into a partial agonist. Interestingly, the Gly to Ala change had no partial agonism-inducing effect on STKR-induced cAMP changes, but the Ala-substituted analogues stimulated both types of response at considerably lower threshold concentrations (Poels *et al.*, 2004b). Based on these observations, a model was proposed that takes into account the existence of at least two active receptor conformations (Poels *et al.*, 2004a,b). In analogy, the existence of multiple conformational states has also been suggested for mammalian NK receptors (Holst *et al.*, 2001; Maggi and Schwartz, 1997; Palanche *et al.*, 2001; Sachon *et al.*, 2002). The fact that NK receptor conformers can differentiate between a number of peptide agonists could prove to be an important mechanism of physiological finetuning (Poels *et al.*, 2004b).

B. MYOKININ RECEPTORS

Kinins were first isolated from the cockroach, *Leucophaea maderae*, by their ability to stimulate cockroach hindgut contractility (and hence, they are also designated as "leucokinins" or "myokinins") (Holman *et al.*, 1986a). Other members of the myokinin peptide family have since then been purified from many other insect species, as well as from crustaceans and mollusks (Cox *et al.*, 1997; Nieto *et al.*, 1998). In addition to their stimulatory action on visceral muscle motility, myokinins are important regulators of fluid and ion secretion in the Malpighian tubules (Hayes *et al.*, 1989). The bioactivity of kinins on Malpighian tubules is most probably mediated via the Ca^{2+} pathway. Insect kinins possess a characteristic C-terminal pentapeptide motif FX_1X_2WGa, with the amidated carboxyl terminus a prerequisite for activity.

The first myokinin receptor was cloned from a mollusk (i.e., the pond snail *Lymnaea stagnalis*). When expressed in CHO cells, receptor activation by the endogenous *Lymnaea* kinin resulted in an increase of intracellular Ca^{2+} ions (Cox et al., 1997). Subsequently, a second kinin receptor was cloned from the cattle tick, *Boophilus microplus* (Holmes et al., 2000). Based on these myokinin receptor sequences, the authors identified a putative *Drosophila* orthologue (CG10626) in the genome sequence database. This prediction was later confirmed when this fruit fly receptor was functionally expressed and characterized as the first insect leucokinin receptor, DLKR (Radford et al., 2002). Surprisingly, DLKR is not only expressed in Malpighian tubules and hindgut but also in the CNS and in the gonads of the fly. These tissues displayed Ca^{2+} stimulation upon leucokinin application. In the genital tract, leucokinin might play a role in the peristaltic transfer of eggs or sperm or regulate fertility (Radford et al., 2002). Although we could not find a clear vertebrate orthologue of the fly, tick, and snail myokinin receptors, a computational BLAST search of the sequenced *C. elegans* genome revealed the presence of a putatively orthologous receptor in nematode worms (NP500725; Fig. 1).

C. SULFAKININ RECEPTORS

Sulfakinins typically contain a C-terminal X(E/D)YGHMRFa motif (X represents one or more additional amino acids). When these peptides were purified, the Tyr-residue (underlined) was most of the times but not always sulfated. In any case, reports exist that show that this sulfation can be critical for activity. The first identified insect sulfakinins originated from cockroaches (hence the name "leucosulfakinins") and were identified by their ability to stimulate hindgut contraction (Nachman et al., 1986). These peptides showed sequence similarity to the hormonally active portion of the vertebrate hormones gastrin and cholecystokinin (CCK), suggesting an evolutionary relationship. Since then, other insect and crustacean family members were identified or purified (Duve et al., 1995; Johnsen et al., 2000; Nichols, 1992a). The precursors of these peptides are expressed in neural and gut tissues. Apart from stimulating hindgut contraction, sulfakinins increase the contraction frequency of the heart in the cockroach, *Periplaneta americana* (Predel et al., 2001). In the desert locust, *Schistocerca gregaria*, they significantly inhibit food uptake in fifth-instar nymphs (Wei et al., 2000), an effect that is also seen in the cockroach, *Blatella germanica* (Maestro et al., 2001).

Recently, DSK-R1, a drosulfakinin receptor was discovered. It was expressed in multiple cell lines and could be activated by the sulfated form of drosulfakinin-I (*Drm*-SK-1S) (Kubiak et al., 2002). The unsulfated peptide was approximately 3000 times less effective than its sulfated counterpart. DSK-R1 stimulated the PLC pathway in a pertussis toxin insensitive manner, suggesting functional coupling to G proteins of the $G_{q/11}$ class. The receptor was not responsive to related vertebrate sulfated peptides, such as human CCK-8 and gastrin-II, at concentrations up to 10^{-5} M.

D. RECEPTORS FOR NPF-LIKE AND NPY-LIKE PEPTIDES

In mammals, neuropeptide Y (NPY) and homologous peptides, peptide YY and pancreatic polypeptide (PP), serve as neurotransmitters and gastrointestinal hormones. These peptides exert their effects via a subfamily of GPCRs termed Y-receptors. Purification of NPY-like peptides from invertebrates revealed that these peptides usually contain a Phe- rather than a Tyr- residue at their C-terminus and hence are termed NPFs. In *Drosophila* the gene and the encoded NPF (*Drm*-NPF) were identified in 1999 (Brown *et al.*, 1999), while a putative *Drosophila* NPY-receptor had already been cloned in 1992 (Li *et al.*, 1992). This GPCR, termed NepYr (CG5811), was expressed in *Xenopus* oocytes and demonstrated to be responsive to vertebrate NPY-like peptides in the following order—YY > NPY ≫ PP. Today the placement of NepYr within the Y-receptor family is questioned because of limited specific sequence similarity of NepYr with human Y_1 and Y_2 receptors (Larhammar *et al.*, 1998). In addition, when expressed in CHO-K1 cells, the receptor was not responsive to NPY, PYY, or *Drm*-NPF, and no specific binding to ^{125}I-*Drm*-NPF was evident (Garczynski *et al.*, 2002). Moreover, this receptor shows more similarity with GRL106, a cardioexcitatory peptide (LyCEP) receptor from mollusks (Tensen *et al.*, 1998a), than with mammalian NPY receptors.

A *Drosophila* receptor for the NPY-related peptide *Drm*-NPF (CG1147 or NPFR1) has been cloned and expressed in CHO-K1 cells (Garczynski *et al.*, 2002). This GPCR specifically interacted with ^{125}I-*Drm*-NPF. In addition, it showed agonist concentration–dependent inhibition of forskolin-stimulated adenylyl cyclase activity, a functional feature that is shared by other members of the NPY receptor family. NPFR1 is related to the Y_2 receptor, as well as to an earlier cloned NPY receptor from mollusks (Tensen *et al.*, 1998b). HEK-293 cells expressing *Drm*-NPFR1 displayed β-arrestin2-GFP translocation in response to *Drm*-NPF at micromolar concentrations (Johnson *et al.*, 2003a). Interestingly, NPFR1 is expressed in the CNS and midgut of *Drosophila* larvae, a pattern that is comparable to *Drm*-NPF. As a consequence, similar to vertebrate NPY family members, both the *Drosophila* NPF receptor and its ligand are well positioned for regulating feeding and digestion (Brown *et al.*, 1999; Garczynski *et al.*, 2002). In an elegant series of experiments, Wu *et al.* (2003) demonstrated the influence of *Drm*-NPF and *Drm*-NPFR1 on the foraging and social behavior of *Drosophila* larvae. The feeding activity of wild-type young third-instar larvae decreases as they mature and become more mobile in search for food-free sites that are appropriate for pupation. Transgenic young larvae with ablated NPF- or NPFR1-neurons thereby exhibit food aversion, hyperexcitation in the presence of food, premature onset of bordering (migration to the periphery of food agar plates), and clumping behavior. These are phenotypes that are normally associated with older third-instar larvae. Conversely, overexpression of *Drm*-NPF in older larvae prolonged feeding and suppressed bordering and clumping. Interestingly, also in *C. elegans*, a putative NPY receptor (NPR-1) plays a role in feeding behavior. Two isoforms of this receptor that differ at a single residue occur in the wild. These isoforms are sufficient to account for two different foraging patterns (de Bono and Bargmann, 1998). Thus, although both organisms display different adaptations, social behavior and food response are controlled by a conserved NPY-like system.

While several NPF-like peptides from invertebrates range in size from 36 to 40 amino acids, others are much shorter but display some C-terminal sequence similarity with NPY and NPF (Spittaels *et al.*, 1996). These peptides are designated as "short NPFs" (sNPFs). In *Drosophila*, a gene (CG13968) encoding four sNPFs has been found by *in silico* analysis of the genome data (Vanden Broeck, 2001a). The predicted precursor protein contains two peptides with a conserved LRLRFa C-terminus (*Drm*-sNPF1 and 2), as well as two additional peptides possessing a C-terminal RLRWa sequence (*Drm*-sNPF3 and 4). A *Drosophila* receptor (NPFR76F, CG7395) that can be activated by insect sNPF-like peptides when expressed in *Xenopus* oocytes was first discovered by the group of Evans (Feng *et al.*, 1999; Reale *et al.*, 2000). This GPCR was shown to be equally responsive to all four fruit fly *Drm*-sNPFs when analyzed in CHO-cells that stably express aequorin and $G_\alpha 16$ (Mertens *et al.*, 2002). In contrast to this report, Feng *et al.* (2003) observed that *Drm*-sNPF3 and 4 were considerably less effective and less potent than *Drm*-sNPF1 and 2. Whether this discrepancy reflects a physiologically relevant phenomenon, rather than just differences in sensitivity or G protein–coupling efficiency of the utilized expression and detection systems (i.e., CHO cells versus *Xenopus* oocytes), is uncertain (Reale *et al.*, 2004). In addition, in *Xenopus* oocytes, it was shown that *Drm*-NPF peptides (NPF-A1 and NPF-A2, consisting of 36 and 28 amino acids, respectively) were considerably less effective than sNPFs in activating NPFR76F-mediated inward currents (Feng *et al.*, 2003).

In insects, sNPFs seem to play a role as neurotransmitters or neuromodulators, and they probably function as regulators of feeding and reproduction. Transcripts of the *Drm*-sNPF receptor are present in ovaries, heads, and to a lesser extent in the bodies of adult flies (Feng *et al.*, 2003; Mertens *et al.*, 2002). The presence of the *Drm*-sNPF receptor thus corresponds with presumed sNPF functions. Very recently, the *Drm*-sNPF2 peptide was localized in the embryonic and larval nervous system and adult brain but not in the gut. By utilizing gain-of-function and loss-of-function *Drm-sNPF* mutants, *Drm*-sNPF was shown to control food intake and body size. In contrast to *Drm*-NPF, *Drm*-sNPF overexpression in wandering larvae did not prolong feeding, suggesting that these peptides are involved in different aspects of feeding behavior in *Drosophila* (Lee *et al.*, 2004).

E. ALLATOSTATIN RECEPTORS

In insects, the rate of juvenile hormone (JH) biosynthesis and release appears to be under neuronal and endocrine control. The activity of the *corpora allata* (CA) is regulated by peptides with a stimulatory (allatotropins or AT) or inhibitory (allatostatins or Ast) effect (Horodyski *et al.*, 2001; Stay, 2000; Tobe and Bendena, 1999). Allatotropin receptors still have to be discovered, whereas receptors for each of the three families of insect allatostatins have recently been identified by "reverse physiology" or "reverse pharmacology" approaches.

Allatostatins isolated from the cockroach, *Diploptera punctata*, belong to a family of brain–gut peptides (i.e., the "AstA family" with a characteristic C-terminal "**YXFGLa**" sequence) that is widely conserved in insects, as well as in other classes of arthropods (Bendena *et al.*, 1999; Duve *et al.*, 1997). In a broad range of insect species, these peptides have an inhibitory effect on the contractility of a variety of visceral muscles. In some species, such as cockroaches and related insects, they also inhibit *in vitro* JH biosynthesis and are present in the nerves leading to the CA, as well as in the CA itself. Moreover, *in vivo* injections of *Dippu*-AstA or of pseudopeptide mimetic analogues into mated female cockroaches (*Diploptera punctata*) can inhibit JH biosynthesis and reduce the growth rate of basal oocytes (Garside *et al.*, 2000). A "reverse physiology" approach based on functional expression in *Xenopus* oocytes of DAR-1 (or DAlstR-1), a *Drosophila* GPCR displaying sequence similarities to mammalian galanin receptors, resulted in the purification of an AstA-like peptide (**SRPYSFGLa**) from *Drosophila* head extracts (Birgül *et al.*, 1999). This peptide is part of a larger precursor encoded in the *Drosophila* genome (Vanden Broeck *et al.*, 2000; Lenz *et al.*, 2000; Vanden Broeck, 2001a; Hewes and Taghert, 2001), which displays similarities with pre-pro-AstA precursors of other species (Belles *et al.*, 1999; Donly *et al.*, 1993; Vanden Broeck *et al.*, 1996). In addition, a second AstA receptor, DAR-2 (or DAlstR-2), closely related to the first one, was discovered in the fruit fly (Larsen *et al.*, 2001; Lenz *et al.*, 2000, 2001). Related AstA receptors were functionally characterized in other insect species, such as the American cockroach (Auerswald *et al.*, 2001) and the silkworm (Secher *et al.*, 2001). The genome sequences of the malaria mosquito, the silkworm, and the honeybee contain orthologues of these AstA receptors (as verified by searching the NCBI website for BLAST analysis of these insect genomes [http://www.ncbi.nlm.nih.gov/BLAST/Genome/Insects.html]).

Members of the second allatostatin (AstB) family reduce JH biosynthesis in the CA of some insect species (e.g., crickets; Lorenz *et al.*, 1995). Based on their more general inhibitory effects on visceral muscle motility, they are also referred to as "myoinhibiting peptides" (MIP). Furthermore, in the lepidopteran species, *Manduca sexta* and *B. mori*, AstB/MIP-related peptides exert a prothoracicostatic activity, since they inhibit ecdysteroid synthesis in the prothoracic glands (Hua *et al.*, 1999). Recently, a *Drosophila* AstB-like peptide has been shown to specifically induce the subcellular translocation of a β-arrestin2-GFP fusion protein in HEK-293 cells expressing CG14484, a previously orphan GPCR with sequence similarity to vertebrate bombesin receptors (Johnson *et al.*, 2003a).

In a number of insects belonging to higher insect orders, a third neuropeptide family (AstC) displays allatostatic activities. Its first member was isolated from the tobacco hornworm, *M. sexta* (Kramer *et al.*, 1991). "Flatline", the *Drosophila* AstC orthologue, is a very strong inhibitor of spontaneous muscle contractions (Price *et al.*, 2002). Two *Drosophila* GPCRs, Drostar-1 and Drostar-2, are each activated by AstC-like agonists, as shown by their functional expression in *Xenopus* oocytes in the presence of a G protein–gated inwardly rectifying potassium channel (GIRK) (Kreienkamp *et al.*, 2002). In addition, both receptors specifically induced translocation of a β-arrestin2-GFP fusion protein in HEK-293 cells, when co-expressed with a GPCR kinase (GRK2) and challenged with AstC (Johnson *et al.*, 2003a). These two receptors are related to vertebrate somatostatin and opioid receptors, as well as to the AstA and galanin receptors, and they are more distantly related to the AstB and bombesin receptors. However, their peptide ligand (AstC) does not show very obvious sequence similarities with AstA, AstB, somatostatin-, opioid-, galanin-, or bombesin-like peptides. The fact that the peptide agonists of all these related receptors do not show clear sequence similarities with each other probably indicates that the domains conferring ligand specificity of these GPCRs may have evolved more rapidly than their overall structure and their signaling properties.

F. RECEPTORS FOR PRXa PEPTIDES: PYROKININ/PBAN, CAP2b, AND ETH

Several insect peptides display a C-terminal -PRXamide motif. Among these are members of the pyrokinin/myotropin/PBAN peptide family (-FXPRLa), the periviscerokinin/Cap2b-like peptides (-FPRXa), and the ecdysis-triggering hormones (-PRXa). These insect peptides display sequence similarities to vertebrate neuromedin U (NMU)-related peptides, which also possess a C-terminal -PRX-amide feature (Park *et al.*, 2002).

The first insect pyrokinin (PK) was isolated from the cockroach, *L. maderae*, and showed myostimulatory activity on cockroach hindgut preparations (Holman *et al.*, 1986b). Since then, several pyrokinin/myotropin-like peptides have been discovered in a variety of insect species, including the fruit fly (Choi *et al.*, 2001; Meng *et al.*, 2002; Vanden Broeck, 2001a). In addition to their effect on visceral muscle contractions, these peptides are regulating physiological activities, such as sex pheromone biosynthesis in female lepidopterans (PBAN: pheromone biosynthesis activating neuropeptide), cuticle coloration, and diapause induction in moths and pupariation in flies. Cardioacceleratory peptide 2b (Cap2b) was first isolated from the tobacco hawkmoth, *M. sexta* (Huesmann *et al.*, 1995). It causes an increase in heart rate in *Manduca*, as well as in *Drosophila*. In addition, it has also been implicated in the regulation of fluid secretion by Malpighian tubules in *Drosophila* (Davies *et al.*, 1995). The insect ecdysis-triggering hormones (ETHs), first identified in the epitracheal glands (Inka cells) of *M. sexta* (Zitnan *et al.*, 1996), are important regulatory peptides that initiate the process of ecdysis (i.e., the shedding of the old cuticle at the end of each molt in insects and other arthropods).

Based on the assumption of ligand–receptor co-evolution, the *Drosophila* genome data were searched for GPCRs related to vertebrate NMU receptors. Some of these receptors were indeed shown to be activated by insect -PRXa–containing peptides (Park *et al.*, 2002). When expressed in *Xenopus* oocytes, the receptor protein CG8795 responded to a set of nonoverlapping -PRXa peptides and showed highest sensitivity to *Drm*-PK2 and *Drm*-hugγ (Park *et al.*, 2002), two *Drosophila* neuropeptides that are both encoded in the *hugin* gene (Meng *et al.*, 2002). The same receptor also showed moderate sensitivity to *Drm*-ETH1 and *Drm*-Cap2b-3/PK1. However, due to rapid desensitization in the oocyte expression system, reliable EC_{50} values could not be determined (Park *et al.*, 2002). CG8784, another GPCR related to vertebrate NMU receptors, was only activated at high concentrations (≥ 10 μM) of hugγ and *Drm*-PK2 in the oocyte expression system (Park *et al.*, 2002). Nevertheless, a second research group has characterized CG8784, together with CG8795, as physiologically relevant *Drm*-PK receptors by expressing them in a mammalian (Chinese hamster ovary) cell line (showing concentration–activity curves with EC_{50} values in the nanomolar range) (Rosenkilde *et al.*, 2003). This discovery nicely illustrated the importance of the expression and functional assay systems employed for receptor deorphanization (Rosenkilde *et al.*, 2003). The same authors also found that *Drm*-PK2 is probably crucial during *Drosophila* embryonic development. Posttranscriptional gene silencing of CG8784 by means of the dsRNA-mediated interference technique (RNAi) caused embryonic lethality, while the knockdown of CG8795 transcripts significantly reduced the survival of embryos and first-instar larvae (Rosenkilde *et al.*, 2003).

Two additional receptors, CG14575 and CG9918, showing similarity to vertebrate NMU receptors, have been studied. Whereas CG9918 could be activated by very high levels (≥ 10 μM) of Cap2b-3/*Drm*-PK1 (-FXPRLa), CG14575 appeared to be more ligand selective, since it responded to submicromolar concentrations of Cap2b-1 and Cap2b-2 (both peptides have a C-terminal -FPRXa motif) (Park *et al.*, 2002). The activation of

CG14575 by Cap2b-1 and Cap2b-2 was confirmed by another research group (Iversen et al., 2002a). In addition to these *Drosophila* receptors, GPCRs for PBAN were recently identified in the pheromone glands of the moth *Helicoverpa zea* (Choi et al., 2003) and of the silkmoth *B. mori* (Hull et al., 2004).

The first insect ETH receptors were discovered in the fruit fly. The *Drosophila* genome sequence database harbors a gene sequence (CG5911) encoding two receptor subtypes that are generated through alternative splicing (CG5911a and CG5911b). Both receptors preferentially respond to ETH peptides (Iversen et al., 2002b; Park et al., 2003). *Drm*-ETH1 was the most potent agonist of these receptors (Iversen et al., 2002b). The EC_{50} values (see Table I) correspond well with the physiological ETH concentrations determined in the hemolymph (Zitnan et al., 1996). The characterization of these *Drosophila* ETH receptors provides the opportunity to clone the orthologues from other research model species, as well as from important pest insects or disease vectors. The malaria mosquito, *A. gambiae*, contains a receptor closely related to the *Drosophila* ETH receptors (Hill et al., 2002; Iversen et al., 2002b), and other orthologous genes are present in the recently sequenced insect genomes (BLAST searches of *Apis* and *Bombyx* genomes).

G. PROCTOLIN RECEPTOR

Proctolin (RYLPT) was the first insect neuropeptide to be sequenced (Starratt and Brown, 1975). Immunolocalization studies suggested a role for this myostimulatory peptide as a neuromodulator in motoneurons and interneurons or as a neurohormone that could be released from neurosecretory terminals (Lange, 2002; Nässel and O'Shea, 1987). The biosynthetic precursor (Proct, CG7105) was recently discovered in the genome of *D. melanogaster*. It was shown to contain a single copy of the pentapeptide sequence (Taylor et al., 2004). Moreover, two different research groups have recently identified the previously orphan GPCR, CG6986, as a functional receptor that can be activated by the pentapeptide in a physiological concentration range (Egerod et al., 2003a; Johnson et al., 2003b). In agreement with these findings, this receptor protein appears to be localized in tissues that were shown to be responsive to proctolin (Johnson et al., 2003b). Interestingly, clear orthologues of this GPCR seem to be absent in the current genome databases of other insect species (mosquito, bee, and silkworm), indicating that it is probably not highly conserved during evolution. However, since the peptide agonist is well conserved in arthropods, this observation probably suggests that other RYLPT-receptor types may exist. The *Drosophila* RYLPT receptor (CG6986) is distantly related to the orphan GPCR, CG16726, and to CG2114 (FMRFaR). The current knowledge of the biosynthesis, receptor interaction, and metabolic inactivation of RYLPT was recently reviewed by Isaac et al. (2004).

H. RECEPTORS FOR FMRFa/FLRFa-LIKE PEPTIDES

The *Drosophila* FMRFa gene codes for a neuropeptide precursor that contains multiple, putative peptide sequences, some of which have been fully confirmed by peptide purification and amino acid sequencing (Nambu *et al.*, 1988; Schneider and Taghert, 1988). Whether all these peptides are functionally redundant and they are all expressed in the same cells is still a matter of discussion (Merte and Nichols, 2002; Taghert, 1999). Several studies in mollusks and vertebrates indicate that FMRFa-like peptides can directly activate ligand-gated ion channels, while other activities appear to be mediated via GPCRs (Green *et al.*, 1994; Jeziorski *et al.*, 2000; Lingueglia *et al.*, 1995; Nishimura *et al.*, 2000). *Drm*-FMRFa peptides can elicit bioluminescent responses in coelenterazine-loaded CHO cells expressing CG2114, a GPCR predicted from *Drosophila* genome sequence data, together with the promiscuous $G\alpha_{16}$ subunit and apoaequorin (Cazzamali and Grimmelikhuijzen, 2002; Meeusen *et al.*, 2002). In addition, a FMRFa-like peptide stimulates translocation of a β-arrestin2-GFP fusion protein toward the plasma membrane in HEK-293 cells expressing CG2114 (Johnson *et al.*, 2003a). The transcript for this GPCR is widely expressed in a variety of tissues and developmental stages of the fruit fly.

Myosuppressins (or FLRFa) display sequence similarity with FMRFa-like peptides and produce myoinhibitory effects on various visceral muscle preparations (Nichols, 2003). Dromyosuppressin (DMS) or TDVDHVFLRFa only differs from cockroach and locust myosuppressins by the N-terminal amino acid residue (Nichols, 1992b). Dromyosuppressin is not derived from the FMRFa gene described earlier, but a single copy of the peptide is present at the C-terminus of CG6440, a relatively short precursor polypeptide (Vanden Broeck, 2001a). Two specific DMS receptors, DMS-R1 and DMS-R2 (CG8985, CG13803), have been identified by reverse pharmacological approaches (Egerod *et al.*, 2003b; Johnson *et al.*, 2003a). When expressed in HEK-293 cells, both receptors attenuate forskolin-induced cyclic AMP accumulation if they are challenged with DMS (Johnson *et al.*, 2003a). In addition, they induce arrestin translocation in these cells (Johnson *et al.*, 2003a) and mediate agonist-dependent luminescent responses in CHO cells expressing $G\alpha_{16}$ and apoaequorin (Egerod *et al.*, 2003b).

Together with a number of orphans, the insect FMRFa, myosuppressin, and proctolin receptors appear to form a subgroup of GPCRs with closer relatives in the nematode worm than in the chordate–vertebrate lineage (Fig. 1).

I. AKH, CRZ, AND CCAP RECEPTORS

During their search for *Drosophila* peptide receptors in the database of the "Berkeley *Drosophila* Genome Project" (BDGP), Hauser *et al.* (1998) discovered a gene coding for a GPCR that displayed clear sequence similarity with vertebrate gonadotropin-releasing hormone receptors (GnRHR). Once the

Drosophila genome project was completed, several groups have predicted the presence of approximately 40 putative neuropeptide receptors belonging to the rhodopsin family (Brody and Cravchik, 2000; Hewes and Taghert, 2001; Vanden Broeck, 2001b). This analysis yielded three GPCRs (CG6111, CG10698, and CG11325) related to vertebrate receptors for GnRH or vasopressin-related peptides. Although the presence of an insect vasopressin-like peptide had been reported in locusts (Proux *et al.*, 1987), searches in the *Drosophila* and *Anopheles* genomes did not reveal sequences similar to these vertebrate peptides in two higher dipteran species (Hewes and Taghert, 2001; Riehle *et al.*, 2002; Park *et al.*, 2002; Vanden Broeck, 2001a). Therefore, other strategies had to be explored to identify the endogenous ligands of these orphan receptors.

1. Adipokinetic Hormone Receptor

The *Drosophila* receptor (CG11325) resembling vertebrate GnRHR was further expressed in a mammalian cell line, and its peptide agonist was chromatographically purified from homogenates of third-instar larvae ("orphan receptor strategy"). This ligand turned out to be identical to *Drosophila* adipokinetic hormone (*Drm*-AKH; pQLTFSPDWa) (Staubli *et al.* 2002). This finding was confirmed by another group that also discovered that CG11325 can be specifically activated by *Drm*-AKH (Park *et al.*, 2002). The silkworm orthologue of the *Drosophila* AKHR was characterized by means of a homology cloning strategy in combination with a "reverse pharmacology" approach. This *Bom*-AKHR could be activated by a moth AKH-like peptide, *Helicoverpa zea* hypertrehalosaemic hormone, at low concentrations (Staubli *et al.*, 2002).

Adipokinetic hormones were identified several decades ago (Stone *et al.*, 1976). They are present in intrinsic glandular cells of the *corpora cardiaca* and play a crucial role in the regulation of energy metabolism by stimulating the mobilization from the fat body of energy substrates, such as carbohydrates and lipids, during flight and locomotion. In addition, they display stimulatory effects on visceral muscle contractility (Scarborough *et al.*, 1984). Adipokinetic hormones belong to a conserved family of arthropod peptides (Gäde, 1996; Gäde *et al.*, 1997). They are structurally related to red pigment–concentrating hormone (RPCH) from crustaceans (Fernlund and Josefsson, 1972; Mordue and Stone, 1977; Stone *et al.*, 1976) and are believed to be distantly related to the cardiostimulatory peptide, corazonin (Veenstra, 1994). The fruit fly AKH-related peptide (*Drm*-AKH) is an eight amino acid member of this family, and its precursor gene shows the same organization as in other insect species (Noyes *et al.*, 1995; Schaffer *et al.*, 1990).

2. Corazonin Receptor

A similar reverse pharmacology strategy also resulted in deorphanization of the second GnRH-like receptor (CG10698). Several groups have identified the endogenous ligand for this receptor as corazonin (CRZ;

pQTFQYSRGWTa) (Cazzamali et al., 2002; Johnson et al., 2003a; Park et al., 2002). The receptor is expressed from embryonic (8 h) to adult stages. In adult flies, both head and body regions contain the receptor transcript. The corazonin and AKH receptor genes appear to be structurally related and have three introns at the same position, with the same intron phasing (Cazzamali et al., 2002). Corazonin and AKHs display some structural similarity (Veenstra, 1989), and their precursor genes have a similar organization (Veenstra, 1994). The fact that both peptides also have similar receptor genes supports the hypotheses that they probably have a common evolutionary origin and that the corresponding ligand–receptor couples have co-evolved in arthropods (Cazzamali et al., 2002; Park et al., 2002).

A CRZ-receptor orthologue (*Mas*CRZR) was recently identified in the tobacco hornworm, *M. sexta* (Kim et al., 2004). High levels of *Mas*CRZR transcript were detected in the epitracheal endocrine glands (Inka cells), which were shown to release pre-ETH (PETH) and ETH in response to CRZ exposure. Corazonin is best known for its cardioactive activity in cockroaches (Veenstra, 1989) and pigment modulation in locusts (Tawfik et al., 1999), but these data, combined with evidence that CRZ circulates in the hemolymph just before pre-ecdysis onset, revealed a new function in the initiation of ecdysis behavior in *Manduca* (Kim et al., 2004).

3. Crustacean Cardioactive Peptide Receptor

CG6111 displays sequence similarity to vertebrate vasopressin and molluskan conopressin receptors and is distantly related to vertebrate GnRHR, as well as to *Drosophila* AKH and CRZ receptors. This receptor was found to be highly specific for crustacean cardioactive peptide (CCAP; PFCNAFTGCa) (Cazzamali et al., 2003). Crustacean cardioactive peptide was first discovered in crustaceans, but it also occurs in identical form in several insect species (Furuya et al., 1993; Lehman et al., 1993; Stangier et al., 1987). Cazzamali et al. (2003) discovered that the ORF of the receptor gene had not been correctly predicted, and they corrected its cDNA sequence. High concentrations of CCAP partly inactivate the receptor, probably due to desensitization. Park et al. (2002) had also identified CG6111 as a putative *Drosophila* CCAP receptor. However, there were major differences between both studies. The much lower affinity (i.e., higher EC_{50}) found by Park et al. (2002) was probably due to the fact that the last exon was missing, corresponding to a portion of the C-terminus of the receptor protein (Cazzamali et al., 2003).

Crustacean cardioactive peptide was initially characterized based on its cardioacceleratory activity on the heart of the shore crab, *Carcinus maenas* (Stangier et al., 1987). The primary structure appears to be strictly conserved in a variety of arthropod species. In addition to its myotropic activities (Nichols et al., 1999; Tublitz and Truman, 1985), the peptide induces

AKH release from *corpora cardiaca* in locusts (Veelaert *et al.*, 1997) and plays a role in the induction of ecdysis behavior (Ewer and Truman, 1996; Gammie and Truman, 1999; Phlippen *et al.*, 2000).

Crustacean cardioactive peptide is a remarkable peptide, since its structure has remained unchanged during the last 500 million years of evolution (since the divergence of insects and crustaceans) (Budd *et al.*, 2001). This would suggest that its receptor may also have remained relatively invariable, and it should be feasible to clone related CCAP receptors in other arthropod species.

J. "LEU-RICH REPEAT-CONTAINING GPCRS" AND BURSICON

The leucine-rich repeat-containing GPCRs (LGRs) form a unique subgroup of receptors that possess a typical, large N-terminal ectodomain involved in ligand recognition. The first insect LGR (fly LGR1, *Fsh*) was cloned by Hauser *et al.* (1997) from the fruit fly *D. melanogaster* and shown to be evolutionarily related to mammalian glycoprotein hormone (FSH/LH/CG/TSH) receptors (type A LGR subclass). The expression of this receptor already starts 8–16 h after oviposition and stays high until after pupation. In adult male flies, the receptor transcript is about six times more highly expressed than in adult females (Hauser *et al.*, 1997). A second *Drosophila* LGR (fly LGR2, *rk*) exhibits structural homology with the subgroup of vertebrate orphan LGRs 4, 5, and 6 (type B LGR subclass) (Eriksen *et al.*, 2000, Nishi *et al.*, 2000) and is mainly expressed in embryos and pupae (Eriksen *et al.*, 2000). Two additional LGRs appear to be encoded in the fruit fly genome, CG5042 and CG4187 (Hewes and Taghert, 2001). These receptors belong to a third subgroup of LGRs (type C LGR subclass), together with the mammalian LGRs 7 and 8, that are activated by relaxin-like peptides (Hsu *et al.*, 2002). Orthologous receptors showing homology with each of the three LGR subclasses have recently been identified in the genome of the malaria mosquito, *A. gambiae* (Hill *et al.*, 2002).

At present, none of these receptors has been fully deorphanized (i.e., the endogenous ligands for insect LGRs remain unknown). However, the *Drosophila* genome encodes proteins showing some sequence similarity to the vertebrate glycoprotein hormone subunits (Vitt *et al.*, 2001), as well as several members of the insulin superfamily (which includes relaxin-like peptides) (Claeys *et al.*, 2002). Based on the phenotype of *Drosophila rickets* mutants, the fly LGR2 was recently suggested to be the putative receptor for the neurohormone bursicon (Baker and Truman, 2002). Flies with a defective LGR2 signaling display defective tanning of their cuticle (Baker and Truman, 2002) as well as abnormal wing spreading after adult eclosion (Kimura *et al.*, 2004). Both processes are known to be triggered by bursicon. Recently, partial peptide sequence information of cockroach bursicon led to the identification of a *Drosophila* gene that appears to be required for these processes, as shown by genetic analyses (Dewey *et al.*, 2004). Interestingly, the putative bursicon protein appears to be very well conserved in other insect genomes (as verified by *in silico* analysis of the sequence databases) and is a cystine knot family member, distantly related to vertebrate glycoprotein hormones, as well as to a large variety of TGF-β receptor antagonists (Vitt *et al.*, 2001).

III. MEMBERS OF OTHER 7TM FAMILIES

A. SECRETIN RECEPTOR (GPCR CLASS B) FAMILY

1. Insect Diuretic Hormone Receptors

In insects, the precise control of salt and water balance is crucial for their survival. This process is regulated by corticotrophin-releasing factor (CRF)-related diuretic hormones (DH), which stimulate the rate of fluid secretion by the Malpighian tubules in a cAMP-dependent manner (Kay *et al.*, 1991). Diuretic hormones are expressed throughout the central nervous system and gut, often within identified neuroendocrine neurons (Coast, 1996). The first insect DH receptor was characterized in the tobacco hornworm, *M. sexta*, by expression cloning in COS-7 cells (Reagan, 1994). This receptor binds *Mas*-DH as well as several other insect CRF-related diuretic peptides, with high affinity and functionally coupled to a cAMP second messenger system (see Table II). The *M. sexta* DH receptor is the first identified insect member of a separate GPCR family, the "secretin receptor" family (also referred to as "family 2" or "B-family"). This receptor family is composed of a multitude of vertebrate receptors for peptide hormones, such as calcitonin, glucagon, parathyroid hormone, secretin, and vasoactive intestinal peptide. Reagan (1996) has also cloned the cricket, *Acheta domesticus*, DH receptor by using degenerate oligonucleotides based on conserved regions in the *M. sexta* DH receptor and the vertebrate CRF receptors. When expressed in COS-7 cells, the *A. domesticus* DH receptor binds *Acd*-DH with high affinity and stimulates adenylate cyclase activity with high potency, in agreement with the *in vivo* activity of this peptide in Malpighian tubules (Coast, 1996). Another related receptor was cloned from Malpighian tubules of the silkmoth, *B. mori* (Ha *et al.*, 2000), but its functional activity has not been described yet.

The *Drosophila* genome contains at least five gene sequences coding for proteins that are very similar to vertebrate B-family receptors for peptide hormones, in particular to calcitonin and CRF receptors. The gene products corresponding to these predicted receptors are—CG13758, CG17415, CG4395, CG8422, and CG12370 (Hewes and Taghert, 2001, Vanden Broeck, 2001b). The first three are most similar to vertebrate calcitonin receptors, whereas the last two are orthologous to *Manduca* and *Acheta*

DH receptors, as well as to vertebrate CRF receptors. Recently, CG8422 has indeed been characterized as a functional DH receptor in *Drosophila* (Johnson et al., 2004). *Drm*-DH stimulated cAMP synthesis in CG8422 expressing cells. At substantially higher doses of *Drm*-DH, CG8422 activation also led to an increase of intracellular Ca^{2+} ions. Furthermore, microarray analysis showed the presence of CG8422 transcripts in *Drosophila* heads. The CG12370 paralogue displays 59% identity with CG8422 in its transmembrane domains. However, this receptor is still to be considered as an orphan GPCR (Johnson et al., 2004).

In addition to the previously mentioned receptors, the *Drosophila* GPCR-B family also contains a GPCR related to the latrophilin receptor (CG8639), two GPCRs (CG11318 and CG15556) that are distantly related to the human HE6-receptor, and a large subset of GPCRs related to "Methuselah" (Mth), a *Drosophila* receptor shown to play a role in the control of longevity (Vanden Broeck, 2001b).

2. Methuselah-Like Receptor Subfamily

The *mth* gene codes for Methuselah, a *Drosophila* GPCR that plays an important role in stress response and biological ageing. Mutant flies, with a heterozygous mutation in the *mth* gene, display an increase in average life span and enhanced resistance to several forms of stress, such as starvation, high temperature, and oxidative damage (Lin et al., 1998). Mutant flies have a defective synaptic transmission, suggesting that Mth and its natural ligand(s) may be part of a nervous control system for stress responses and ageing (Song et al., 2002). Methuselah is a member of the secretin receptor family, which has a large N-terminal ectodomain that constitutes the putative ligand-binding site. Moreover, the *Drosophila* genome codes for approximately 12 Mth-like orphan receptors (Brody and Cravchik, 2000; Vanden Broeck, 2001b). West et al. (2001) analyzed the crystal structure of the ectodomain of Mth in relation to predicted Mth homologues and potential ligand-binding features. Recently, Stunted A and B (Sun A and Sun B), the first endogenous peptide ligands of Mth, were purified and characterized via a "reverse physiology" approach (Cvejic et al., 2004). Both ligands specifically induce a rise of intracellular Ca^{2+} ions in HEK cells expressing Mth or its splicing variant MthB. Binding of Sun A or Sun B to Mth was confirmed by competition with a labeled N-terminal peptide fragment of Sun. Flies with mutations in the gene encoding these ligands (*sun*), also show an increase in life span and resistance to oxidative stress (Cvejic et al., 2004).

B. THE FRIZZLED 7TM RECEPTOR FAMILY

The frizzled-like proteins (Fz) are a small group of receptors that have the typical structure of a GPCR but show little sequence identity with members of other GPCR families. Fz proteins have a large extracellular N-terminus containing a conserved cysteine-rich domain, followed by a hydrophobic linker, 7TM regions, and a short cytoplasmatic C-terminus (Wodarz and Nusse, 1998). Genetic investigations have shown the involvement of the frizzled (fz) gene in the establishment or maintenance of planar polarity (Gubb and Garcia-Bellido, 1982). Fz proteins act as receptors for Wingless (Wg) and other signaling molecules of the Wnt gene family. The first identification came from the discovery that *Drosophila* S2 cells transfected with Dfz2 respond to added Wingless protein by elevating the level of the Armadillo protein. Moreover, Dfz2 confers cell-surface binding of Wg in both homologous (S2) and heterologous (HEK-293) cell systems. Fz itself was also shown to confer Wg-binding activity (Bhanot *et al.*, 1996). Although these binding studies are consistent with a function for Fz proteins as Wnt receptors, they do not strictly prove a physiological ligand–receptor relationship. Several genetic studies provide additional evidence for this. Overexpression of either Fz or Fz2 in *Drosophila* causes phenotypes associated with ectopic Wg signaling (Cadigan and Nusse, 1997; Tomlinson *et al.*, 1997; Zhang and Carthew, 1998). Elimination of both Fz and Fz2 activity causes an absolute loss of Wg transduction in all of the Wg signaling processes that were assessed. In contrast, the absence of either Fz or Fz2 activity alone has little if any detectable effect on Wg signaling. This shows that Fz and Fz2 are functionally redundant—the presence of either protein is sufficient to confer Wg transducing activity on most or all cells throughout development (Chen and Struhl, 1999). Sivasankaran *et al.* (2000) identified a third member of the *Drosophila* Frizzled family, Dfz3. Dfz-3 is also a target gene of Wg signaling, but, in contrast to Dfz2, it is activated rather than repressed by Wg signaling in imaginal discs. Experiments have shown that this Dfz3 is not required for viability but is necessary for optimal Wg signaling at the wing margin.

Another member of the Fz-related GPCR family is Smoothened (Smo). Experiments with Smo mutant embryos indicate a role for Smo in Hedgehog (Hh) signaling, which is important in many developmental processes (Alcedo *et al.*, 1996). Hh is bound and transduced by a receptor complex that includes two multispan transmembrane proteins Patched (Ptc) and Smo. In the absence of Hh, Smo activity is blocked by Ptc, whereas the reception of a Hh signal overcomes Ptc inhibition of Smo (Chen and Struhl 1996; Stone *et al.*, 1996), allowing Smo to signal downstream. The Smo C-terminal tail appears essential for its function. Moreover, Smo transduces Hh signal by physically interacting with Costal2 (Cos2) and Fused (Fu) through its C-tail (Jia *et al.*, 2003). Genome sequence data predict the presence of a few additional genes coding for members of the Fz/Smo 7TM family in insects (Janson *et al.*, 2001; Vanden Broeck, 2001b).

IV. SINGLE TRANSMEMBRANE RECEPTORS

This highly diverse and heterogenous category of signal-transducing receptors is characterized by the presence of only a single transmembrane segment. Most of these receptors have an extracellular domain responsible for ligand binding and a large cytoplasmic part that possesses an intrinsic enzyme activity, as well as interaction sites for cellular effector proteins, which mediate downstream signaling events. The activation mechanism depends on the receptor class but usually requires a receptor dimerization or oligomerization step initiated by interaction with the extracellular ligand. In this review, the focus is on three evolutionary conserved families of single TM receptors, which are known to mediate the cellular responses to a variety of peptide hormones or paracrines in Metazoans: (1) receptor tyrosine kinase (RTK), (2) TGF-β receptor, and (3) receptor guanylyl cyclase (rGC) families. However, it is worth mentioning that several other membrane receptor families do occur in insects. For instance, receptor tyrosine phosphatases, which are probably activated by cell–cell adhesion events, regulate axon guidance and synaptogenesis in the nervous system (Schindelholz *et al.*, 2001). Another example of a receptor involved in cell–cell contact signaling is notch, which initiates a highly conserved signaling pathway that determines cell fate and regulates pattern formation in animals (Lai, 2004). Insects also possess several receptors that resemble vertebrate cytokine receptors and function in early development and/or in the innate immune system (e.g., Toll [Hoffmann, 2003], Wengen [Kanda *et al.*, 2002], and Domeless/Mom [Boulay *et al.*, 2003; Brown *et al.*, 2001]).

A. RECEPTOR TYROSINE KINASES

1. General Properties

Many protein hormones and growth factors regulate cellular processes through the activation of single transmembrane segment containing proteins, such as receptor tyrosine kinases (RTKs). The RTKs form a large family of integral membrane proteins with highly divergent extracellular domains coupled to a conserved intracellular protein tyrosine kinase motif. This enzyme activity can control a diversity of cellular processes, including differentiation, migration, growth, and viability (Van Der Geer *et al.*, 1994). The transmembrane segment anchors the receptor in the plasma membrane, while the extracellular domains interact with protein hormones or growth factors. Characteristically, the extracellular domains possess one or more identifiable structural motifs, such as cysteine-rich regions, fibronectin III-like domains, immunoglobulin-like domains, EGF-like domains, cadherin-like domains, kringle-like domains, factor VIII-like domains, glycine-rich regions, leucine-rich regions, acidic regions and discoidin-like domains.

The binding of extracellular ligands initiates a complex machinery of downstream regulators, which finally can lead to changes in gene expression by the regulation of transcription factors. Detailed analyses of signal transduction processes initiated from activated RTKs have led to the identification of a number of key regulators that are highly conserved components of RTK signaling pathways. The strong evolutionary conservation of the RTK pathways implies the usefulness of small invertebrate model organisms, such as *D. melanogaster* and other insect species, to further unravel the mode of action of the different mediators, as well as their signaling network interactions. Table IIIA lists different insect members of the RTK family.

Different signaling pathways can be activated by receptor tyrosine kinases. These activation events include the phosphatidylinositol-3-OH-kinase (PI3K), 70-kDa S6 kinase (S6K), mitogen-activated protein kinase (MAPK), phospholipase C-γ, and the JAK-STAT pathways. The MAPK pathway is well studied, and a combination of genetic (e.g., with *Drosophila* mutants) and biochemical data has led to the following model of RTK signal transduction via MAPK signaling.

Upon activation by its ligand, the RTK dimer (monomer in inactivated state) becomes autophosphorylated at specific tyrosyl residues. The adaptor protein, growth factor receptor bound protein 2 (Grb2; in *Drosophila* also designated as "downstream of receptor kinase" [Drk]) then binds the activated receptor via its Src-homology-2 (SH2) domain, while it can associate through its Src-homology-3 (SH3) domain with "Son-of-Sevenless" (Sos), a ubiquitously expressed guanine nucleotide exchange factor of Ras, a small G protein (Olivier *et al.*, 1993; Simon *et al.*, 1993). The formation of the Drk:Sos complex results in the relocalization of Sos to the plasma membrane where it promotes the exchange of GDP to GTP on Ras (Gaul *et al.*, 1992), thereby inducing a conformational change that activates Ras. Ras activation leads to the activation of a protein kinase cascade that includes Raf, a serine/threonine-specific protein kinase, MAPKK (MEK) a dual serine tyrosine/threonine kinase, and the serine/threonine kinase MAPK (ERK). Once activated, MAPK homodimerizes and imported into the nucleus (Fukuda *et al.*, 1997; Khokhlatchev *et al.*, 1998) where it can phosphorylate nuclear target proteins that regulate transcription, the ultimate goal of the signaling pathway.

The *Drosophila* FGF receptor homologue, Breathless (Shishido *et al.*, 1993), shares this downstream effector pathway with Torso (Sprenger *et al.*, 1989), Sevenless (Basler and Hafen, 1988), and EGF-R/Torpedo (Livneh *et al.*, 1985). Thus, the "all purpose" Ras–Raf pathway is induced by each of these receptors. This raises the question—"How are specific downstream targets regulated when the signaling pathway is redundant?" Marshall *et al.* (1995) described a model for the possible regulation of different cellular responses via RTK signaling, emphasizing the critical importance of the cellular context in understanding the diversity of signaling processes. These authors propose that when cells make decisions about proliferation versus differentiation through RTK signaling, they do this by differences in the duration of MAPK/ERK activation. Such considerations highlight the importance in development of restricted expression of receptors and, especially, of their corresponding ligands.

2. Signaling Through the Insulin Receptor

a. *The Insulin Receptor*

As stated earlier, RTKs mediate the effects of extracellular signals through well conserved signaling processes. This is also the case for the insulin receptor (IR), a member of the RTK family (see Table IIIA). Unlike other RTKs, members of the IR subfamily are composed of disulfide-linked polypeptide chains forming an $\alpha_2\beta_2$ receptor complex. Insulin binding to the extracellular domain of the IR induces a rearrangement in the quaternary heterotetrameric structure that leads to increased autophosphorylation of the cytoplasmic domain. The same characteristic features are also present in the IGF receptor. Interestingly, IGF 2 also appears to interact specifically with a cation-independent mannose-6-phosphate receptor type.

Several IRs have already been identified in insects. The *Drosophila* IR (DIR or DInR) displays similarities with human insulin and insulin-like growth factor (IGF) receptors. Nevertheless, it is expected to be a larger protein, since it has approximately 300 additional amino acids at the NH$_2$- and C-termini (Ruan *et al.*, 1995). The overlapping portions of the α- and β-subunits between DIR and human IR show approximately 30 and 40% sequence identity, respectively. *Drosophila* IR is capable of binding human insulin with lower affinity than human IR, but its signaling capacity is similar, suggesting that essential functions of the β-subunit are conserved from insects to mammals (Fernandez *et al.*, 1995; Yamaguchi *et al.*, 1995).

Fernandez *et al.* (1995) reported that the processing of the *Drosophila* proreceptor is somewhat different from that of mammalian insulin and insulin-like growth factor receptor precursors. The DIR β-subunit contains an extra domain at the carboxyterminal side of the tyrosine kinase, in the form of a 60-kDA extension. In some *D. melanogaster* cell types, the C-terminal extension is cleaved from the receptor, resulting in the existence of two DIR isoforms. As a result, one isoform resembles mammalian receptors, while the other contains a 368 amino acid C-terminal extension. DAF-2, the insulin receptor homologue of *C. elegans*, also contains a long C-terminal extension that may function analogously to mammalian insulin receptor substrate proteins (IRS) (Kimura *et al.*, 1997). Besides DIR, a few other insect insulin receptors were characterized. Graf *et al.* (1997) cloned the mosquito (*Aedes aegypti*) insulin receptor (MIR) homologue from ovarian mRNA. In addition, an ovarian insulin receptor has also been cloned from the silkmoth, *B. mori* (Lindstrom-Dinnetz and Iatrou, unpublished data, database accession number AF025542). Unraveling the genome

TABLE III. Functionally Characterized Members of the RTK and TGFβ Receptor Families in Insects

Receptor type	Aliases	Species	Synonyms	Accession	Agonist	Reference
A. Receptor tyrosine kinase						
Fibroblast growth factor receptor homolog 1, fgf-r homolog 1: Heartless	DFR-1 DFGF-R2	D. melanogaster	Fibroblast growth factor receptor homolog 1 precursor (Heartless protein)	Q07407 A49120	Thisbe (ths, FGF8-like 1) Pyramus (pyr, FGF8-like 2)	Gryzik and Muller (2004); Shishido et al. (1993)
Fibroblast growth factor receptor homolog 2, fgf-r homolog 2: Breathless	DFGFR-1; Dtk2, fgf-r, DFR2;	D. melanogaster	Fibroblast growth factor receptor homolog 2 precursor (Breathless protein) (dFGF-R1)	Q09147 A44065 CAA51340	Branchless	Shishido et al. (1993); Sutherland et al. (1996)
Eph receptor tyrosine kinase;		D. melanogaster	D. melanogaster Eph receptor tyrosine kinase (dek) mRNA, complete cds	AF132028 AAD38508 AAC99308 AAD30170 T13039 AAM48432	Ephrin	Scully et al. (1999)
Epidermal growth factor receptor	DER, torpedo, ellipse	D. melanogaster	Epidermal growth factor receptor precursor (Egfr) (Gurken receptor) (Torpedo protein) (Drosophila relative of ERBB).	P04412 AAD26134	Gurken, Spitz, Argos	Livneh et al. (1985)
			Epidermal growth factor receptor isoform II	AAC08535		Clifford and Schupbach (1994)
Insulin-like receptor	DIR; InR	D. melanogaster	Insulin receptor	P09208 A56081 AAC47458 S57245 AAA28644 AAA28645	Dilp 1–7	Fernandez et al. (1995)
		B. mori	Insulin receptor-like protein precursor	AAF21243	Bombyxin	Lindstrom-Dinnetz and Iatrou, unpublished

PDGF- and VEGF-receptor related	Pvr	*D. melanogaster*	PDGF/VEGF receptor	CAC39205 AAL91092 AAL91089 AAL91090	PDGF/VEGF factor 1 (pvf1)	Heino et al. (2001)
Sevenless-Boss receptor	sev	*D. melanogaster*	Sevenless protein	P13368 TVFF7L CAB55310	BOSS	Basler and Hafen (1988)
Torso	tor	*D. melanogaster*	Putative receptor tyrosine kinase	CAA33247	Torso-like Thrunk	Sprenger et al. (1989)
B. Serine/threonine kinase receptor						
Baboon	Atr-1 *Babo*	*D. melanogaster*	CG8224-PA RE55648p Serine/threonine kinase Atr-1 CG8224-PB, isoform B	NP_477000 AAQ22445 A55921 AAM71094		Brummel et al. (1999)
Punt	put	*D. melanogaster*	Activin receptor II precursor	A48678	Decapentaplegic-like ligands: Dpp, Screw, Gbb (60A)	Childs et al. (1993)
Saxophone		*D. melanogaster*	Dpp receptor SAX precursor	I45712 2017214B AAA18208 A55921	Decapentaplegic-like ligands: Dpp, Screw, Gbb (60A)	Brummel et al. (1994)
Thick veins	Dtfr, Tkv	*D. melanogaster*	DTFR	BAA06330 AAA61948 NP_787991 NP_787989 NP_787990	Decapentaplegic-like ligands: Dpp, Screw, Gbb (60A)	Okano et al. (1994)
Wishful thinking	wit	*D. melanogaster*	Wishful thinking	AAL16073 AAF60175		Aberle et al. (2002)

sequence of several insects has also led to the annotation of possible insulin-like peptide receptors in *A. gambiae* and *Apis mellifera*. The characterized insect insulin receptors are listed together with other RTK in Table IIIA.

Multiple alignment of the amino acid sequences of insulin (or insulin-like peptide) receptors of different organisms ranging from lower invertebrates to humans, reveals a strong sequence conservation, particularly in receptor parts that are responsible for ATP-binding and kinase activity.

b. The Conserved Insulin-Signaling Pathway

Upon binding of insulin to its receptor, the IR β-subunits undergo autophosphorylation, due to the activation of their intrinsic tyrosine kinase activity. Unlike other receptor tyrosine kinases, the insulin receptor acts through a family of soluble scaffolding adaptor molecules, known as insulin receptor substrates (IRS-1–4), to initiate its signaling program (White *et al.*, 1998). Interaction of IRS with the tyrosine-phosphorylated cytoplasmic tail of IR is facilitated by the cytoskeleton (Whitehead *et al.*, 2000) and results in phosphorylation of the IRS molecule at specific tyrosine-residues (Yenush *et al.*, 1996). Activated IRS mediates the insulin signal toward downstream molecules. Numerous tyrosine phosphorylation sites on IRS act as docking sites for SH_2 domain-containing molecules (White *et al.*, 1998). Two very important SH_2-domain proteins can interact with IRS, namely Grb2 and PI3K. Each of these proteins can initiate a specific pathway—the MAPK and the PI3K-PKB (protein kinase B) pathway, respectively (Blenis, 1993; Shepherd *et al.*, 1998). Gbr2 recruitment results in the activation of the Ras pathway, as described earlier (see Section IV.A.1.). Following tyrosine phosphorylation of IRS on key sites, PI3K (p110) can be recruited (via the p85 SH_2 adaptor subunit of PI3K). The cytoskeleton targets the activated IRS-PI3K complex to specific microdomains within the plasma membrane. Phosphatidyl-inositol-(3,4,5)-trisphosphate (PIP_3) is generated at these specific sites, serving to recruit phosphoinositide-dependent protein kinases (PDK 1 and PDK 2) and downstream kinases, such as Akt 2 (PKB) and PKC. Phosphoinositide-dependent protein kinase mediates their activation by phosphorylation (Shepherd *et al.*, 1998). Activated Akt/PKB and PKC each can phosphorylate downstream effectors.

Orthologues of components involved in the mammalian insulin or IGF signaling pathways have also been identified in *D. melanogaster* (see Fig. 2). In addition to the insulin receptor (DIR) (Ruan *et al.*, 1995), fruit fly homologues of IRS (CHICO) (Böhni *et al.*, 1999; Kunlansky Poltilove *et al.*, 2000), Grb2 (Drk) (Raabe *et al.*, 1995), MEK (DSORT), ERK (ERK-A) (Biggs *et al.*, 1994), PDK (Clyde and Bownes, 2000; MacDougall *et al.*, 1999), Sos (Park *et al.*, 1998), PI3K (Dp110) (Leevers *et al.*, 1996) and Akt/PKB (Stavely *et al.*, 1998) have been discovered. However, the existence of structural differences between DIR and mammalian insulin receptors raised the possibility of a major difference between mammalian and insect signaling pathways. The DIR-tail shows similarity to a region in IRS-1 that is required for insulin-stimulated DNA-synthesis. Moreover, this DIR-tail contains three YXXM-motifs (Tyr-1941, Tyr-1957, and Tyr-1978), which align with tyrosine phosphorylation sites in IRS-1 and can serve as docking sites for signaling proteins containing SH_2-domains, such as PI3K (Fernandez et al., 1995). This C-terminal domain is functional, since expression of a chimeric receptor consisting of the extracellular domain of the human IR, and the intracellular domain of DIR in murine 32D cells that lack endogenous IRS-1 can partially activate mammalian PI3K. In contrast, the ability of the human IR to activate PI3K depends on co-expression of IRS-1. These findings, together with the identification of CHICO, suggest that DIR can mediate the insulin signal in two distinct ways, one utilizes docking sites in the C-terminal tail of DIR and the other employs docking sites in CHICO, an IRS homologue (Yenush et al., 1996).

All seven Drosophila insulin-like peptides (DILPs) encoded by the fly genome are believed to act through the insulin receptor orthologue DIR, leading to activation of the PI3-kinase homologue Dp110. Dp110 and its antagonist PTEN are central to this pathway, because they regulate the levels of phosphoinositide PIP3, which appears to function as an activator of multiple downstream kinases, namely the serine/threonine kinases Pdk1 and Akt (see Fig. 2).

Recently, other molecules that are involved in the control of overall development, cell growth, and size in flies seem to participate in the insulin-signaling pathway. Tuberous sclerosis complex proteins 1 and 2 (TSC1 and 2), target of rapamycin (TOR), and 40 S ribosomal protein S6 kinase (S6K) function downstream of the insulin receptor. 40 S ribosomal protein S6 kinase can be modulated both through Akt and TOR (Brown et al., 1995; Han et al., 1995; Weng et al., 1995), forming an integrating cross-point between two cascades. The signaling function of TOR appears to be regulated in response to nutrient levels, particularly those of amino acids (nitrogen supply) (Jacinto and Hall, 2003). How TOR activity responds to nutrient conditions needs further investigation, but the proteins of the TSC complex (TSC1 and TSC 2) have been identified as potential upstream regulators of TOR and may play a role in nutrient sensing (Gao et al., 2002; McManus and Alessi, 2002). The convergence of these two signaling pathways can be considered as a nutrient-dependent checkpoint on growth factor signaling. Communication between PI3K and TOR is likely to play an important role in coordinating cell growth with other metabolic pathways.

B. TGF-β RECEPTOR FAMILY

The TGF-β superfamily of growth and differentiation factors includes the TGF-βs, activins, inhibins, bone morphogenetic proteins (BMPs), and Müllerian-inhibiting substance (Massagué et al., 1997). Members of the

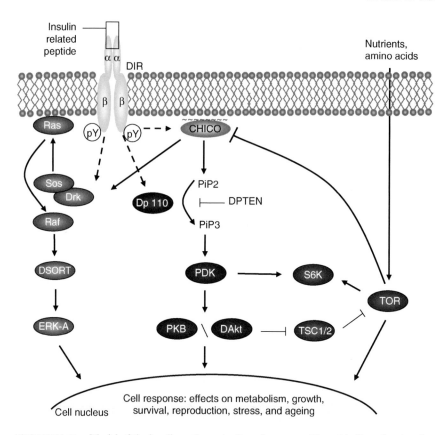

FIGURE 2. Model of the insulin pathway in *D. melanogaster*. Upon binding of an insulin-like peptide (dilp), the activated insulin receptor initiates different pathways, namely the Ras/MAPK – and the PI3K/TOR pathway (Blenis, 1993; Shepherd *et al.*, 1998). Directly or indirectly (via CHICO), the activated insulin receptor induces the conversion of the membrane lipid phosphatidylinositol 4,5-bisphosphate (PIP2) into the second messenger phosphatidylinositol 3,4,5-trisphosphate (PIP3) through PI3K (Dp110) (Shepherd *et al.*, 1998). Increased levels of PIP3 cause translocation of the serine/threonine kinases PDK and DAkt to the cell membrane through interactions between PIP3 and the pleckstrin homology (PH) domains of these proteins. PDK then promotes activation of both DAkt (PKB) and S6K through phosphorylation of their activation loops. 40 S ribosomal protein S6 kinase can also be activated through the serine/threonine kinase TOR (Brown *et al.*, 1995). The signaling function of TOR appears to be activated in response to nutrient levels, particularly those of amino acids (Jacinto and Hall, 2003). Recent studies suggest that a link between PI3K and TOR may occur through DAkt-mediated phosphorylation of TSC-2 (tuberous sclerosis complex protein 2), which was found to disrupt and inactivate the TSC1/TSC2 complex (Dan *et al.*, 2002; Inoki *et al.*, 2002, Potter *et al.*, 2002). An additional level of cross-talk between the PI3K and TOR pathways occurs through a negative feedback loop involving TOR-mediated inhibition of IRS (CHICO), an adaptor protein required for the PI3K activation by the insulin receptor. Activation of TOR results in phosphorylation and subsequent proteasomal degradation of CHICO, leading to reduced PI3K signaling (Haruta *et al.*, 2000). The parallel PI3K/TOR pathways are likely to play an important role in coordinating cell growth (and other cell responses) with other metabolic programs. Binding of

TGF-β family exert a wide range of physiological effects (e.g., they regulate cell growth, differentiation, and morphogenesis). In embryonic development, they play a crucial role in (fruit fly) pattern formation.

TGF-β family members initiate their cellular action by binding to membrane receptors with intrinsic serine/threonine kinase activity, known as type I and II receptors (Heldin et al., 1997). These classes are structurally similar, with small cysteine-rich extracellular regions and intracellular parts consisting mainly of the kinase domains. Type I but not type II receptors have regions rich in glycine and serine residues (GS domain) in the juxtamembrane domain. Type II receptors possess a constitutively active intracellular kinase activity and, in the case of activins and TGF-βs, are the primary determinant of ligand recruitment. On ligand binding, the type II receptor associates with an appropriate type I partner whose intracellular domain serves as a substrate for the type II-dependent phosphorylation (Wrana et al., 1994). Upon receptor activation, downstream components, such as the receptor-regulated class of Smads (R-Smad) signaling proteins, are recruited and activated through phosphorylation. Upon phosphorylation, the R-Smads can associate with the common class of Smads (co-Smads). After nuclear translocation of the Smad signaling complex, gene activity can be affected.

The unraveling of the genetic information of *D. melanogaster* has led to the identification of representative molecules from all the different classes of TGF-β–like signaling components. At the ligand level, three BMP-type factors, *Decapentaplegic* (*Dpp*), *Screw* (*Scw*), and *Glass bottom boat* (*Gbb*, also known as *60A*), have been identified and functionally characterized via loss of function mutations (Arora et al., 1994; Chen et al., 1996; Khalsa et al., 1998; Padgett et al., 1987). These ligands act through a receptor complex composed of heterodimeric combinations of two receptor types. Three type I receptors, *thick veins* (*tkv*) (Okano et al., 1994), *saxophone* (*sax*) (Brummel et al., 1994), *baboon* (*babo*) (Brummel et al., 1999), and two type II receptors, *punt* (*put*) (Childs et al., 1993) and *wishful thinking*

ILP to the receptor also leads to the formation of the Drk/Sos complex and subsequently to the activation of membrane associated Ras-protein. Ras-activation triggers the activation of a kinase cascade that includes Raf, MAPKK (MEK, DSORT), and MAPK (ERK, ERK-A). Once activated MAPK homodimerizes and is imported into the nucleus where it phosphorylates target proteins that regulate transcription, the ultimate goal of the signaling pathway.

Abbreviations: Dilp, "*Drosophila* insulin-like peptide;" DIR, "*Drosophila* insulin receptor;" Drk, "Downstream of RTK" (Grb2); DSORT, orthologue of MEK, "Mitogen-activated ERK-activating kinase"; ERK-A, orthologue of ERK, "Extracellular-signal-regulated kinase"; IRS, "Insulin receptor substrate"; MAPK, "Mitogen-activated protein kinase"; PDK, "Phosphatidylinositol-dependent protein kinase"; PI3K, "Phosphatidylinositol-3-OH kinase"; PKB, "Protein kinase B"; PTEN, phosphatidylinositide phosphatase; S6K, "Ribosomal protein S6 kinase"; Sos, "Son of Sevenless"; TOR, "Target of rapamycin;" TSC1/TSC2, "Tuberous sclerosis complex" protein 1 and 2. (See Color Insert.)

(*wit*) (Aberle *et al.*, 2002), interact with either two R-Smads, *mothers against dpp* (*mad*) or *dSmad2* (Kretzschmar *et al.*, 1997; Liu *et al.*, 1997). In general, *tkv* transmits a *dpp*/BMP signal through *mad*, while *babo* transmits an activin signal through *dSmad2*. Different insect members of the TGF-β receptor family are summarized in Table IIIB.

Ongoing investigations on the regulation of fundamental, cellular processes, such as proliferation, differentiation, development, and apoptosis, are suggesting a crucial coordinating role of the TGF-β signaling pathways (e.g., the *dpp*/BMP and activin pathways) in these processes. Recent studies on neuronal remodeling (i.e., the outgrowth of dendrites and axons), in the *Drosophila* larval brain confirm the importance of TGF-β signaling. *Tkv* and *babo* are both linked to the regulation of functional neural circuits. Yang *et al.* (2004) have identified target genes for the *Tkv* and *babo* pathways by employing microarray techniques using activated forms of the receptors expressed in the brain. Twenty-seven genes seem to be co-regulated through the *Tkv* and *babo* pathways. Among the regulated genes is ultraspiracle, the heterodimeric partner of the nuclear ecdysone receptor. These observations led to the model that *Drosophila* activin signaling results in the induction of the ecdysone receptor complex, leading to transcriptional regulation of downstream target genes that control the neuronal circuits in the brain of the fruit fly (Zheng *et al.*, 2003).

Stem cells are responsible for replacing damaged or dying cells in various adult tissues throughout a lifetime. Therefore, they form an ideal model for studying cell differentiation. *Drosophila* testis has become one of the premier stem cell systems for these investigations. BMP signals from somatic cells surrounding germline stem cells (GSC) of the testis are essential for maintaining GSC in the *Drosophila* testis. Somatic cells express two BMP molecules, Gbb and Dpp, which seem to be involved in the regulation of the GSC reservoir (Kawase *et al.*, 2004). Dpp signaling is also known to be essential for maintaining GSCs in the *Drosophila* ovary (Song *et al.*, 2004). This study further suggests that both *Drosophila* male and female GSCs use BMP signals to maintain GSCs.

TGF-β signaling has also been implicated in synapse assembly and plasticity in both vertebrate and invertebrate systems. Recently, *wishful thinking*, a *Drosophila* BMP type II receptor, has been shown to be required for the normal function and development of the neuromuscular junction (NMJ). Furthermore, Rawson *et al.* (2003) demonstrated that *saxophone* and downstream transcription factor *mad* are also essential for normal structural and functional development of the *Drosophila* NMJ. These findings suggest that a TGF-β ligand activates a signaling cascade involving type I and II receptors and Smad family of transcription factors to orchestrate the assembly of the NMJ, a synapse that displays activity-dependent plasticity.

C. GUANYLYL CYCLASE RECEPTOR FAMILY

1. Guanylyl Cyclases

Guanylyl cyclases (GC) comprise a family of enzymes, which catalyze the conversion of GTP to cyclic GMP. Based on the cellular location and activation mechanism, GC can be subdivided in two major classes—the cytoplasmically located soluble GC (sGC) and the membrane-bound receptor GC (rGC). While sGC exist as heterodimeric proteins with a heme prosthetic group, which binds free radicals, such as nitric oxide (NO), conventional rGC are homo-oligomers with an extracellular ligand-binding site for (peptide) hormones. After GC-activation, increasing cGMP-levels will initiate downstream signaling by effector molecules, including cyclic nucleotide–gated ion channels as well as cGMP-dependent kinases and phosphodiesterases (Lucas *et al.*, 2000).

For many years, reports on GC mainly focused on vertebrates, indicating a regulatory role for the second messenger cGMP in diverse physiological processes, such as retinal phototransduction, water balance, blood pressure, smooth muscle motility, and neuronal plasticity. Sequencing of several invertebrate genomes, however, not only led to the prediction of numerous additional GC sequences but also to a growing interest in cGMP-mediated signaling pathways in invertebrates. As a result, a plethora of GC has been identified in invertebrates (Morton, 2004a), many of which are not conforming to the "classic" sGC and rGC. In support of this, two studies have shown that atypical soluble GC of *C. elegans* and *D. melanogaster* are regulated in response to low or changing oxygen levels, rather than by nitric oxide (Grey *et al.*, 2004; Morton, 2004b).

2. Identification and Characterization of Guanylyl Cyclase Receptors

a. Discovery of Guanylyl Cyclase Receptors

In 1988, the first rGC was cloned from a sea urchin testis cDNA library (Singh *et al.*, 1988). Based on this sequence information, the natriuretic peptide receptors, guanylyl cyclase A (GC-A) and B (GC-B), were cloned from mammalian tissues (Chang *et al.*, 1989; Chinkers *et al.*, 1989; Lowe *et al.*, 1989; Schulz *et al.*, 1989), facilitating the identification of five additional rGC sequences (GC-C to GC-G) by degenerate RT-PCR (Lucas *et al.*, 2000). All rGC have a single transmembrane region, separating the extracellular ligand-binding domain from three distinct intracellular domains. Next to the membrane is the protein kinase homology domain, which is connected to the catalytic domain by a hinge domain. In addition, the GC-C, -D, -E, -F isoforms also possess a C-terminal tail without significant homology to other known proteins. Although all mammalian rGC share an extracellular ligand-binding domain, ligands for only three rGC have been identified so far. While the atrial and brain natriuretic peptides (ANP and BNP, respectively) activate GC-A and GC-B, which in turn regulate blood pressure/volume, the

C-type natriuretic peptide is thought to act locally as an autocrine/paracrine endothelian factor via GC-B (Kuhn, 2003; Lucas et al., 2000). Besides GC-A and GC-B, a third receptor binds the three natriuretic peptides with similar affinity. But although the extracellular domain of this receptor is related to GC-A and GC-B, it has only 37 intracellular amino acids and thus lacks guanylate cyclase activity. Instead this "clearance" receptor (NPR-C) serves for cellular internalization and degradation of NPs and can inhibit cAMP synthesis by interaction with a Gi-protein (Patel, 2004). Guanylyl cyclase C was characterized as the docking site for bacterial enterotoxins, leading to secretory diarrhea (Schulz et al., 1990), and as the receptor for two intestinal mammalian peptides, guanylin and uroguanylin (Currie et al., 1992; Hamra et al., 1993). In contrast, GC-D, -E, -F, and -G are orphan receptors for which ligands remain to be identified. However, the existence of ligands for retinal GC-E and GC-F is still very controversial, since the activation of these receptors by intracellular Ca^{2+}-binding proteins (GCAP-1–3) in vertebrate rods during visual processes appears to be independent from the presence of an extracellular ligand-binding domain (Sokal et al., 2003).

b. Arthropod Guanylyl Cyclase Receptors

Shortly after the identification of the mammalian natriuretic peptide receptors, Gigliotti et al. (1993) reported the cloning of the first insect rGC (Gyc32E) from D. melanogaster. The deduced primary sequence is most related to GC-A, revealing the same domain organization as encountered in vertebrates. A second Drosophila rGC (Gyc76C) was identified, but in contrast to Gyc32E, the encoded protein contains a 430 C-terminal amino acid tail in addition to the "classic" rGC-domains. Furthermore, while the Gyc32E-mRNA was mainly detected in the ovaries, the Gyc76C-transcript is widely distributed in adult tissues, such as the CNS (Liu et al., 1995; McNeil et al., 1995). From the silk moth, B. mori, a receptor guanylyl cyclase, named BmGC-I, has been cloned that is also widely expressed in many different tissues, including the nervous system (Tanoue et al., 2001). Again, no biochemical characterization was reported, but the expression of this cyclase was found to be regulated in flight muscle in a circadian fashion (Tanoue and Nishioka, 2001). More recently, an insect rGC (MsGC-II) was cloned from nervous tissue of M. sexta, revealing highest sequence homology with vertebrate retinal rGC, which are known to be activated by GCAPs in the presence of low Ca^{2+} levels (Morton and Nighorn, 2003). Interestingly, this study also showed that guanylyl cyclase activity of transiently expressed MsGC-II was inhibited by high levels of Ca^{2+} in the absence of exogenous Ca^{2+}-binding proteins, whereas exogenous mammalian GCAP-2 or frequenin, a Drosophila Ca^{2+}-binding protein, further enhanced the activity of the receptor in the presence of low Ca^{2+}. Altogether, these data suggest that both the primary structure as well as the activation mechanism of MsGC-II is similar to the vertebrate retinal guanylyl cyclases. Finally, a cDNA encoding a putative membrane form GC (PcGC-M2) from the muscle of the crayfish, *Procambarus clarkii*, was identified (Liu *et al.*, 2004).

Based on the available genome sequences of several insects, including *D. pseudoobscura* and *Apis mellifera*, several putative rGC can be predicted. In *D. melanogaster*, there are at least four predicted rGC in addition to the previously reported receptors (see Fig. 3). *Anopheles* is predicted to have an orthologue to each of them, while, at present, "only" four orthologues have been found in the (partial) genome of *D. pseudoobscura*. Also in the preliminarily annotated *Apis mellifera* genome, at least four rGC have been predicted. In Table IV these annotated proteins are represented, as well as additional orthologues in other insects. It is worth mentioning that MsGC-II and its predicted orthologues in *D. melanogaster* and *A. gambiae*, all of which are most similar to the vertebrate retinal rGC-F, lack an extracellular ligand-binding domain, conforming to an activation mechanism by intracellular Ca^{2+}-binding proteins rather than by an extracellular peptide ligand. Apart from the previously mentioned insect rGC, guanylyl cyclases (e.g., MsGC-I Simpson *et al.*, 1999) have been identified that lack an extracellular and transmembrane domain, while the catalytic and dimerization domain (hinge region) resemble the corresponding domains in rGC (Morton, 2004a). Furthermore, transient co-expression of MsGC-I and MsGC-II *in vitro* reduced the amount of MsGC-I in the soluble fraction, suggesting

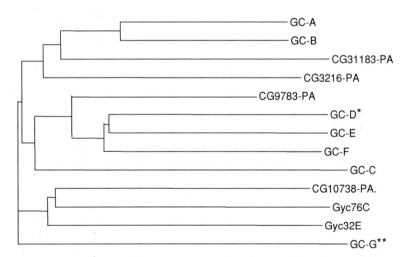

FIGURE 3. Dendrogram of a sequence comparison of mammalian and *Drosophila* receptor guanylyl cyclases (rGC). Characterized and orphan receptor guanylyl cyclases (rGC) of mammalian or insect origin were compared by CLUSTALW analysis. The different receptors are discussed in the text and related receptors from other organisms are summarized in Table IV. (*) All mammalian rGC are derived from *H. sapiens*, except for GC-D, which is from *Rattus norvegicus*. (**) Predicted gene (Accession number: XP_497249). (See Color Insert.)

TABLE IV. Members of the Receptor Guanylyl Cyclase (rGC) Family

D. melanogaster		A. gambiae		D. pseudoobscura	A. mellifera	Other Arthropods	H. sapiens
Gene	Protein	Gene[a]	Protein	Protein	Protein	Protein	Protein
CG3216 RA	Q9w2p1	00000011655	EAA10186	EAL26478	XP_396555[b]		P16066 (NPR1 or GC-A)
CG3216 RB	Q8mlx0	00000004849	EAA14802	EAL33862			P16066 (NPR1 or GC-A)
CG33114 (Gyc32E)	Q07553						
CG8742 (Gyc76C)	Q24051	0000006858	EAA00176		XP_393152	AAQ74970 (PcGC-2)	P16066 (NPR1 or GC-A)
CG10738 RA	Q9VU79	00000006859	EAA00177	EAL30246	ENSAPMP00000 1198[e]	AAM94353 B. dorsalis[d]	P20594 (NPR2 or GC-B)
CG10738 RB	Q81QK2						
CG31183	Q9VF17	00000010445	EAA03567	EAL28393	ENSAPMG00000 10873[e] XP_394065	BAB32672 (BmGC-1)	P16066 (NPR1 or GC-A)
CG9783	Q9VN32[c]	00000010835	EAA03699[c]			AAN16469[c] (MsGC-II)	P51841 (GUCY2F or GC-F)

[a] Full Gene ID = ENSANGG followed by the number code: http://www.ensembl.org/Anopheles_gambiae/.
[b] Incomplete GC-doman.
[c] No ANP-binding domain.
[d] Bactrocera dorsalis (oriental fruit fly).
[e] Protein ID: Ensembl Honeybee Genome Server.

that both proteins could heterodimerize *in vivo* (Morton and Nighorn, 2003). Consistent with this observation, both CNS specific GC show an overlapping expression pattern in neurons of the optic and antennal lobes, as well as of the mushroom bodies (Nighorn *et al.*, 2001). If this hypothesis proves to be true, it underlines the complexity of cGMP signaling cascades in the nervous system.

c. Guanylyl Cyclase Receptors and Insect Physiology

Apparently no close homologues of the mammalian natriuretic peptides are present in insects. Moreover, no other insect rGC-ligands have been identified so far. As a result, the physiological relevance of cGMP signaling cascades, which are activated by rGC, remains poorly documented as yet. Nevertheless, it has been reported that crustacean hyperglycemic hormone (CHH) induces cGMP synthesis through a membrane-bound GC in lobster muscle (Goy, 1990). The fact that the PcGC-M2 transcript is expressed in several CHH target tissues, including muscle, hepatopancreas, heart, ovary, testis, and gill (Liu *et al.*, 2004), suggests that PcGC-M2 may be activated by CHH. Furthermore, it was shown that a neurohemal gland extract, containing a CHH-like factor, activates a rGC in the crab stomatogastric nervous system (Scholz *et al.*, 1996). Since CHH-like peptides have also been identified in insects, it can be speculated that these peptides (the so-called ion transport peptides) are possible ligands for the orphan insect rGC.

From head extracts of the beetle, *Tenebrio molitor*, a potent antidiuretic factor was purified, which inhibits fluid secretion by the Malpighian tubules via cGMP (Eigenheer *et al.*, 2002). Neither NO-donors nor inhibitors of NO-synthase affect the cGMP synthesis induced by this antidiuretic factor, conforming to an rGC-mediated response (Eigenheer *et al.*, 2002). In addition, it was shown that cAMP levels in *Tenebrio*, stimulated by endogenous CRF-related diuretic hormone (see Section III.A.1.), drop when the antidiuretic factor is added to the Malpighian tubules (Eigenheer *et al.*, 2002; Wiehart *et al.*, 2002), suggesting the involvement of a cGMP-dependent PDE, downstream of the presumed rGC. Altogether, these observations indicate that diuretic and antidiuretic peptides regulate fluid secretion in *T. molitor* via antagonistic second messengers (cAMP and cGMP, respectively). However, since capa peptides (related to CAP2b, see Section II.F.) have been shown to stimulate fluid secretion from Malpighian tubules in the intestinal lumen via soluble GC and cGMP-synthesis in several dipteran species (Kean *et al.*, 2002; Pollock *et al.*, 2004), mechanisms regulating homeostasis appear to differ between insect orders. Finally, it is worth considering that also in mammals, regulation of intestinal fluid secretion is mediated by peptides, which bind an rGC (GC-C) (Lucas *et al.*, 2000).

Although its ligand is yet not characterized, a role for the *Drosophila* rGC, Gyc76C, in repulsive signaling in the developing embryonic system has been established by using a genetic screening approach (Ayoob *et al.*, 2004). During neural development, axons, which are initially arranged as parallel bundles, will branch off (defasciculate) to establish (neuromuscular) connectivity with target tissues. Semaphore-1a (Sema-1a) is present on embryonic motor axons and acts as a repellent, by interacting with the PlexA receptor, which targets individual neurons away from the central axon bundle (Bashaw, 2004). Since homozygous *gyc76C* mutants show an identical phenotype as Sema-1a mutants (i.e., a lack of axon defasciculation), leading to a failure of muscle target innervation, proper Gyc76C receptor activity is critical for normal Sema-1a repulsive guidance of motor axons. Interestingly, local production of cAMP seems to antagonize Sema-1a-mediated signaling via the so-called Nervy protein, which couples a cAMP-dependent protein kinase A to the PlexA receptor (Terman and Kolodkin, 2004).

During phototransduction in vertebrate photoreceptor cells, cGMP levels are balanced by the antagonistical enzyme activities of PDE and retinal rGC (Sokal *et al.*, 2003). Although in *Drosophila* similar rGC are predicted, the response to photons leads to the activation of PLC, which generates IP_3 and DAG as second messengers (Hardie *et al.*, 2002). In *Limulus* the visual transduction process is highly complicated, in line with both high sensitivity and a broad dynamic range. In contrast to *Drosophila*, it was shown recently that a GC enzyme is implicated in the cascade, which couples the light-induced Ca^{2+} elevation to the production of cGMP. Based on its pharmacological profile, this GC is most likely a membrane-bound enzyme, indicating that in this ancient arthropod, phototransduction signaling is mediated by rGC (Garger *et al.*, 2004).

V. EMERGING CONCEPTS AND FUTURE DEVELOPMENTS

A. ELEMENTS OF DISCUSSION

1. Deorphanization of the Remaining Putative Peptide Receptors

During the past years, an impressive list of peptide receptor discoveries (see Tables) has substantially revolutionized the field of insect neurobiology and endocrinology. The current situation in the area of insect peptide receptor characterizations starts resembling the one encountered in mammals, where the remaining number of orphan, putative peptide receptors is much larger than the number of "orphan peptides," (i.e., known peptides for which the receptor still has to be characterized). A combination of powerful methods has been very instrumental to achieve the present situation. In addition to the more traditional binding studies (Reagan, 1994), these methods also include a number of very elegant functional receptor assays, such as: (1) electrophysiological analysis on *Xenopus* oocytes (Birgül *et al.*, 1999), (2) detection of cellular Ca^{2+} changes in multiwell plate format by aequorin bioluminescence (Fig. 4) or indicator dye fluorescence (Torfs *et al.*, 2000,

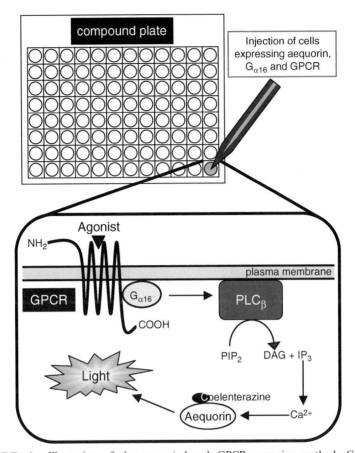

FIGURE 4. Illustration of the aequorin-based GPCR screening method. Cells co-expressing apoaequorin, $G_{\alpha 16}$ and a GPCR of interest are loaded with coelenterazine and subsequently injected into the wells of a compound plate that is mounted in a luminometer. The promiscuous G protein, $G_{\alpha 16}$, couples most GPCRs to phospholipase-C-β (PLCβ) resulting in the generation of inositol 1,4,5-trisphosphate (IP$_3$) and diacylglycerol (DAG) from phosphatidylinositol 4,5-bisphosphate (PIP$_2$). IP$_3$ is responsible for the subsequent release of Ca^{2+}-ions from the intracellular stores. In the presence of Ca^{2+}-ions, aequorin catalyses the oxidation of coelenterazine to coelenteramide, with a concomitant emission of light.

2002c), and (3) transfluor technology analyzing cellular βarrestin2-GFP translocation (Johnson et al., 2003a). A closer view at Tables I and II clearly shows that it is very difficult to predict which method is the most appropriate deorphanization tool for a given receptor. Each method has its disadvantages and restrictions (e.g., Kostenis, 2001). Therefore, it can be expected that all these methods, which are rather complementary, will remain very useful in the continuing search for the natural ligands of the remaining orphan receptors. With more receptors being characterized, deorphanization of the remaining ones will perhaps require novel peptide or protein purifications from insect tissue extracts ("reverse physiology"), rather than using libraries of chemically synthesized peptides ("reverse pharmacology approach"). Nevertheless, various "orphan" insect peptide families still exist, such as, pigment dispersing hormones (PDH), LF/SIFa-like peptides, calcitonin-like diuretic hormones (DH_{31}), ion transport peptides (ITP), eclosion hormones (EH), amnesiac, and sex peptides (SP). In addition, orthologues of several insect peptides were not (yet) found in the *Drosophila* genome but occur in other species (e.g., prothoracicotropic hormones [PTTH], neuroparsins, and ovary ecdysteroidogenic hormones [OEH], the locust ovary maturating parsin [OMP], the locust vasopressin-like peptide, baratin, orcokinins [discovered in crustaceans; according to our own unpublished observations, orcokinin precursors are also encoded in recently sequenced insect genomes], allatotropins, and various myotropic peptides).

The identification of larger peptide and protein ligands obviously causes a number of problems related to the availability, purity, and conformational stability of such ligands. Therefore, growth factor and protein hormone receptor characterizations are mainly based on genetic studies, performed with mutant or transgenic fruit flies. For many of these larger ligand–receptor couples direct evidence on their molecular interaction is still lacking. Biochemical studies showing these molecular interactions represent an important challenge needed to complement and confirm the data from genetic studies. For such analyses, the availability of a sufficient amount of pure ligand(s) is required. In contrast to small peptides, which can be produced via chemical synthesis, most of these larger peptides and protein ligands have to be produced in highly efficient biological expression systems and need further purification by chromatographic techniques. These procedures take a lot of time and impose several difficulties that will have to be overcome in the future in order to biochemically and pharmacologically characterize the receptors for growth factors and protein hormones.

2. Molecular Pharmacology and Signaling Properties of Insect Peptide Receptors

The data summarized in Table I illustrate the existence of important discrepancies (e.g., in the EC_{50} values) between different analyses performed on a single type of receptor. Only part of these remarkable differences can be explained by corrections to the predicted receptor cDNA sequence (Cazzamali *et al.*, 2003). Another possible source of experimental variability probably resides in the quality (purity) and the exact quantification of the chemically synthesized peptide ligands used in different studies; however, this is difficult to verify. In addition, the cellular context provided by the expression system, as well as the type of assay employed for detecting receptor activity constitute major causes of variability in pharmacological analyses. The functional response of a GPCR, as well as the EC_{50} values, can be influenced by the type of expression system (also observed with amine receptors; Poels et al., 2001). Moreover, in particular for receptors that display dual coupling properties, an increasing number of studies provide substantial evidence for the occurrence of multiple conformational receptor states, even within a single clonal cell line (see Poels et al., 2004b). Therefore, molecular pharmacological studies of insect peptide receptors should preferably be based on the comparison of data obtained with the same expression and second messenger detection systems.

So far, most studies utilize heterologous, vertebrate derived-expression systems for the analysis of the properties of receptor proteins from insect origin. These systems are probably not very suitable for a physiologically relevant analysis of the functional coupling properties of insect-derived receptor proteins. Therefore, it would seem more logical to employ more homologous expression systems that provide a cellular context that is closer to the natural physiological situation (Vanden Broeck et al., 1998). In recent years, several permanent insect cell lines have been established, and expression vectors have been developed that can be employed for these purposes. For instance, an insect cell-based receptor assay system for analysis of Ca^{2+}-dependent aequorin bioluminescence (Fig. 4) has been developed (Torfs et al., 2002c).

B. EMERGING CONCEPTS ON RECEPTOR–LIGAND CO-EVOLUTION

The availability of Metazoan genome sequence data represents an important development that has undoubtedly been very fruitful with respect to comparative and evolutionary analyses of receptor–ligand couples. The evolutionary origin of several peptide and receptor families precedes the divergence of the important animal Phyla, suggesting that selective pressure to preserve ligand–receptor matching has conserved structural features of the peptides during co-evolution with their receptor(s). However, this is not a strict rule. As nicely shown for the subgroup of allatostatin (AstA, AstB, and AstC), bombesin, galanin, opioid, and somatostatin GPCRs, receptor conservation does not necessarily imply that the corresponding ligand is strongly conserved. This may be an indication for the existence of "fast co-evolutionary changes" of the ligand–receptor interaction and/or of an evolutionary change of the partnership. That such "unfaithful partnership changes" can occur during evolution is suggested by the fact that some peptide families utilize receptors belonging to multiple, very different classes. Several Metazoan peptide families were reviewed that use more than one category of transmembrane receptors—FMRFa-like peptides can exert their effects via ligand gated ion channels (as shown in mollusks and vertebrates) or via GPCRs; members of the insulin superfamily can activate GPCRs or RTKs; members of the cystine knot family of protein hormones are agonists for GPCRs (LGRs) or function as antagonists of receptors belonging to the TGF-β receptor family; members of the Wnt family of secreted signaling molecules are ligands for the Fz heptahelical receptors, as well as for Derailed, an atypical RTK (Yoshikawa et al., 2003). It is well established that small mutational changes in the ligand (e.g., single amino acid changes) or in the receptor can already have very drastic consequences for ligand–receptor interaction and biological activity. Furthermore, a single receptor can have more than one natural ligand and can even show agonist-dependent differences in functional coupling specificity. This phenomenon could represent an important, new mechanism for physiological finetuning (see also Poels et al., 2004b).

Another mechanism that can contribute to physiological finetuning emerged from the discovery that several families of soluble peptide ligand-binding proteins are evolutionary conserved in Metazoans. These binding proteins, such as "IGF-binding proteins" (IGF-BP; Claeys et al., 2003) and "CRF-binding proteins;" (CRF-BP; Jahn et al., 2002; according to our own unpublished observations, related proteins are encoded in the recently sequenced insect genomes), are believed to have a modulatory effect on peptide ligand availability by acting upstream of the receptor activation process (see also Simonet et al., 2004).

C. FUTURE PROSPECTS

Insect genome sequencing projects have revealed the existence of many different genes that are believed to be involved in peptide or protein ligand–induced signaling processes. Future challenges in the field of insect peptides and their receptors are to study the regulation, the biological functions, and the biodynamic interplay of this large set of genes. In addition to the molecular data that have been gathered in recent years, more functional data are needed in order to unravel the physiological processes in which different peptide–receptor couples are involved. In this postgenomic research era, some very powerful methods have now become available for functional genomic analyses, in addition to the more traditional insect physiology and the fruitful combination of *Drosophila* genetics and germline transformation technology. For instance, the use of the RNA interference (RNAi) technique, which provides a multitude of new opportunities to study the phenotypic effects of specific posttranscriptional gene silencing, has been successfully applied *in vitro* as well as *in vivo* (Denell and Shippy, 2001; Kuttenkeuler and Boutros, 2004; Rosenkilde et al., 2003). In addition, more broadly applicable genetic transformation systems have recently been established for several insect species (Wimmer, 2003). Another methodological development allows for the study of the regulation of multiple transcripts by real-time quantitative RT-PCR. Moreover, this can be done at a genome-wide scale by microarray analysis (e.g., Yang et al., 2004). Both methods are complementary to each other and may prove to become very useful for the identification of peptide induced downstream signaling events leading to changes in gene expression at the transcriptional level. Although similar effector proteins and transcription factors may be involved, the regulation of such downstream processes can show important differences between vertebrates and insects (Poels and Vanden Broeck, 2004; Poels et al., 2004c). Therefore, it will be of interest to investigate cellular signaling processes in insects and compare them with the situation in vertebrates. This comparative study of peptide signaling pathways will probably reveal well-conserved aspects, as well as insect-specific ones. More insight in these pathways will perhaps result in the discovery of pharmacological targets for the development of novel insect control agents and/or lead to new methods and applications (Vanden Broeck et al., 1997).

Note Added in Proof

Since this manuscript was submitted for publication on the 30th of November 2004, very recent discoveries on insect peptide receptors were not incorporated.

In March 2005, two different research groups (Luo et al., 2005; Mendive et al., 2005) simultaneously revealed that bursicon, the neurohormone which induces post-ecdysial cuticular tanning, is a heterodimer and the natural agonist of the *rickets* receptor, DLGR2. In the fruit fly, bursicon is composed of two different cystine knot subunits, CG13419 and CG15284.

ACKNOWLEDGMENTS

This study was supported by the Belgian Program on Interuniversity Poles of Attraction (IUAP/PAI P5/30, Belgian Science Policy Programming). Also supported by grants from the "Fonds voor Wetenschappelijk Onderzoek" (FWO) and the "K.U.Leuven Onderzoeksfonds." Ilse Claeys, Vanessa Franssens and Tom Van Loy obtained a PhD fellowship from the "Instituut voor de aanmoediging van Innovatie door Wetenschap en Technologie in Vlaanderen" (IWT). Jeroen Poels and Gert Simonet obtained a postdoctoral fellowship from the "K.U.Leuven Onderzoeksfonds."

REFERENCES

A letter of comment regarding this discovery was published by Vassart et al. (2005) [*EMBO Rep.* **6**(7), 592].

Aberle, H., Haghighi, A. P., Fetter, R. D., McCabe, B. D., Magalhaes, T. R., and Goodman, C. S. (2002). Wishful thinking encodes a BMP type II receptor that regulates synaptic growth in *Drosophila*. *Neuron* **33**, 545–558.

Adams, M. D., Celniker, S. E., Holt, R. A., *et al.* (2000). The genome sequence of *Drosophila melanogaster*. *Science* **287**, 2185–2195.

Alcedo, J., Ayzenzon, M., Von Ohlen, T., Noll, M., and Hooper, J. E. (1996). The *Drosophila smoothened* gene encodes a seven-pass membrane protein, a putative receptor for the Hedgehog signal. *Cell* **86**, 221–232.

Anastasi, A., and Erspamer, V. (1962). Occurrence and some properties of eledoisin in extracts of posterior salivary glands of *Eledone*. *Br. Pharm. Chemother.* **19**, 326–336.

Arora, J. M., Levine, M., and O'Connor, M. (1994). The *screw* gene encodes a ubiquitously expressed member of the TGF-beta family required for specification of dorsal cell fates in the *Drosophila* embryo. *Genes Dev.* **8**, 2588–2601.

Auerswald, L., Birgül, N., Gäde, G., Kreienkamp, H. J., and Richter, D. (2001). Structural, functional, and evolutionary characterization of novel members of the allatostatin receptor family from insects. *Biochem. Biophys. Res. Commun.* **282**, 904–909.

Ayoob, J. C., Yu, H. H., Terman, J. R., and Kolodkin, A. L. (2004). The *Drosophila* receptor guanylyl cyclase Gyc76C is required for semaphorin-1a-plexin A-mediated axonal repulsion. *J. Neurosci.* **24**, 6639–6649.

Baker, J. D., and Truman, J. W. (2002). Mutations in the *Drosophila* glycoprotein hormone receptor, rickets, eliminate neuropeptide-induced tanning and selectively block a stereotyped behavioral program. *J. Exp. Biol.* **205**, 2555–2565.

Bashaw, G. J. (2004). Semaphorin signaling unplugged: A nervy AKAP cAMP(s) out on plexin. *Neuron* **42**, 363–366.

Basler, K., and Hafen, E. (1988). Control of photoreceptor cell fate by the sevenless protein requires a functional tyrosine kinase domain. *Cell* **54**, 299–311.

Belles, X., Graham, L. A., Bendena, W. G., Ding, Q. I., Edwards, J. P., Weaver, R. J., and Tobe, S. S. (1999). The molecular evolution of the allatostatin precursor in cockroaches. *Peptides* **20**, 11–22.

Bendena, W. G., Donly, B. C., and Tobe, S. S. (1999). Allatostatins: A growing family of neuropeptides with structural and functional diversity. *Ann. N.Y. Acad. Sci.* **897**, 311–329 (Review).

Bhanot, P., Brink, M., Samos, C. H., Hsieh, J. C., Wang, Y., Macke, J. P., Andrew, D., Nathans, J., and Nusse, R. (1996). A new member of the frizzled family from *Drosophila* functions as a Wingless receptor. *Nature* **382**, 225–230.

Biggs, W. H. III, Zavitz, K. H., Dickson, B., Vander Straten, A., Brunner, D., Hafen, E., and Zipursky, S. L. (1994). The *Drosophila* rolled locus encodes a MAP kinase required in the sevenless signal transduction pathway. *EMBO J.* **13**, 1628–1635.

Birgül, N., Weise, C., Kreienkamp, H. J., and Richter, D. (1999). Reverse physiology in *Drosophila*: Identification of a novel allatostatin-like neuropeptide and its cognate receptor structurally related to the mammalian somatostatin/galanin/opioid receptor family. *EMBO J.* **18**, 5892–5900.

Blenis, J. (1993). Signal transduction via the MAP kinases: Proceed at your own RSK. *Proc. Natl. Acad. Sci. USA* **90**, 5889–5892.

Bockaert, J. (1991). G proteins and G protein-coupled receptors: Structure, function and interactions. *Curr. Opin. Neurobiol.* **1**, 32–42 (Review).

Böhni, R., Riergo-Escovar, J., Oldham, S., Brogiolo, W., Stocker, H., Andruss, B. F., Beckingham, K., and Hafen, E. (1999). Autonomous control of cell and organ size by CHICO, a *Drosophila* homologue of vertebrate IRS-1-4. *Cell* **97**, 865–875.

Boulay, J. L., O'Shea, J. J., and Paul, W. E. (2003). Molecular phylogeny within type I cytokines and their cognate receptors. *Immunity* **19**, 159–163 (Review).

Brody, T., and Cravchik, A. (2000). *Drosophila melanogaster* G protein-coupled receptors. *J. Cell Biol.* **150**, F83–F88 (Review).

Brown, E. J., Beal, P. A., Keith, C. T., Chen, J., Shin, T. B., and Schreiber, S. L. (1995). Control of p70 S6 kinase by kinase activity of FRAP *in vivo*. *Nature* **377**, 441–446.

Brown, M. R., Crim, J. W., Arata, R. C., Cai, H. N., Chun, C., and Shen, P. (1999). Identification of a *Drosophila* brain-gut peptide related to the neuropeptide Y family. *Peptides* **20,** 1035–1042.

Brown, S, Hu, N., and Hombria, J. C. (2001). Identification of the first invertebrate interleukin JAK/STAT receptor, the *Drosophila* gene domeless. *Curr. Biol.* **11,** 1700–1705.

Brummel, T. J., Twombly, V., Marques, G., Wrana, J. L., Newfeld, S. J., Attisano, L., Massague, J., O'Connor, M. B., and Gelbart, W. M. (1994). Characterization and relationship of Dpp receptors encoded by the saxophone and thick veins genes in *Drosophila*. *Cell* **78,** 251–261.

Brummel, T., Abdollah, S., Haerry, T. E., Shimell, M. J., Merriam, J., Raftery, L., Wrana, J. L., and O'Connor, M. B. (1999). The *Drosophila* activin receptor baboon signals through dSmad2 and controls cell proliferation but not patterning during larval development. *Genes Dev.* **13,** 98–111.

Budd, G. E., Butterfield, N. J., and Jensen, S. (2001). Crustaceans and the "Cambrian explosion." *Science* **294,** 2047.

Cadigan, K. M., and Nusse, R. (1997). Wnt signalling: A common theme in animal development. *Genes Dev.* **11,** 3286–3305.

Cazzamali, G., and Grimmelikhuijzen, C. J. (2002). Molecular cloning and functional expression of the first insect FMRFamide receptor. *Proc. Natl. Acad. Sci. USA* **99,** 12073–12078.

Cazzamali, G., Saxild, N., and Grimmelikhuijzen, C. J. P. (2002). Molecular cloning and functional expression of a *Drosophila* corazonin receptor. *Biochem. Biophys. Res. Commun.* **298,** 31–36.

Cazzamali, G., Hauser, F., Kobberup, S., Williamson, M., and Grimmelikhuijzen, C. J. P. (2003). Molecular identification of a *Drosophila* G protein-coupled receptor specific for crustacean cardioactive peptide. *Biochem. Biophys. Res. Commun.* **303,** 146–152.

Champagne, D. E., and Ribeiro, J. M. (1994). Sialokinin I and II: Vasodilatory tachykinins from the yellow fever mosquito *Aedes aegypti*. *Proc. Natl. Acad. Sci. USA* **91,** 138–142.

Chang, M. S., Lowe, D. G., Lewis, M., Hellmiss, R., Chen, E., and Goeddel, D. V. (1989). Differential activation by atrial and brain natriuretic peptides of two different receptor guanylate cyclases. *Nature* **341,** 68–72.

Chen, C. M., and Struhl, G. (1999). Wingless transduction by the Frizzled and Frizzled2 proteins of *Drosophila*. *Development* **126,** 5441–5452.

Chen, X., Rubock, M. J., and Whitman, M. (1996). A transcriptional partner for MAD proteins in TGF-beta signaling. *Nature* **383,** 691–696.

Chen, Y., and Struhl, G. (1996). Dual roles for patched in sequestering and transducing Hedgehog. *Cell* **87,** 553–563.

Childs, S. R., Wrana, J. L., Arora, K., Attisano, L., O'Connor, M. B., and Massague, J. (1993). Identification of a *Drosophila* activin receptor. *Proc. Natl. Acad. Sci. USA* **90,** 9475–9479.

Chinkers, M., Garbers, D. L., Chang, M. S., Lowe, D. G., Chin, H. M., Goeddel, D. V., and Schulz, S. (1989). A membrane form of guanylate cyclase is an atrial natriuretic peptide receptor. *Nature* **338,** 78–83.

Choi, M. Y., Rafaeli, A., and Jurenka, R. A. (2001). Pyrokinin/PBAN-like peptides in the central nervous system of D*rosophila melanogaster*. *Cell Tissue Res.* **306,** 459–465.

Choi, M. Y., Fuerst, E. J., Rafaeli, A., and Jurenka, R. (2003). Identification of a G protein-coupled receptor for pheromone biosynthesis activating neuropeptide from pheromone glands of the moth *Helicoverpazea*. *Proc. Natl. Acad. Sci. USA* **100,** 9721–9726.

Claeys, I., Simonet, G., Poels, J., Van Loy, T., Vercammen, L., De Loof, A., and Vanden Broeck, J. (2002). Insulin-related peptides and their conserved signal transduction pathway. *Peptides* **23,** 807–816.

Claeys, I., Simonet, G., Van Loy, T., De Loof, A., and Vanden Broeck, J. (2003). cDNA cloning and transcript distribution of two novel members of the neuroparsin family in the desert locust, *Schistocerca gregaria*. *Insect Mol. Biol.* **12,** 473–481.

Clifford, R., and Schupbach, T. (1994). Molecular analysis of the *Drosophila* EGF receptor homolog reveals that several genetically defined classes of alleles cluster in subdomains of the receptor protein. *Genetics* **137**, 531–550.

Clyde, D., and Bownes, M. (2000). The Dstpk61 locus of *Drosophila* produces multiple transcripts and protein isoforms, suggesting it is involved in multiple signalling pathways. *J. Endocrinol.* **167**, 391–401.

Coast, G. M. (1996). Neuropeptides implicated in the control of diuresis in insects. *Peptides* **17**, 327–336.

Cox, K. J., Tensen, C. P., Van der Schors, R. C., Li, K. W., van Heerikhuizen, H., Vreugdenhil, E., Geraerts, W. P., and Burke, J. F. (1997). Cloning, characterization, and expression of a G-protein-coupled receptor from *Lymnaea stagnalis* and identification of a leucokinin-like peptide, PSFHSWSamide, as its endogenous ligand. *J. Neurosci.* **17**, 1197–1205.

Currie, M. G., Fok, K. F., Kato, J., Moore, R. J., Hamra, F. K., Duffin, K. L., and Smith, C. E. (1992). Guanylin: An endogenous activator of intestinal guanylate cyclase. *Proc. Natl. Acad. Sci. USA* **89**, 947–951.

Cvejic, S., Zhu, Z., Felice, S. J., Berman, Y., and Huang, X. Y. (2004). The endogenous ligand Stunted of the GPCR Methuselah extends lifespan in *Drosophila*. *Nat. Cell Biol.* **6**, 540–546.

Dan, H. C., Sun, M., Yang, L., Feldman, R. I., Sui, X. M., Ou, C. C., Nellist, M., Yeung, R. S., Halley, D. J., Nicosia, S. V., Pledger, W. J., and Cheng, J. Q. (2002). Phosphatidylinositol 3-kinase/Akt pathway regulates tuberous sclerosis tumor suppressor complex by phosphorylation of tuberin. *J. Biol. Chem.* **277**, 35364–35370.

Davies, S. A., Huesmann, G. R., Maddrell, S. H., O'Donnell, M. J., Skaer, N. J., Dow, J. A., and Tublitz, N. J. (1995). CAP2b, a cardioacceleratory peptide, is present in *Drosophila* and stimulates tubule fluid secretion via cGMP. *Am. J. Physiol.* **269**, R1321–R1326.

de Bono, M., and Bargmann, C. I. (1998). Natural variation in neuropeptide Y receptor homolog modifies social behavior and food response in *C. elegans*. *Cell* **94**, 679–689.

Denell, R, and Shippy, T. (2001). Comparative insect developmental genetics: Phenotypes without mutants. *Bioessays* **23**, 379–382 (Review).

Dewey, E. M., McNabb, S. L., Ewer, J., Kuo, G. R., Takanishi, C. L., Truman, J. W., and Honegger, H. W. (2004). Identification of the gene encoding bursicon, an insect neuropeptide responsible for cuticle sclerotization and wing spreading. *Curr. Biol.* **14**, 1208–1213.

Donly, B. C., Ding, Q., Tobe, S. S., and Bendena, W. G. (1993). Molecular cloning of the gene for the allatostatin family of neuropeptides from the cockroach *Diploptera punctata*. *Proc. Natl. Acad. Sci. USA* **90**, 8807–8811.

Duve, H., Thorpe, A., Scott, A. G., Johnsen, A. H., Rehfeld, J. F., Hines, E., and East, P. D. (1995). The sulfakinins of the blowfly *Calliphora vomitoria*. Peptide isolation, gene cloning and expression studies. *Eur. J. Biochem.* **232**, 633–640.

Duve, H., Johnsen, A. H., Maestro, J. L., Scott, A. G., Jaros, P. P., and Thorpe, A. (1997). Isolation and identification of multiple neuropeptides of the allatostatin superfamily in the shore crab *Carcinus maenas*. *Eur. J. Biochem.* **250**, 727–734.

Egerod, K., Reynisson, E., Hauser, F., Williamson, M., Cazzamali, G., and Grimmelikhuijzen, C. J. (2003a). Molecular identification of the first insect proctolin receptor. *Biochem. Biophys. Res. Commun.* **306**, 437–442.

Egerod, K., Reynisson, E., Hauser, F., Cazzamali, G., Williamson, M., and Grimmelikhuijzen, C. J. (2003b). Molecular cloning and functional expression of the first two specific insect myosuppressin receptors. *Proc. Natl. Acad. Sci. USA* **100**, 9808–9813.

Eigenheer, R. A., Nicolson, S. W., Schegg, K. M., Hull, J. J., and Schooley, D. A. (2002). Identification of a potent antidiuretic factor acting on beetle Malpighian tubules. *Proc. Natl. Acad. Sci. USA* **99**, 84–89.

Eriksen, K. K., Hauser, F., Schiott, M., Pedersen, K. M., Sondergaard, L., and Grimmelikhuijzen, C. J. (2000). Molecular cloning, genomic organization, developmental regulation, and a knock-out mutant of a novel leu-rich repeats-containing G protein-coupled receptor (DLGR-2) from *Drosophila melanogaster. Genome Res.* **10,** 924–938.

Ewer, J., and Truman, J. W. (1996). Increases in cyclic $3'$, $5'$-guanosine monophosphate (cGMP) occur at ecdysis in an evolutionarily conserved crustacean cardioactive peptide-immunoreactive insect neuronal network. *J. Comp. Neurol.* **370,** 330–341.

Feng, G., Reale, V., Kennedy, K., Chatwin, H. M., Evans, P. D., and Hall, L. M. (1999). Cloning and functional characterization of a novel neuropeptide F-like receptor from *Drosophila melanogaster. Soc. Neurosci. Abstr.* **25,** 183.

Feng, G., Reale, V., Chatwin, H., Kennedy, K., Venard, R., Ericsson, C., Yu, K., Evans, P. D., and Hall, L. M. (2003). Functional characterization of a neuropeptide F-like receptor from *Drosophila melanogaster. Eur. J. Neurosci.* **18,** 227–238.

Fernandez, R., Tabarini, D., Azpiazu, N., Frasch, M., and Schlessinger, J. (1995). The *Drosophila* insulin receptor homologue: A gene essential for embryonic development encodes two receptor isoforms with different signaling potential. *EMBO J.* **14,** 3373–3384.

Fernlund, P., and Josefsson, L. (1972). Crustacean color-change hormone: Amino acid sequence and chemical synthesis. *Science* **177,** 173–175.

Fujisawa, J., Muneoka, Y., Takahashi, T., Takao, T., Shimonishi, Y., Kubota, I., Ikeda, T., Minakata, H., Nomoto, K., Kiss, T., and Hiripi, L. (1994). An invertebrate-type tachykinin isolated from the freshwater bivalve mollusc *Anodonta cygnea*. *In* "Peptide Chemistry 1993" (Y. Okada, Ed.), Osaka Protein Research, Foundation, 161–164.

Fukuda, M., Gotoh, Y., and Nishida, E. (1997). Interaction of MAP kinase with MAP kinase kinase: Its possible role in the control of nucleocytoplasmatic transport of MAP kinase. *EMBO J.* **16,** 1901–1908.

Furuya, K., Liao, S., Reynolds, S. E., Ota, R. B., Hackett, M., and Schooley, D. A. (1993). Isolation and identification of a cardioactive peptide from *Tenebrio molitor* and *Spodoptera eridania. Biol. Chem. Hoppe Seyler* **374,** 1065–1074.

Gao, X., Zhang, Y., Arrazola, P., Hino, O., Kobayashi, T., Yeung, R. S., Ru, B., and Pan, D. (2002). Tsc tumor suppressor proteins antagonize amino acid-TOR signaling. *Nat. Cell Biol.* **12,** 332–339.

Gäde, G. (1996). The revolution in insect neuropeptides illustrated by the adipokinetic hormone/red pigment-concentrating hormone family of peptides. *Z. Naturforsch.* **51,** 607–617.

Gäde, G., Hoffmann, K. H., and Spring, J. H. (1997). Hormonal regulation in insects: Facts, gaps, and future directions. *Physiol. Rev.* **77,** 963–1032.

Gammie, S. C., and Truman, J. W. (1999). Eclosion hormone provides a link between ecdysis-triggering hormone and crustacean cardioactive peptide in the neuroendocrine cascade that controls ecdysis behavior. *J. Exp. Biol.* **202,** 343–352.

Garczynski, S. F., Brown, M. R., Shen, P., Murray, T. F., and Crim, J. W. (2002). Characterization of a functional neuropeptide F receptor from *Drosophila melanogaster. Peptides* **23,** 773–780.

Garger, A. V., Richard, E. A., and Lisman, J. E. (2004). The excitation cascade of *Limulus* ventral photoreceptors: Guanylate cyclase as the link between InsP3-mediated Ca^{2+} release and the opening of cGMP-gated channels. *BMC Neurosci.* **5,** 7.

Garside, C. S., Nachman, R. J., and Tobe, S. S. (2000). Injection of Dip-allatostatin or Dip-allatostatin pseudopeptides into mated female *Diploptera punctata* inhibits endogenous rates of JH biosynthesis and basal oocyte growth. *Insect Biochem. Mol. Biol.* **30,** 703–710.

Gaul, U., Mardon, G., and Rubin, G. M. (1992). A putative Ras GTPase activating protein acts as a negative regulator of signaling by the sevenless receptor tyrosine kinase. *Cell* **68,** 1007–1019.

Gigliotti, S., Cavaliere, V., Manzi, A., Tino, A., Graziani, F., and Malva, C. (1993). A membrane guanylate cyclase *Drosophila* homologue gene exhibits maternal and zygotic expression. *Dev. Biol.* **159,** 450–461.

Goy, M. F. (1990). Activation of membrane guanylate cyclase by an invertebrate peptide hormone. *J. Biol. Chem.* **265,** 20220–20227.

Graf, R., Neuenschwander, S., Brown, M. R., and Ackermann, U. (1997). Insulin-mediated secretion of ecdysteroids from mosquito ovaries and molecular cloning of the insulin receptor homologue from ovaries of bloodfed *Aedes aegypti. Insect Mol. Biol.* **6**, 151–163.

Green, K. A., Falconer, S. W., and Cottrell, G. A. (1994). The neuropeptide Phe-Met-Arg-Phe-NH2 (FMRFamide) directly gates two ion channels in an identified *Helix* neurone. *Pflugers Arch.* **428**, 232–240.

Grey, J. M., Karow, D. S., Lu, H., Chang, A. J., Chang, J. S., Ellis, R. E., Marletta, M. A., and Bargmann, C. I. (2004). Oxygen sensation and social feeding mediated by a *C. elegans* guanylate cyclase homologue. *Nature* **430**, 317–322.

Gryzik, T., and Muller, H. A. (2004). *FGF8-like1* and *FGF8-like2* encode putative ligands of the FGF receptor Htl and are required for mesoderm migration in the *Drosophila* gastrula. *Curr. Biol.* **14**, 659–667.

Gubb, D., and Garcia-Bellido, A. (1982). A genetic analysis of the determination of cuticular polarity during development in *Drosphila melanogaster. J. Embryol. Exp. Morphol.* **68**, 37–57.

Guerrero, F. D. (1997). Cloning of a cDNA from stable fly which encodes a protein with homology to a *Drosophila* receptor for tachykinin-like peptides. *Ann. N.Y. Acad. Sci.* **814**, 310–311.

Ha, S. D., Kataoka, H., Suzuki, A., Kim, B. J., Kim, H. J., Hwang, S. H., and Kong, J. Y. (2000). Cloning and sequence analysis of cDNA for diuretic hormone receptor from *Bombyx mori. Mol. Cells* **10**, 13–17.

Hamra, F. K., Forte, L. R., Eber, S. L., Pidhorodeckyj, N. V., Krause, W. J., Freeman, R. H., Chin, D. T., Tompkins, J. A., Fok, K. F., and Smith, C. E. (1993). Uroguanylin: Structure and activity of a second endogenous peptide that stimulates intestinal guanylate cyclase. *Proc. Natl. Acad. Sci. USA* **90**, 10464–10468.

Han, J. W., Pearson, R. B., Dennis, P. B., and Thomas, G. (1995). Rapamycin, wortmannin, and the methylxanthine SQ200006 inactivate p70s6k by inducing dephosphorylation of the same subset of sites. *J. Biol. Chem.* **270**, 21396–21403.

Hardie, R. C., Martin, F., Cochrane, G. W., Juusola, M., Georgiev, P., and Raghu, P. (2002). Molecular basis of amplification in *Drosophila* phototransduction: Roles for G protein, phospholipase C, and diacylglycerol kinase. *Neuron* **36**, 689–701.

Haruta, T., Uno, T., Kawahara, J., Takano, A., Egawa, K., Sharma, P. M., Olefsky, J. M., and Kobayashi, M. (2000). A rapaycin-sensitive pathway down-regulates insulin signaling via phosphorylation and proteasomal degradation of insulin receptor substrate-1. *Mol. Endocrinol.* **14**, 782–794.

Hauser, F., Nothacker, H. P., and Grimmelikhuijzen, C. J. (1997). Molecular cloning, genomic organization, and developmental regulation of a novel receptor from *Drosophila melanogaster* structurally related to members of the thyroid-stimulating hormone, follicle-stimulating hormone, luteinizing hormone/choriogonadotropin receptor family from mammals. *J. Biol. Chem.* **272**, 1002–1010.

Hauser, F., Sondergaard, L., and Grimmelikhuijzen, C. J. P. (1998). Molecular cloning, genomic organization and developmental regulation of a novel receptor from *Drosophila melanogaster* structurally related to gonadotropin-releasing hormone receptors for vertebrates. *Biochem. Biophys. Res. Commun.* **249**, 822–828.

Hayes, T. K., Pannabecker, T. L., Hinckley, D. J., Holman, G. M., Nachman, R. J., Petzel, D. H., and Beyenbach, K. W. (1989). Leucokinins, a new family of ion transport stimulators and inhibitors in insect Malpighian tubules. *Life Sci.* **44**, 1259–1266.

Heino, T. I., Karpanen, T., Wahlstrom, G., Pulkkinen, M., Eriksson, U., Alitalo, K., and Roos, C. (2001). The *Drosophila* VEGF receptor homolog is expressed in hemocytes. *Mech. Dev.* **109**, 69–77.

Heldin, C.H, Miyazono, K, and Ten Dijke, P. (1997). TGF-beta signalling from cell membrane to nucleus through SMAD proteins. *Nature* **390**, 465–471 (Review).

Hewes, R. S., and Taghert, P. H. (2001). Neuropeptides and neuropeptide receptors in the *Drosophila melanogaster* genome. *Genome Res.* **11**, 1126–1142.

Hill, C. A., Fox, A. N., Pitts, R. J., Kent, L. B., Tan, P. L., Chrystal, M. A., Cravchik, A., Collins, F. H., Robertson, H. M., and Zwiebel, L. J. (2002). G protein-coupled receptors in *Anopheles gambiae*. *Science* **298**, 176–178.

Hoffmann, J. A. (2003). The immune response of *Drosophila*. *Nature* **426**, 33–38 (Review).

Holman, G. M., Cook, B. J., and Nachman, R. J. (1986a). Isolation, primary structure and synthesis of two neuropeptides from *Leucophaea maderae*: Members of a new family of cephalomyotropins. *Comp. Biochem. Physiol. C* **84**, 205–211.

Holman, G. M., Cook, B. J., and Nachman, R. J. (1986b). Primary structure and synthesis of a blocked myotropic neuropeptide isolated from the cockroach, *Leucophaea maderae*. *Comp. Biochem. Physiol. C* **85**, 219–224.

Holmes, S. P., He, H., Chen, A. C., Ivie, G. W., and Pietrantonio, P. V. (2000). Cloning and transcriptional expression of a leucokinin-like peptide receptor from the southern cattle tick, *Boophilus microplus* (Acari: Ixodidae). *Insect Mol. Biol.* **9**, 457–465.

Holst, B., Hastrup, H., Raffetseder, U., Martini, L., and Schwartz, T. W. (2001). Two active molecular phenotypes of the tachykinin NK1 receptor revealed by G-protein fusions and mutagenesis. *J. Biol. Chem.* **276**, 19793–19799.

Horodyski, F. M., Bhatt, S. R., and Lee, K. Y. (2001). Alternative splicing of transcripts expressed by the *Manduca sexta* allatotropin (Mas-AT) gene is regulated in a tissue-specific manner. *Peptides* **22**, 263–269.

Hsu, S. Y., Nakabayashi, K., Nishi, S., Kumagai, J., Kudo, M., Sherwood, O. D., and Hsueh, A. J. (2002). Activation of orphan receptors by the hormone relaxin. *Science* **295**, 671–674.

Hua, Y. J., Tanaka, Y., Nakamura, K., Sakakibara, M., Nagata, S., and Kataoka, H. (1999). Identification of a prothoracicostatic peptide in the larval brain of the silkworm, *Bombyx mori*. *J. Biol. Chem.* **274**, 31169–31173.

Huesmann, G. R., Cheung, C. C., Loi, P. K., Lee, T. D., Swiderek, K. M., and Tublitz, N. J. (1995). Amino acid sequence of CAP2b, an insect cardioacceleratory peptide from the tobacco hawkmoth *Manduca sexta*. *FEBS Lett.* **371**, 311–314.

Hull, J. J., Ohnishi, A., Moto, K., Kawasaki, Y., Kurata, R., Suzuki, M. G., and Matsumoto, S. (2004). Cloning and characterization of the pheromone biosynthesis activating neuropeptide receptor from the Silkmoth, *Bombyx mori* : Significance of the carboxyl terminus in receptor internalization. *J. Biol. Chem.* [Epub ahead of print].

Ikeda, T., Minakata, H., Nomoto, K., Kubota, I., and Muneoka, Y. (1993). 2 Novel tachykinin-related neuropeptides in the echiuroid worm, *Urechis unicinctus*. *Biochem. Biophys. Res. Commun.* **192**, 1–6.

Inoki, K., Li, Y., Zhu, T., Wu, J., and Guan, K. L. (2002). TSC2 is phosphorylated and inhibited by Akt and supresses mTOR signalling. *Nat. Cell Biol.* **4**, 648–657.

Isaac, R. E., Taylor, C. A., Hamasaka, Y., Nassel, D. R., and Shirras, A. D. (2004). Proctolin in the post-genomic era: New insights and challenges. *Invert. Neurosci.* **5**, 51–64.

Iversen, A., Cazzamali, G., Williamson, M., Hauser, F., and Grimmelikhuijzen, C. J. (2002a). Molecular cloning and functional expression of a *Drosophila* receptor for the neuropeptides capa-1 and -2. *Biochem. Biophys. Res. Commun.* **299**, 628–633.

Iversen, A., Cazzamali, G., Williamson, M., Hauser, F., and Grimmelikhuijzen, C. J. (2002b). Molecular identification of the first insect ecdysis triggering hormone receptors. *Biochem. Biophys. Res. Commun.* **299**, 924–931.

Jacinto, E., and Hall, M. N. (2003). Tor signaling in bugs, brain and brawn. *Nat. Rev. Mol. Biol.* **4**, 117–126.

Jahn, O., Eckart, K., Brauns, O., Tezval, H., and Spiess, J. (2002). The binding protein of corticotropin-releasing factor: Ligand-binding site and subunit structure. *Proc. Natl. Acad. Sci. USA* **99**, 12055–12060.

Janson, K., Cohen, E. D., and Wilder, E. L. (2001). Expression of DWnt6, DWnt10, and DFz4 during *Drosophila* development. *Mech. Dev.* **103**, 117–120.

Jeziorski, M. C., Green, K. A., Sommerville, J., and Cottrell, G. A. (2000). Cloning and expression of a FMRFamide-gated Na(+) channel from *Helisoma trivolvis* and comparison with the native neuronal channel. *J. Physiol.* **526**, 13–25.

Jia, J., Tong, C., and Jiang, J. (2003). Smoothened transduces Hedgehog signal by physically interacting with Costal2/Fused complex through its C-terminal tail. *Genes Dev.* **17**, 2709–2720.

Johard, H. A., Muren, J. E., Nichols, R., Larhammar, D. S., and Nassel, D. R. (2001). A putative tachykinin receptor in the cockroach brain: Molecular cloning and analysis of expression by means of antisera to portions of the receptor protein. *Brain Res.* **919**, 94–105.

Johnsen, A. H., Duve, H., Davey, M., Hall, M., and Thorpe, A. (2000). Sulfakinin neuropeptides in a crustacean. Isolation, identification and tissue localization in the tiger prawn *Penaeus monodon*. *Eur. J. Biochem.* **267**, 1153–1160.

Johnson, E. C., Bohn, L. M., Barak, L. S., Birse, R. T., Nassel, D. R., Caron, M. G., and Taghert, P. H. (2003a). Identification of *Drosophila* neuropeptide receptors by G protein-coupled receptors-beta-arrestin2 interactions. *J. Biol. Chem.* **278**, 52172–52178.

Johnson, E. C., Garczynski, S. F., Park, D., Crim, J. W., Nassel, D. R., and Taghert, P. H. (2003b). Identification and characterization of a G protein-coupled receptor for the neuropeptide proctolin in *Drosophila melanogaster*. *Proc. Natl. Acad. Sci. USA* **100**, 6198–6203.

Johnson, E. C., Bohn, L. M., and Taghert, P. H. (2004). *Drosophila* CG8422 encodes a functional diuretic hormone receptor. *J. Exp. Biol.* **207**, 743–748.

Kage, R., McGregor, G. P., Thim, L., and Conlon, J. M. (1988). Neuropeptide-gamma - a peptide isolated from rabbit intestine that is derived from gamma-preprotachykinin. *J. Neurochem.* **50**, 1412–1417.

Kanda, A., Iwakoshi-Ukena, E., Takuwa-Kuroda, K., and Minakata, H. (2003). Isolation and characterization of novel tachykinins from the posterior salivary gland of the common octopus *Octopus vulgaris*. *Peptides* **24**, 35–43.

Kanda, H., Igaki, T., Kanuka, H., Yagi, T., and Miura, M. (2002). Wengen, a member of the *Drosophila* tumor necrosis factor receptor superfamily, is required for Eiger signaling. *J. Biol. Chem.* **277**, 28372–28375.

Kangawa, K., Minamino, N., Fukuda, A., and Matsuo, H. (1983). Neuromedin-K - a novel mammalian tachykinin identified in porcine spinal-cord. *Biochem. Biophys. Res. Commun.* **114**, 533–540.

Kawada, T., Masuda, K., Satake, H., Minakata, H., Muneoka, Y., and Nomoto, K. (2000). Identification of multiple urechistachykinin peptides, gene expression, pharmacological activity, and detection using mass spectrometric analyses. *Peptides* **21**, 1777–1783.

Kawada, T., Furukawa, Y., Shimizu, Y., Minakata, H., Nomoto, K., and Satake, H. (2002). A novel tachykinin-related peptide receptor. Sequence, genomic organization, and functional analysis. *Eur. J. Biochem.* **269**, 4238–4246.

Kawase, E., Wong, M. D., Ding, B. C., and Xie, T. (2004). Gbb/BMP signaling is essential for maintaining germline stem cells and for repressing *bam* transcription in the *Drosophila* testis. *Development* **131**, 1365–1375.

Kay, I., Coast, G. M., Cusinato, O., Wheeler, C. H., Totty, N. F., and Goldworthy, G. J. (1991). Isolation and characterization of a diuretic peptide from *Acheta domesticus*. Evidence for a family of insect diuretic peptides. *Biol. Chem. Hoppe Seyler* **372**, 505–512.

Kean, L., Cazenave, W., Costes, L., Broderick, K. E., Graham, S., Pollock, V. P., Davies, S. A., Veenstra, J. A., and Dow, J. A. (2002). Two nitridergic peptides are encoded by the gene capability in *Drosophila melanogaster*. *Am. J. Physiol. Regul. Integr. Comp. Physiol.* **282**, 1297–1307.

Khalsa, O., Yoon, J., Torres-Schumann, S., and Wharton, K. A. (1998). TGF-beta/BMP superfamily members, Gbb-60A and Dpp, cooperate to provide pattern information and establish cell identity in the *Drosophila* wing. *Development* **125**, 2723–2734.

Khokhlatchev, A. V., Canagarajah, B., Wilsbacher, J., Robinson, M., and Atkinson, M. (1998). Phosphorylation of the MAP kinase ERK2 promotes its homodimerization and nuclear translocation. *Cell* **93**, 605–613.

Kim, Y. J., Spalovska-Valachova, I., Cho, K. H., Zitnanova, I., Park, Y., Adams, M. E., and Zitnan, D. (2004). Corazonin receptor signaling in ecdysis initiation. *Proc. Natl. Acad. Sci. USA* **101**, 6704–6709.

Kimura, K., Tissenbaum, H. A., Lui, Y., and Ruvkun, G. (1997). DAF-2, an insulin receptor-like gene that regulates longevity and diapause in *Caenorhabditis elegans*. *Science* **277**, 942–946.

Kimura, K., Kodama, A., Hayasaka, Y., and Ohta, T. (2004). Activation of the cAMP/PKA signaling pathway is required for post-ecdysial cell death in wing epidermal cells of *Drosophila melanogaster*. *Development* **131**, 1597–1606.

Kostenis, E. (2001). Is Galpha16 the optimal tool for fishing ligands of orphan G-protein-coupled receptors? *Trends Pharmacol. Sci.* **22**, 560–564 (Review).

Kuhn, M. (2003). Structure, regulation, and function of mammalian membrane guanylyl cyclase receptors, with a focus on guanylyl cyclase-A. *Circ. Res.* **93**, 700–709.

Kramer, S. J., Toschi, A., Miller, C. A., Kataoka, H., Quistad, G. B., Li, J. P., Carney, R. L., and Schooley, D. A. (1991). Identification of an allatostatin from the tobacco hornworm *Manduca sexta*. *Proc. Natl. Acad. Sci. USA* **88**, 9458–9462.

Kreienkamp, H. J., Larusson, H. J., Witte, I., Roeder, T., Birgül, N., Honck, H. H., Harder, S., Ellinghausen, G., Buck, F., and Richter, D. (2002). Functional annotation of two orphan G-protein-coupled receptors, Drostar-1 and -2, from *Drosophila melanogaster* and their ligands by reverse pharmacology. *J. Biol. Chem.* **277**, 39937–39943.

Kretzschmar, M., Liu, F., Hata, A., Doody, J., and Massagué, J. (1997). The TGF-b family mediator Smad1 is phosphorylated directly and activated functionally by the BMP receptor kinase. *Genes Dev.* **11**, 984–995.

Kubiak, T. M., Larsen, M. J., Burton, K. J., Bannow, C. A., Martin, R. A., Zantello, M. R., and Lowery, D. E. (2002). Cloning and functional expression of the first *Drosophila* melanogaster sulfakinin receptor DSK-R1. *Biochem. Biophys. Res. Commun.* **291**, 313–320.

Kunlansky Poltilove, R. M., Jacobs, A. R., Hafts, C. R., Xu, J., and Taylor, S. I. (2000). Characterization of *Drosophila* Insulin receptor substrate. *J. Biol. Chem.* **275**, 23346–23354.

Kurtz, M. M., Wang, R., Clements, M. K., Cascieri, M. A., Austin, C. P., Cunningham, B. R., Chicchi, G. G., and Liu, Q. (2002). Identification, localization and receptor characterization of novel mammalian substance P-like peptides. *Gene* **296**, 205–212.

Kuttenkeuler, D, and Boutros, M. (2004). Genome-wide RNAi as a route to gene function in *Drosophila*. *Brief Funct. Genomic. Proteomic.* **3**, 168–176 (Review).

Lai, E. C. (2004). Notch signaling: Control of cell communication and cell fate. *Development* **131**, 965–973 (Review).

Lange, A. B. (2002). A review of the involvement of proctolin as a cotransmitter and local neurohormone in the oviduct of the locust, *Locusta migratoria*. *Peptides* **23**, 2063–2070 (Review).

Larhammar, D., Soderberg, C., and Lundell, I. (1998). Evolution of the neuropeptide Y family and its receptors. *Ann. N. Y. Acad. Sci.* **839**, 35–40.

Larsen, M. J., Burton, K. J., Zantello, M. R., Smith, V. G., Lowery, D. L., and Kubiak, T. M. (2001). Type A allatostatins from *Drosophila melanogaster* and *Diplotera puncata* activate two *Drosophila* allatostatin receptors, DAR-1 and DAR-2, expressed in CHO cells. *Biochem. Biophys. Res. Commun.* **286**, 895–901.

Lee, K., You, K., Choo, J., Han, Y., and Yu, K. (2004). *Drosophila* short neuropeptide F regulates food intake and body size. *J. Biol. Chem.* **279**, 50781–50789.

Leevers, S. J., Weinkove, D., MacDougall, L. K., Hafen, E., and Waterfield, M. D. (1996). The *Drosophila* phosphoinositide 3-kinase Dp110 promotes cell growth. *EMBO J.* **15**, 6584–6594.

Lehman, H. K., Murgiuc, C. M., Miller, T. A., Lee, T. D., and Hildebrand, J. G. (1993). Crustacean cardioactive peptide in the sphinx moth, *Manduca sexta*. *Peptides* **14**, 735–741.

Lenz, C., Williamson, M., and Grimmelikhuijzen, C. J. (2000). Molecular cloning and genomic organization of a second probable allatostatin receptor from *Drosophila melanogaster*. *Biochem. Biophys. Res. Commun.* **273**, 571–577.

Lenz, C., Williamson, M., Hansen, G. N., and Grimmelikhuijzen, C. J. (2001). Identification of four *Drosophila* allatostatins as the cognate ligands for the *Drosophila* orphan receptor DAR-2. *Biochem. Biophys. Res. Commun.* **286**, 1117–1122.

Li, X. J., Wolfgang, W., Wu, Y. N., North, R. A., and Forte, M. (1991). Cloning, heterologous expression and developmental regulation of a *Drosophila* receptor for tachykinin-like peptides. *EMBO J.* **10**, 3221–3229.

Li, X. J., Wu, Y. N., North, R. A., and Forte, M. (1992). Cloning, functional expression, and developmental regulation of a neuropeptide Y receptor from *Drosophila melanogaster*. *J. Biol. Chem.* **267**, 9–12.

Lin, Y. J., Seroude, L., and Benzer, S. (1998). Extended life-span and stress resistance in the *Drosophila* mutant *methuselah*. *Science* **282**, 943–946.

Lingueglia, E., Champigny, G., Lazdunski, M., and Barbry, P. (1995). Cloning of the amiloride-sensitive FMRFamide peptide-gated sodium channel. *Nature* **378**, 730–733.

Liu, H. F., Lai, C. Y., Watson, R. D., and Lee, C. Y. (2004). Molecular cloning of a putative membrane form guanylyl cyclase from the crayfish *Procambarus clarkii*. *J. Exp. Zoolog. Part A Comp. Exp. Biol.* **301**, 512–520.

Liu, W., Moon, J., Burg, M., Chen, L., and Pak, W. L. (1995). Molecular characterization of two *Drosophila* guanylate cyclases expressed in the nervous system. *J. Biol. Chem.* **270**, 12418–12427.

Liu, X., Sun, Y., Constantinescu, S. N., Karam, E., Weinberg, R. A., and Lodish, H. F. (1997). Transforming growth factor beta-induced phosphorylation of Smad3 is required for growth inhibition and transcriptional induction in epithelial cells. *Proc. Natl. Acad. Sci. USA* **94**, 10669–10674.

Livneh, E., Glazer, L., Segal, D., Schlessinger, J., and Shilo, B. Z. (1985). The *Drosophila* EGF receptor gene homolog: Conservation of both hormone binding and kinase domains. *Cell* **40**, 599–607.

Lorenz, M. W., Kellner, R., and Hoffmann, K. H. (1995). A family of neuropeptides that inhibit juvenile hormone biosynthesis in the cricket, *Gryllus bimaculatus*. *J. Biol. Chem.* **270**, 21103–21108.

Lowe, D. G., Chang, M. S., Hellmiss, R., Chen, E., Singh, S., Garbers, D. L., and Goeddel, D. V. (1989). Human atrial natriuretic peptide receptor defines a new paradigm for second messenger signal transduction. *EMBO J.* **8**, 1377–1384.

Lucas, K. A., Pitari, G. M., Kazerounian, S., Ruiz-Stewart, I., Park, J., Schulz, S., Chepenik, K. P., and Waldman, S. A. (2000). Guanylyl cyclases and signaling by cyclic GMP. *Pharmacol. Rev.* **52**, 375–413.

Luo, C. W., Dewey, E. M., Sudo, S., Ewer, J., Hsu, S. Y., Honegger, H. W., and Hsueh, A. J. (2005). Bursicon, the insect cuticle-hardening hormone, is a heterodimeric cystine knot protein that activates G protein-coupled receptor LGR2. *Proc. Natl. Acad. Sci. USA* **102**, 2820–2825.

MacDougall, C. N., Clyde, D., Wood, T., Todman, M., Harbison, D., and Bownes, M. (1999). Sex-specific transcripts of the Dstpk61 serine/threonine kinase gene in *Drosophila melanogaster*. *Eur. J. Biochem.* **262**, 456–466.

Maestro, J. L., Aguilar, R., Pascual, N., Valero, M. L., Piulachs, M. D., Andreu, D., Navarro, I., and Belles, X. (2001). Screening of antifeedant activity in brain extracts led to the identification of sulfakinin as a satiety promoter in the German cockroach. Are arthropod sulfakinins homologous to vertebrate gastrins-cholecystokinins? *Eur. J. Biochem.* **268**, 5824–5830.

Maggi, C. A., and Schwartz, T. W. (1997). The dual nature of the tachykinin NK1 receptor. *Trends Pharmacol. Sci.* **18**, 351–355.

Marshall, C. J. (1995). Specificity of receptor tyrosine kinase signaling: Transient versus sustained extracellular signal-regulated kinase activation. *Cell* **80**, 179–185.

Massagué, J., Hata, A., and Liu, F. (1997). TGF-b signaling through the Smad pathway. *Trends Cell Biol.* **7,** 187–192 (Review).

Masu, Y., Nakayama, K., Tamaki, H., Harada, Y., Kuno, M., and Nakanishi, S. (1987). cDNA cloning of bovine substance-K receptor through oocyte expression system. *Nature* **329,** 836–838.

McManus, E. J., and Alessi, D. R. (2002). TSC1-TSC2: A complex tale of PKB-mediated S6K regulation. *Nat. Cell Biol.* **4,** E214–E216.

McNeil, L., Chinkers, M., and Forte, M. (1995). Identification, characterization and developmental regulation of a receptor guanylyl cyclase expressed during early stages of *Drosophila* development. *J. Biol. Chem.* **270,** 7189–7196.

Meeusen, T., Mertens, I., Clynen, E., Baggerman, G., Nichols, R., Nachman, R. J., Huybrechts, R., De Loof, A., and Schoofs, L. (2002). Identification in *Drosophila melanogaster* of the invertebrate G protein-coupled FMRFamide receptor. *Proc. Natl. Acad. Sci. USA* **99,** 15363–15368.

Mendive, F., Van Loy, T., Claeysen, S., Poels, J., Williamson, M., Hauser, F., Grimmelikhuijzen, C., Vassart, G., and Vanden Broeck, J. (2005). *Drosophila* molting neurohormone bursicon is a heterodimer and the natural agonist of the orphan receptor DLGR2. *FEBS Lett.* **579,** 2171–2176.

Meng, X., Wahlstrom, G., Immonen, T., Kolmer, M., Tirronen, M., Predel, R., Kalkkinen, N., Heino, T. I., Sariola, H., and Roos, C. (2002). The *Drosophila* hugin gene codes for myostimulatory and ecdysis-modifying neuropeptides. *Mech. Dev.* **117,** 5–13.

Merte, J, and Nichols, R. (2002). *Drosophila melanogaster* FMRFamide-containing peptides: Redundant or diverse functions? *Peptides* **23,** 209–220 (Review).

Mertens, I., Meeusen, T., Huybrechts, R., De Loof, A., and Schoofs, L. (2002). Characterization of the short neuropeptide F receptor from *Drosophila melanogaster*. *Biochem. Biophys. Res. Commun.* **297,** 1140–1148.

Monnier, D., Colas, J. F., Rosay, P., Hen, R., Borrelli, E., and Maroteaux, L. (1992). NKD, a developmentally regulated tachykinin receptor in *Drosophila*. *J. Biol. Chem.* **267,** 1298–1302.

Mordue, W., and Stone, J. V. (1977). Relative potencies of locust adipokinetic hormone and prawn red pigment-concentrating hormone in insect and crustacean systems. *Gen. Comp. Endocrinol.* **33,** 103–108.

Morton, D. B. (2004a). Invertebrates yield a plethora of atypical guanylyl cyclases. *Mol. Neurobiol.* **29,** 97–115.

Morton, D. B. (2004b). Atypical soluble guanylyl cyclases in *Drosophila* can function as molecular oxygen sensors. *J. Biol. Chem.* **279,** 50651–50653.

Morton, D. B., and Nighorn, A. (2003). MsGC-II, a receptor guanylyl cyclase isolated from the CNS of *Manduca sexta* that is inhibited by calcium. *J. Neurochem.* **84,** 363–372.

Nachman, R. J., Holman, G. M., Haddon, W. F., and Ling, N. (1986). Leucosulfakinin, a sulfated insect neuropeptide with homology to gastrin and cholecystokinin. *Science* **234,** 71–73.

Nachman, R. J., Moyna, G., Williams, H. J., Zabrocki, J., Zadina, J. E., Coast, G. M., and Vanden Broeck, J. (1999). Comparison of active conformations of the insect atachykinin/tachykinin and insect kinin/Tyr-W-MIF-1 neuropeptide family pairs. *Ann. N. Y. Acad. Sci.* **897,** 388–400.

Nakajima, Y., Tsuchida, K., Negishi, M., Ito, S., and Nakanishi, S. (1992). Direct linkage of three tachykinin receptors to stimulation of both phosphatidylinositol hydrolysis and cyclic AMP cascades in transfected Chinese hamster ovary cells. *J. Biol. Chem.* **267,** 2437–2442.

Nambu, J. R., Murphy-Erdosh, C., Andrews, P. C., Feistner, G. J., and Scheller, R. H. (1988). Isolation and characterization of a *Drosophila* neuropeptide gene. *Neuron* **1,** 55–61.

Nässel, D. R. (1999). Tachykinin-related peptides in invertebrates: A review. *Peptides* **20,** 141–158 (Review).

Nässel, D. R. (2002). Neuropeptides in the nervous system of *Drosophila* and other insects: Multiple roles as neuromodulators and neurohormones. *Prog. Neurobiol.* **68,** 1–84.

Nässel, D. R., and O'Shea, M. (1987). Proctolin-like immunoreactive neurons in the blowfly central nervous system. *J. Comp. Neurol.* **265**, 437–454.

Nawa, H., Doteuchi, M., Igano, K., Inouye, K., and Nakanishi, S. (1984). Substance-K—a novel mammalian tachykinin that differs from substance-P in its pharmacological profile. *Life Sci.* **34**, 1153–1160.

Nichols, R. (1992a). Isolation and expression of the *Drosophila* drosulfakinin neural peptide gene-product, DSK-I. *Mol. Cell. Neurosci.* **3**, 342–347.

Nichols, R. (1992b). Isolation and structural characterization of *Drosophila* TDVDHVFLRF amide and FMRFamide-containing neural peptides. *J. Mol. Neurosci.* **3**, 213–218.

Nichols, R. (2003). Signaling pathways and physiological functions of *Drosophila melanogaster* FMRFamide-related peptides. *Annu. Rev. Entomol.* **48**, 485–503 (Review).

Nichols, R., Kaminski, S., Walling, E., and Zornik, E. (1999). Regulating the activity of a cardioacceleratory peptide. *Peptides* **20**, 1153–1158.

Nieto, J., Veelaert, D., Derua, R., Waelkens, E., Cerstiaens, A., Coast, G., Devreese, B., Van Beeumen, J., Calderon, J., De Loof, A., and Schoofs, L. (1998). Identification of one tachykinin- and two kinin-related peptides in the brain of the white shrimp, *Penaeus vannamei*. *Biochem. Biophys. Res. Commun.* **248**, 406–411.

Nighorn, A., Simpson, P. J., and Morton, D. B. (2001). The novel guanylyl cyclase MsGC-I is strongly expressed in higher order neuropils in the brain of *Manduca sexta*. *J. Exp. Biol.* **204**, 305–314.

Nishi, S., Hsu, S. Y., Zell, K., and Hsueh, A. J. (2000). Characterization of two fly LGR (leucine-rich repeat-containing, G protein-coupled receptor) proteins homologous to vertebrate glycoprotein hormone receptors: Constitutive activation of wild-type fly LGR1 but not LGR2 in transfected mammalian cells. *Endocrinology* **141**, 4081–4090.

Nishimura, M., Ohtsuka, K., Takahashi, H., and Yoshimura, M. (2000). Role of FMRFamide-activated brain sodium channel in salt-sensitive hypertension. *Hypertension* **35**, 443–450.

Noyes, B. E., Katz, F. N., and Schaffer, M. H. (1995). Identification and expression of the *Drosophila* adipokinetic hormone gene. *Mol. Cell. Endocrinol.* **109**, 133–141.

Okano, H., Yoshikawa, S., Suzuki, A., Ueno, N., Kaizu, M., Okabe, M., Takahashi, T., Matsumoto, M., Sawamoto, K., and Mikoshiba, K. (1994). Cloning of a *Drosophila melanogaster* homologue of the mouse type-I bone morphogenetic proteins-2/-4 receptor: A potential decapentaplegic receptor. *Gene* **148**, 203–209.

Olivier, J. P., Raabe, T., Henkemeyer, M., Dickson, B., Mbamalu, G., Margolis, B., Schlessinger, J., Hafen, E., and Pawson, T. (1993). Biochemical and genetic analysis of Drk SH2/SH3 adapter protein implicated in coupling the sevenless tyrosine kinase to a activator of Ras guanine nucleotide exchange, Sos. *Cell* **73**, 179–191.

Padgett, R. W., Johnston, R. D. S., and Gelbart, W. M. (1987). A transcript from a *Drosophila* pattern gene predicts a protein homologues to the transforming growth factor-(beta) family. *Nature* **325**, 81–84.

Page, N. M., Bell, N. J., Gardiner, S. M., Manyonda, I. T., Brayley, K. J., Strange, P. G., and Lowry, P. J. (2003). Characterization of the endokinins: Human tachykinins with cardiovascular activity. *Proc. Natl. Acad. Sci. USA* **100**, 6245–6250.

Palanche, T., Ilien, B., Zoffmann, S., Reck, M. P., Bucher, B., Edelstein, S. J., and Galzi, J. L. (2001). The neurokinin A receptor activates calcium and cAMP responses through distinct conformational states. *J. Biol. Chem.* **276**, 34853–34861.

Park, C., Choi, Y., and Yun, Y. (1998). Son of Sevenless binds to the SH3 domain of scr-type tyrosine kinase. *Moll. Cells* **8**, 518–523.

Park, Y., Kim, Y. J., and Adams, M. E. (2002). Identification of G protein-coupled receptors for *Drosophila* PRXamide peptides, CCAP, corazonin, and AKH supports a theory of ligand-receptor coevolution. *Proc. Natl. Acad. Sci. USA* **99**, 11423–11428.

Park, Y., Kim, Y. J., Dupriez, V., and Adams, M. E. (2003). Two subtypes of ecdysis-triggering hormone receptor in *Drosophila melanogaster*. *J. Biol. Chem.* **278**, 17710–17715.

Patel, T. B. (2004). Single transmembrane spanning heterotrimeric G protein-coupled receptors and their signaling cascades. *Pharmacol. Rev.* **56,** 371–385.
Pennefather, J. N., Lecci, A., Candenas, M. L., Patak, E., Pinto, F. M., and Maggi, C. A. (2004). Tachykinins and tachykinin receptors: A growing family. *Life Sci.* **74,** 1445–1463.
Phlippen, M. K., Webster, S. G., Chung, J. S., and Dircksen, H. (2000). Ecdysis of decapod crustaceans is associated with a dramatic release of crustacean cardioactive peptide into the haemolymph. *J. Exp. Biol.* **203,** 521–536.
Poels, J., and Vanden Broeck, J. (2004). Insect basic leucine zipper proteins and their role in cyclic AMP dependent regulation of gene expression. *Int. Rev. Cytol.* **241,** 277–309.
Poels, J., Suner, M. M., Needham, M., Torfs, H., De Rijck, J., De Loof, A., Dunbar, S. J., and Vanden Broeck, J. (2001). Functional expression of a locust tyramine receptor in murine erythroleukaemia cells. *Insect Mol. Biol.* **10,** 541–548.
Poels, J., Nachman, R. J., Åkerman, K. E., Oonk, H. B., Guerrero, F., De Loof, A., Janecka, A. E., Torfs, H., and Vanden Broeck, J. (2004a). Pharmacology of stomoxytachykinin receptor depends on second messenger system. *Peptides* **26,** 109–114.
Poels, J., Van Loy, T., Franssens, V., Detheux, M., Nachman, R. J., Oonk, H. B., Akerman, K. E., Vassart, G., Parmentier, M., De Loof, A., Torfs, H., and Vanden Broeck, J. (2004b). Substitution of conserved glycine residue by alanine in natural and synthetic neuropeptide ligands causes partial agonism at the stomoxytachykinin receptor. *J. Neurochem.* **90,** 472–478.
Poels, J., Franssens, V., Van Loy, T., Martinez, A., Suner, M.-M., Dunbar, S. J., De Loof, A., and Vanden Broeck, J. (2004c). Isoforms of cyclic AMP response element binding proteins (CREB) in *Drosophila* S2 cells. *Biochem. Biophys. Res. Commun.* **320,** 318–324.
Pollock, V. P., McGettigan, J., Cabrero, P., Maudlin, I. M., Dow, J. A., and Davies, S. A. (2004). Conservation of capa peptide-induced nitric oxide signalling in Diptera. *J. Exp. Biol.* **207,** 4135–4145.
Potter, C. J., Pedraza, L. G., and Xu, T. (2002). Akt regulates growth by directly phosphorylating Tsc2. *Nat. Cell Biol.* **4,** 658–665.
Predel, R., Rapus, J., and Eckert, M. (2001). Myoinhibitory neuropeptides in the American cockroach. *Peptides* **22,** 199–208.
Price, M. D., Merte, J., Nichols, R., Koladich, P. M., Tobe, S. S., and Bendena, W. G. (2002). Drosophila melanogaster flatline encodes a myotropin orthologue to *Manduca sexta* allatostatin. *Peptides* **23,** 787–794.
Proux, J. P., Miller, C. A., Li, J. P., Carney, R. L., Girardie, A., Delaage, M., and Schooley, D. A. (1987). Identification of an arginine vasopressin-like diuretic hormone from *Locusta migratoria*. *Biochem. Biophys. Res. Commun.* **149,** 180–186.
Raabe, T., Olivier, J. P., Dickson, B., Liu, X., Gish, G. D., Pawson, T., and Hafen, E. (1995). Biochemical and genetic analysis of Drk SH2/SH3 adaptor protein of *Drosophila*. *EMBO J.* **14,** 2509–2518.
Radford, J. C., Davies, S. A., and Dow, J. A. (2002). Systematic G-protein-coupled receptor analysis in Drosophila melanogaster identifies a leucokinin receptor with novel roles. *J. Biol. Chem.* **277,** 38810–38817.
Rawson, M., Lee, M, Kennedy, E. L., and Selleck, S. B. (2003). *Drosophila* neuromuscular synapse assembly and function require the TGF-beta type I receptor saxophone and the transcription factor Mad. *J. Neurobiol.* **55,** 134–150.
Reagan, J. D. (1994). Expression cloning of an insect diuretic hormone receptor. A member of the calcitonin/secretin receptor family. *J. Biol. Chem.* **269,** 9–12.
Reagan, J. D. (1996). Molecular cloning and functional expression of a diuretic hormone receptor from the house cricket, *Acheta domesticus*. *Insect Biochem. Mol. Biol.* **26,** 1–6.
Reale, V., Chatwin, H. M., Hall, L. M., and Evans, P. D. (2000). The action of endogenous Neuropeptide F-like peptides on a cloned Neuropeptide F-like receptor from *Drosophila melanogaster*. *Soc. Neurosci. Abstr.* **26,** 915.

Reale, V., Chatwin, H. M., and Evans, P. D. (2004). The activation of G-protein gated inwardly rectifying K^+ channels by a cloned *Drosophila melanogaster* neuropeptide F-like receptor. *Eur. J. Neurosci.* **19**, 570–576.

Riehle, M. A., Garczynski, S. F., Crim, J. W., Hill, C. A., and Brown, M. R. (2002). Neuropeptides and peptide hormones in *Anopheles gambiae*. *Science* **298**, 172–175.

Rosenkilde, C., Cazzamali, G., Williamson, M., Hauser, F., Sondergaard, L., De Lotto, R., and Grimmelikhuijzen, C. J. (2003). Molecular cloning, functional expression, and gene silencing of two *Drosophila* receptors for the *Drosophila* neuropeptide pyrokinin-2. *Biochem. Biophys. Res. Commun.* **309**, 485–494.

Ruan, Y., Chen, C., Cao, Y., and Garofalo, R. S. (1995). The *Drosophila* insulin receptor contains a novel carboxyl-terminal extension likely to play an important role in signal transduction. *J. Biol. Chem.* **270**, 4236–4243.

Sachon, E., Girault-Lagrange, S., Chassaing, G., Lavielle, S., and Sagan, S. (2002). Analogs of Substance P modified at the C-terminus which are both agonist and antagonist of the NK-1 receptor depending on the second messenger pathway. *J. Pept. Res.* **59**, 232–240.

Satake, H., Kawada, T., Nomoto, K., and Minakata, H. (2003). Insight into tachykinin-related peptides, their receptors, and invertebrate tachykinins: A review. *Zoolog. Sci.* **20**, 533–549.

Scarborough, R. M., Jamieson, G. C., Kalish, F., Kramer, S. J., McEnroe, G. A., Miller, C. A., and Schooley, D. A. (1984). Isolation and primary structure of two peptides with cardioacceleratory and hyperglycemic activity from the corpora cardiaca of *Periplaneta americana*. *Proc. Natl. Acad. Sci. USA* **81**, 5575–5579.

Schaffer, M. H., Noyes, B. E., Slaughter, C. A., Thorne, G. C., and Gaskell, S. J. (1990). The fruit fly *Drosophila melanogaster* contains a novel charged adipokinetic-hormone-family peptide. *Biochem. J.* **269**, 315–320.

Schindelholz, B., Knirr, M., Warrior, R., and Zinn, K. (2001). Regulation of CNS and motor axon guidance in *Drosophila* by the receptor tyrosine phosphatase DPTP52F. *Development* **128**, 4371–4382.

Schneider, L. E., and Taghert, P. H. (1988). Isolation and characterization of a *Drosophila* gene that encodes multiple neuropeptides related to Phe-Met-Arg-Phe-NH2 (FMRFamide). *Proc. Natl. Acad. Sci. USA* **85**, 1993–1997.

Scholz, N. L., Goy, M. F., Truman, J. W., and Graubard, K. (1996). Nitric oxide and peptide neurohormones activate cGMP synthesis in the crab stomatogastric nervous system. *J. Neurosci.* **16**, 1614–1622.

Schoofs, L., Holman, G. M., Hayes, T. K., Nachman, R. J., and De Loof, A. (1990). Locustatachykinin-I and Locustatachykinin-II, 2 novel insect neuropeptides with homology to peptides of the vertebrate tachykinin family. *FEBS Lett.* **261**, 397–401.

Schulz, S., Green, C. K., Yuen, P. S., and Garbers, D. L. (1990). Guanylyl cyclase is a heat stable enterotoxin receptor. *Cell* **63**, 941–948.

Schulz, S., Singh, S., Bellet, R. A., Singh, G., Tubb, D. J., Chin, H., and Garbers, D. L. (1989). The primary structure of a plasma membrane guanylate cyclase demonstrates diversity within this new receptor family. *Cell* **58**, 1155–1162.

Scully, A. L., McKeown, M, and Thomas, J. B. (1999). Isolation and characterization of Dek, a *Drosophila* eph receptor protein tyrosine kinase. *Mol. Cell. Neurosci.* **13**, 337–347.

Secher, T., Lenz, C., Cazzamali, G., Sorensen, G., Williamson, M., Hansen, G. N., Svane, P., and Grimmelikhuijzen, C. J. (2001). Molecular cloning of a functional allatostatin gut/brain receptor and an allatostatin preprohormone from the silkworm *Bombyx mori*. *J. Biol. Chem.* **276**, 47052–47060.

Shepherd, P. R., Withers, D. J., and Siddle, K. (1998). Phosphoinositide 3-kinase: The key switch mechanism in insulin signaling. *Biochem. J.* **333**, 471–490.

Shigemoto, R., Yokota, Y., Tsuchida, K., and Nakanishi, S. (1990). Cloning and expression of a rat neuromedin K receptor cDNA. *J. Biol. Chem.* **265**, 623–628.

Shishido, E., Higashijima, S., Emori, Y., and Saigo, K. (1993). Two FGF-receptor homologues of *Drosophila*: One is expressed in mesodermal primordium in early embryos. *Development* **117**, 751–761.
Simon, M. A., Dodson, G. S., and Rubin, G. M. (1993). An SH3-SH2-SH3 protein is required for p21Ras1 activation and binds to sevenless and SOS proteins *in vitro*. *Cell* **73**, 169–177.
Simonet, G., Poels, J., Claeys, I., Van Loy, T., Franssens, V., De Loof, A., and Vanden Broeck, J. (2004). Neuro-endocrinological and molecular aspects of insect reproduction. *J. Neuroendocrinol.* **16**, 649–659.
Simpson, P. J., Nighorn, A., and Morton, D. B. (1999). Identification and characterization of a novel guanylyl cyclase that is related to receptor guanylyl cyclases, but lacks extracellular and transmembrane domains. *J. Biol. Chem.* **274**, 4440–4446.
Singh, S., Lowe, D. G., Thorpe, D. S., Rodriguez, H., Kuang, W. J., Dangott, L. J., Chinkers, M., Goeddel, D. V., and Garbers, D. L. (1988). Membrane guanylate cyclase is a cell-surface receptor with homology to protein kinases. *Nature* **334**, 708–712.
Sivasankaran, R., Calleja, M., Morata, G., and Basler, K. (2000). The Wingless target gene Dfz3 encodes a new member of the *Drosophila* Frizzled family. *Mech. Dev.* **91**, 427–431.
Siviter, R. J., Coast, G. M., Winther, A. M., Nachman, R. J., Taylor, C. A., Shirras, A. D., Coates, D., Isaac, R. E., and Nassel, D. R. (2000). Expression and functional characterization of a *Drosophila* neuropeptide precursor with homology to mammalian preprotachykinin A. *J. Biol. Chem.* **275**, 23273–23280.
Sokal, I., Alekseev, A., and Palczewski, K. (2003). Photoreceptor guanylate cyclase variants: cGMP production under control. *Acta Biochim. Pol.* **50**, 1075–1095.
Song, W., Ranjan, R., Dawson-Scully, K., Bronk, P., Marin, L., Seroude, L., Lin, Y. J., Nie, Z., Atwood, H. L., Benzer, S., and Zinsmaier, K. E. (2002). Presynaptic regulation of neurotransmission in *Drosophila* by the G protein-coupled receptor methuselah. *Neuron* **36**, 105–119.
Song, X., Wong, M. D., Kawase, E., Xi, R., Ding, B. C., McCarthy, J. J., and Xie, T. (2004). Bmp signal from niche directly repress transcription of a differentiation-promoting gene, *bag of marbles*, in the germline stem cells of the *Drosophila* ovary. *Development* **131**, 1353–1364.
Spittaels, K., Verhaert, P., Shaw, C., Johnston, R. N., Devreese, B., Van Beeumen, J., and De Loof, A. (1996). Insect neuropeptide F (NPF)-related peptides: Isolation from Colorado potato beetle (*Leptinotarsa decemlineata*) brain. *Insect Biochem. Mol. Biol.* **26**, 375–382.
Sprenger, F., Stevens, L. M., and Nusslein-Volhard, C. (1989). The *Drosophila* gene torso encodes a putative receptor tyrosine kinase. *Nature* **338**, 478–483.
Stangier, J., Hilbich, K., Beyreuther, R., and Keller, R. (1987). Unusual Cardioactive Peptide (CCAP) from Pericardial Organs of the Shore Crab *Carcinus maenas*. *Proc. Natl. Acad. Sci. USA* **84**, 575–579.
Starratt, A. N., and Brown, B. E. (1975). Structure of the pentapeptide proctolin, a proposed neurotransmitter in insects. *Life Sci.* **17**, 1253–1256.
Staubli, F., Jorgensen, T. J., Cazzamali, G., Williamson, M., Lenz, C., Sondergaard, L., Roepstorff, P., and Grimmelikhuijzen, C. J. P. (2002). Molecular identification of the insect adipokinetic hormone receptors. *Proc. Natl. Acad. Sci. USA* **99**, 3446–3451.
Stavely, B. E., Ruel, L., Jin, J., Stambolic, V., Mastronardi, F. G., Heitzlen, P., Woodgett, J. R., and Manoukian, A. S. (1998). Genetic analysis of protein kinase B (AKT) in *Drosophila*. *Curr. Biol.* **8**, 599–602.
Stay, B. (2000). A review of the role of neurosecretion in the control of juvenile hormone synthesis: A tribute to Berta Scharrer. *Insect Biochem. Mol. Biol.* **30**, 653–662 (Review).
Stone, D. M., Hynes, M., Armanini, M., Swanson, T. A., Gu, Q., Johnson, R. L., Scott, M. P., Pennica, D., Goddard, A., Phillips, H., Noll, M., Hooper, J. E., de Sauvage, F., and Rosenthal, A. (1996). The tumour-suppressor gene patched encodes a candidate receptor for sonic hedgehog. *Nature* **384**, 129–134.

Stone, J. V., Mordue, W., Batley, K. E., and Morris, H. R. (1976). Structure of locust adipokinetic hormone, a neurohormone that regulates lipid utilisation during flight. *Nature* **263**, 207–211.

Sutherland, D., Samakovlis, C., and Krasnow, M. A. (1996). Branchless encodes a *Drosophila* FGF homolog that controls tracheal cell migration and the pattern of branching. *Cell* **87**, 1091–1101.

Taghert, P. H. (1999). FMRFamide neuropeptides and neuropeptide-associated enzymes in *Drosophila*. *Microsc. Res. Tech.* **45**, 80–95 (Review).

Takeuchi, H., Yasuda, A., Yasuda-Kamatani, Y., Kubo, T., and Nakajima, T. (2003). Identification of a tachykinin-related neuropeptide from the honeybee brain using direct MALDI-TOF MS and its gene expression in worker, queen and drone heads. *Insect Mol. Biol.* **12**, 291–298.

Tanoue, S., and Nishioka, T. (2001). A receptor type guanylyl cyclase expression is regulated under circadian clock in peripheral tissues of the silk moth. Light-induced shifting of the expression rhythm and correlation with eclosion. *J. Biol. Chem.* **276**, 46765–46769.

Tanoue, S., Sumida, S., Suetsugu, T., Endo, Y., and Nishioka, T. (2001). Identification of a receptor type guanylyl cyclase in the antennal lobe and antennal sensory neurons of the silkmoth, *Bombyx mori*. *Insect Biochem. Mol. Biol.* **31**, 971–979.

Tatemoto, K., Lundberg, J. M., Jornvall, H., and Mutt, V. (1985). Neuropeptide-K—isolation, structure and biological activities of a novel brain tachykinin. *Biochem. Biophys. Res. Commun.* **128**, 947–953.

Tawfik, A. I., Tanaka, S., De Loof, A., Schoofs, L., Baggerman, G., Waelkens, E., Derua, R., Milner, Y., Yerushalmi, Y., and Pener, M. P. (1999). Identification of the gregarization-associated dark-pigmentotropin in locusts through an albino mutant. *Proc. Natl. Acad. Sci. USA* **96**, 7083–7087.

Taylor, C. A., Winther, A. M., Siviter, R. J., Shirras, A. D., Isaac, R. E., and Nassel, D. R. (2004). Identification of a proctolin preprohormone gene (Proct) of *Drosophila melanogaster*: Expression and predicted prohormone processing. *J. Neurobiol.* **58**, 379–391.

Tensen, C. P., Cox, K. J., Smit, A. B., van der Schors, R. C., Meyerhof, W., Richter, D., Planta, R. J., Hermann, P. M., van Minnen, J., Geraerts, W. P., Knol, J. C., Burke, J. F., Vreugdenhil, E., and van Heerikhuizen, H. (1998a). The *Lymnaea* cardioexcitatory peptide (LyCEP) receptor: A G-protein-coupled receptor for a novel member of the RFamide neuropeptide family. *J. Neurosci.* **18**, 9812–9821.

Tensen, C. P., Cox, K. J., Burke, J. F., Leurs, R., Van der Schors, R. C., Geraerts, W. P., Vreugdenhil, E., and van Heerikhuizen, H. (1998b). Molecular cloning and characterization of an invertebrate homologue of a neuropeptide Y receptor. *Eur. J. Neurosci.* **10**, 3409–3416.

Terman, J. R., and Kolodkin, A. L. (2004). Nervy links protein kinase A to Plexin-mediated Semaphorin repulsion. *Science* **303**, 1204–1207.

Tobe, S.S, and Bendena, W. G. (1999). The regulation of juvenile hormone production in arthropods. Functional and evolutionary perspectives. *Ann. N.Y. Acad. Sci.* **897**, 300–310 (Review).

Tomlinson, A., Strapp, W. R., and Heemskerk, J. (1997). Linking Frizzled and the Wnt signaling in *Drosophila* development. *Development* **124**, 4515–4521.

Torfs, H., Shariatmadari, R., Guerrero, F., Parmentier, M., Poels, J., Van Poyer, W., Swinnen, E., De Loof, A., Akerman, K., and Vanden Broeck, J. (2000). Characterization of a receptor for insect tachykinin-like peptide agonists by functional expression in a stable *Drosophila* Schneider 2 cell line. *J. Neurochem.* **74**, 2182–2189.

Torfs, H., Oonk, H. B., Vanden Broeck, J., Poels, J., Van Poyer, W., De Loof, A., Guerrero, F., Meloen, R. H., Akerman, K., and Nachman, R. J. (2001). Pharmacological characterization of STKR, an insect G protein-coupled receptor for tachykinin-like peptides. *Arch. Insect Biochem. Physiol.* **48**, 39–49.

Torfs, H., Akerman, K. E., Nachman, R. J., Oonk, H. B., Detheux, M., Poels, J., Van Loy, T., De Loof, A., Meloen, R. H., Vassart, G., Parmentier, M., and Vanden Broeck, J. (2002a).

Functional analysis of synthetic insectatachykinin analogs on recombinant neurokinin receptor expressing cell lines. *Peptides* **23**, 1999–2005.
Torfs, H., Detheux, M., Oonk, H. B., Akerman, K. E., Poels, J., Van Loy, T., De Loof, A., Vassart, G., Parmentier, M., and Vanden Broeck, J. (2002b). Analysis of C-terminally substituted tachykinin-like peptide agonists by means of aequorin-based luminescent assays for human and insect neurokinin receptors. *Biochem. Pharmacol.* **63**, 1675–1682.
Torfs, H., Poels, J., Detheux, M., Dupriez, V., Van Loy, T., Vercammen, L., Vassart, G., Parmentier, M., and Vanden Broeck, J. (2002c). Recombinant aequorin as a reporter for receptor-mediated changes of intracellular Ca^{2+}-levels in *Drosophila* S2 cells. *Invert. Neurosci.* **4**, 119–124.
Tublitz, N. J., and Truman, J. W. (1985). Identification of neurones containing cardioacceleratory peptides (CAPs) in the ventral nerve cord of the tobacco hawkmoth, *Manduca sexta*. *J. Exp. Biol.* **116**, 395–410.
Vanden Broeck, J. (1996). G protein-coupled receptors in insect cells. *Int. Rev. Cytol.* **164**, 189–268.
Vanden Broeck, J. (2001a). Neuropeptides and their precursors in the fruit fly, *Drosophila melanogaster*. *Peptides* **22**, 241–254.
Vanden Broeck, J. (2001b). Insect G protein-coupled receptors and signal transduction. *Arch. Insect Biochem. Physiol.* **48**, 1–12.
Vanden Broeck, J., Veelaert, D., Bendena, W., Tobe, S. S., and De Loof, A. (1996). Molecular cloning of the precursor cDNA for schistostatins, locust allatostatin-like peptides with myoinhibiting properties. *Mol. Cell. Endocrinol.* **122**, 191–198.
Vanden Broeck, J., Schoofs, L., and De Loof, A. (1997). Insect neuropeptides and their receptors: New leads for medical and agricultural applications. *Trends Endocrinol. Metab.* **8**, 321–326.
Vanden Broeck, J., Poels, J., Simonet, G., Dickens, L., and De Loof, A. (1998). Identification of G protein-coupled receptors in insect cells. *Ann. N.Y. Acad. Sci.* **839**, 123–128.
Vanden Broeck, J., Torfs, H., Poels, J., Van Poyer, W., Swinnen, E., Ferket, K., and De Loof, A. (1999). Tachykinin-like peptides and their receptors, a review. *Ann. N.Y. Acad. Sci.* **897**, 374–387.
Vanden Broeck, J., Schoofs, L., and De Loof, A. (2000). Evolution of developmental peptide hormones and their receptors. *In* "Reproductive biology of invertebrates: Progress in developmental endocrinology" (K. G. Adiyodi, R. G. Adiyodi, and A. Dorn, Eds.), vol. X, part B, pp. 41–69. Wiley-Liss. (ISBN 0 471 49465 8).
Van Der Geer, P., Hunter, T., and Lindberg, R. A. (1994). Receptor protein-tyrosine kinases and their signal transduction pathways. *Annu. Rev. Cell Biol.* **10**, 251–337 (Review).
Van Roey, K., Derks, M., Poels, J., and Vanden Broeck, J. (2004). Genomics and Evolution of Metazoan Ga proteins. *In* "Focus on Genomes Research," pp. 125–160. Nova Science (*ISBN* 1-59033-960-6).
Veelaert, D., Passier, P., Devreese, B., Vanden Broeck, J., Van Beeumen, J., Vullings, H., Diederen, J., Schoofs, L., and De Loof, A. (1997). Isolation and characterization of a locust adipokinetic hormone release inducing factor, the crustacean cardio-active peptide CCAP. *Endocrinology* **138**, 138–142.
Veelaert, D., Oonk, H. B., Vanden Eynde, G., Torfs, H., Meloen, R. H., Schoofs, L., Parmentier, M., De Loof, A., and Vanden Broeck, J. (1999). Immunolocalization of a tachykinin-receptor-like protein in the central nervous system of *Locusta migratoria migratorioides* and *Neobellieria bullata*. *J. Comp. Neurol.* **407**, 415–426.
Veenstra, J. A. (1989). Isolation and structure of two gastrin/CCK-like neuropeptides from the American cockroach homologous to the leucosulfakinins. *Neuropeptides* **14**, 145–149.
Veenstra, J. A. (1994). Isolation and structure of the *Drosophila* corazonin gene. *Biochem. Biophys. Res. Commun.* **204**, 292–296.
Vitt, U. A., Hsu, S. Y., and Hsueh, A. J. (2001). Evolution and classification of cystine knot-containing hormones and related extracellular signaling molecules. *Mol. Endocrinol.* **15**, 681–694 (Review).

Von Euler, U. S., and Gaddum, J. H. (1931). An unidentified depressor substance in certain tissue extracts. *J. Physiol. London* **72**, 74–86.
Wei, Z., Baggerman, G., Nachman, J., Goldsworthy, G., Verhaert, P., De Loof, A., and Schoofs, L. (2000). Sulfakinins reduce food intake in the desert locust, *Schistocerca gregaria*. *J. Insect Physiol.* **46**, 1259–1265.
Weng, Q. P., Andrabi, K., Klippel, A., Kozlowski, M. T., Williams, L. T., and Avruch, J. (1995). Phosphatidylinositol 3-kinase signals activation of p70 s6 kinase *in situ* through site-specific p70 phosphorylation. *Proc. Natl. Acad. Sci. USA* **92**, 5744–5748.
West, A. P., Llamas, L. L., Snow, P. M., Benzer, S., and Bjorkman, P. J. (2001). Crystal structure of the ectodomain of Methuselah, a *Drosophila* G protein-coupled receptor associated with extended lifespan. *Proc. Natl. Acad. Sci. USA* **98**, 3744–3749.
White, F. M. (1998). The IRS-signaling system: A network of docking proteins that mediates insulin action. *Mol. Cell Biochem.* **182**, 3–11.
Whitehead, J. P., Clark, S. F., Urso, B., and James, D. E. (2000). Signaling through the insulin receptor. *Curr. Opinion Cell Biol.* **12**, 222–228.
Wiehart, U. I., Nicolson, S. W., Eigenheer, R. A., and Schooley, D. A. (2002). Antagonistic control of fluid secretion by the Malpighian tubules of *Tenebrio molitor*: Effects of diuretic and antidiuretic peptides and their second messengers. *J. Exp. Biol.* **205**, 493–501.
Wimmer, E. A. (2003). Innovations: Applications of insect transgenesis. *Nat. Rev. Genet.* **4**, 225–232 (Review).
Winther, A. M. E., Siviter, R. J., Isaac, R. E., Predel, R., and Nassel, D. R. (2003). Neuronal expression of tachykinin-related peptides and gene transcript during postembryonic development of *Drosophila*. *J. Comp. Neurol.* **464**, 180–196.
Wodarz, A., and Nusse, R. (1998). Mechanisms of Wnt signalling in development. *Annu. Rev. Cell. Dev. Biol.* **14**, 59–88.
Wrana, J. L., Attisano, L., Wieser, R., Ventura, F., and Massagué, J. (1994). Mechanism of activation of he TGF-beta receptor. *Nature* **370**, 341–347.
Wu, Q., Wen, T., Lee, G., Park, J. H., Cai, H. N., and Shen, P. (2003). Developmental control of foraging and social behavior by the *Drosophila* neuropeptide Y-like system. *Neuron* **39**, 147–161.
Yamaguchi, T., Fernandez, R., and Roth, R. A. (1995). Comparison of the signaling abilities of the *Drosophila* and human insulin receptors in mammalian cells. *Biochemistry* **34**, 4962–4968.
Yang, M., Nelson, D., Funakoshi, Y., and Padgett, R. W. (2004). Genome-wide microarray analysis of TGFb signaling in the *Drosophila* brain. *BMC Dev. Biol.* **4**, 14.
Yenush, L., Makati, K., Smith-Hall, J., Ishibaski, O., Myers, M., and White, M. F. (1996). The pleckstrin homology domain is the principal linkage between the insulin receptor and IRS-1. *J. Biol. Chem.* **271**, 24300–24306.
Yokota, Y., Sasai, Y., Tanaka, K., Fujiwara, T., Tsuchida, K., Shigemoto, R., Kakizuka, A., Ohkubo, H., and Nakanishi, S. (1989). Molecular characterization of a functional cDNA for rat substance P receptor. *J. Biol. Chem.* **264**, 17649–17652.
Yoshikawa, S., McKinnon, R. D., Kokel, M., and Thomas, J. B. (2003). Wnt-mediated axon guidance via the *Drosophila* Derailed receptor. *Nature* **422**, 583–588.
Zhang, J., and Carthew, R. W. (1998). Interactions between Wingless and DFz2 during *Drosophila* wing development. *Development* **125**, 3075–3085.
Zhang, Y., Lu, L., Furlonger, C., Wu, G. E., and Paige, C. J. (2000). Hemokinin is a hemopoietic-specific tachykinin that regulates B lymphopoiesis. *Nat. Immunol.* **1**, 392–397.
Zheng, X., Wang, J., Haerry, T., Wu, A., Martin, J., O'Connor, M. B., Lee, C. H., and Lee, T. (2003). TGF-b signaling activates steroid hormone receptor expression during neuronal remodeling in the *Drosophila* brain. *Cell* **112**, 303–315.
Zitnan, D., Kingan, T. G., Hermesman, J. L., and Adams, M. E. (1996). Identification of ecdysis-triggering hormone from an epitracheal endocrine system. *Science* **271**, 88–91.

INDEX

Page numbers followed by f and t indicate figures and tables, respectively.

A

6a-ethyl-chenodeoxycholic acid (6ECDCA), 120
Abdominal ganglion (AG), 21f
Activation domain (AD), 88f
Adipokinetic hormone receptor, 239
Adipokinetic hormones, 239
Adrenodoxin reductase (AR), 179. *See also* Ecdysteroid biosynthesis
Allatostatin receptors, 234–235
Amnesiac, 262
  analysis, 243
  Anodonta cygnea (*Anc*-TK), 230
  arthropod guanylyl cyclase receptors, 256–257
Apolysis, 3
  post-apolysis process, 148
ATP-binding cassette (ABC), 143

B

5(H)-reductase, 39–41
Berkeley Drosophila Genome Project (BDGP), 239
Black box, 39
Broad-Complex (BR-C) gene, 181
Bursicon, 13, 23–24

C

Calcitonin-like diuretic hormones (DH31), 262
  cellular actions, 253

Calmodulin (CaM), *see* juvenile hormone, gene regulation
cAMP, 13
Campesterol, 33
Cells
  27(A), 24
  704(A), 22f, 24
  brain corazonin, 8f
  inka (IC), 8, 12, 14, 18–19, 21f, 22f, 25
  neurosecretory cells, 177
cGMP, 21f, 22
Chemosterilizing activity, 141
Cholecystokinin (CCK), 231
Cholesterols, 33, 34f
Choristoneura hormone receptor 75 (CHR75), 140
Chromafenozides, 133, 136, 145, 148
Circadian timing, 15–16, 16f
Cis-5β(H)-3β-ol, 39
CNS, 7–9, 11, 15
Co-factors interaction groove, 71
Comparative molecular field analysis (CoMFA), 136, 157
Constitutive androstane receptor (CAR), 120
Corazonin, 12, 18–19, 18f
  receptor, 239–240
Co-repressors, 77
Corpora allata (CA), 234. *See also* Allatostatin receptors

Corticotrophin-releasing factor (CRF), 242.
    See also Insect diuretic hormone
    receptors
CRF-binding proteins, 264
Crustacean cardioactive peptide (CCAP), 7, 9,
    11–14, 20, 22–25
Crustacean cardioactive peptide
    receptor, 240–241
    discovery, 255–256
    D. melanogaster insulin receptor (DIR)
        and, 250–251, 252*f*
Cuticular tanning, 23–24
Cyclic-AMP (cAMP), 177. *See also* PTTH
CYP. *See* Cytochrome P450 enzymes
Cytochrome P450 enzymes, 34*f*, 35, 38–41,
    49, 53

D

DAH, *see* Diacylhydrazines
dare gene, 179–180. *See also* Ecdysteroid
    biosynthesis
3-dehydrecdysone (3DE), 40–41
3-dehydro-6-one-7-ene-14a-ol, 39
7dC. *See* 7-dehydrocholesterol
7-dehydrocholesterol, 35, 37
7-dehydrocholesterol, 35, 37
Descending inhibitor (DI), 22*f*
Desmosterol, 35
DGHR2 (Drosophila hormone
    receptor 2), 13
Diacylhydrazine congeners. *See*
    Tebufenozides; Methoxyfenozide;
    Halofenozide; Chromafenozide
Diacylhydrazines (DAH), 131, 162
    neurotoxic effects of, 152
    synthesis, 134–136, 135*f*
dib gene, 42–45, 44*f*
Diethyl maleate (DEM), 144
disembodied (dib), 180
Diuretic hormones (DH), 242. *See also* Insect
    diuretic hormone receptors
DLKR, 231
dromyosuppressin (DMS) and, 238
DiVerential display technique, 193
DNA binding domain (DBD), 102
    ecdysteroid receptors (EcR), 74–75
DOPA Decarboxylase (DDC), 150
Drosophila E75A, 181
Drosophila melanogaster, life cycle, 6, 9–13,
    15, 17, 19–20, 22, 24, 176–177
Drosophila pachea, 35

Drosophila tachykinin receptor
    (DTKR), 229. *See also* TRP
Drug discovery, 82
DSK-R1, 231–232
dsRNAmediated interference technique
    (RNAi), 236

E

20E. *See* 20-hydroxyecdysone
Ecd. *See* Ecdysone
ecd-1, 179. *See also* Ecdysteroid biosynthesis
Ecdysis or melanization, 140
Ecdysis, 2, 15, 16*f*, 25, 33
    control, 18*f*
    motor program, 9, 15, 17*f*, 20, 22–24
Ecdysis, phases, 4*f*, 5*f*, 17–19, 21*f*, 22–23, 25.
    See also Pre-ecdysis I; Pre-ecdysis II;
    Post-ecdysis
Ecdysis, sequence, 6–7, 14. *See also* Eclosion
    hormone
Ecdysis-triggering hormones (ETHs), 236–237
Ecdysone (Ecd), 33, 35–36, 41, 45, 47, 49, 54,
    178, 180
Ecdysone receptors (EcR), molecular
    biology, 187–189
    DHR3, 188
    dSin3A, 189
    ecdysone response element (EcRE), 187
    EcR/USP receptor and, 187, 189
    SMRTER binding to, 189, 191
Ecdysone response elements (EcREs), 186
Ecdysone synthesis, 178–179, 181
Ecdysone-induced transcription factor
    (CHR3), 141. *See also* RH-5849
Ecdysoneless mutation, *ecd-1*, 180
Ecdysone-mimicking effect, 148
Ecdysteroid biosynthesis, 179–180
Ecdysteroid receptor (EcR), 60, 85, 88*f*, 91,
    101, 134, 155
    cellular locale of, 61–62, 62*f*
    ligand binding domain in. *See* Ligand
        binding domain
    DNA binding domain (DBD) and hinge
        region of. *See* DNA binding
        domain; hinge region
    sequence in synthesis of, 76–77
Ecdysteroidogenesis, 178. *See also* PTTH
Ecdysteroids, 33, 35, 46*f*, 60, 62*f*, 84*f*, 102,
    104. *See also* 20E
Eclosion hormone (EH), 6–9, 7*f*, 18*f*, 19–21,
    23, 25, 262

EcR ligand-binding domains
(LBDs), 102–103
  crystal structure, 109, 111*f*, 112*f*, 113–117,
    113*f*, 116*f*, 121*f*, 122*f*
  dimeric arrangement, 105–107, 106*f*, 108*f*
  sequence comparison of invertebrate, 104
  EcR protein, 120
EcR. *See* Ecdysteroid receptor
EcR/USP
  heterodimeric complexes, 102, 105–107,
    106*f*, 108*f*, 109, 124
  interactions of, 110*t*
EcR/USP heterodimerization, 153–154, 155*t*
EcResponse elements (EcRE), 60
EH gene, 6
Endocrine and neuroendocrine systems
  interaction, 2. *See also* Ecdysis phases
Estrogen receptor (ER), 60, 107
ETH, 9–11, 14–15, 17–20, 25
Expressed sequence tags (EST), 219

F

Farnesoid X receptor (FXR), 119
FLRFA like peptides
  receptors for, 238
FMRFA-like peptides
  receptors for, 238, 263
Frizzled 7TM receptor family. *See*
    Frizzled-like proteins (Fz);
    Wingless (Wg) signaling;
    Smoothened (Smo) experiments
Frizzled-like proteins (Fz), 243
  functional analyses of, 42–43
Functional genomics, 82, 90
Functionally characterized peptide receptors
  of rhodopsin family, 224*t*–228*t*
Fusion proteins, 71, 81*f*

G

GC6111, 12
Gene regulation systems, 82, 87. *See also*
  Gene switches; Medicinal applications
Gene switch systems, 104
Gene switches, 60–64, 71–72, 76–77, 83, 158
  agricultural applications of, 84*f*, 87, 88*f*, 89
  EcR based, 78–80, 79*f*, 89
  medicinal applications of, 82–83
  use in medicine, 78
Gene Therapy, 82

Germline stem cells (GSC), 254
GFP, 45, 50. *See also* Halloween genes
giant gene, 181
Gland protein, 184
Glucocortecoid receptor (GR), 60
Gonadotropin-releasing hormone receptors
  (GnRHR), 238
GPCR CG6986, 237
Guanylyl cyclases (GC) receptors, 255

H

Halloween gene family, 49, 54. *See also*
  disembodied (dib); shadow (sad);
  shade (shd)
Halofenozides, 132–133, 138, 141,
  143–145, 152
Heat-shock protein (HSP), 79*f*. *See also* EcR
  based gene switches
Heterodimerization, 60–61, 64*f*, 69–70, 70*f*
Hinge region, Ecdysteroid receptors
  (EcR), 75
Holo-conformation, 71–72
20-hydroxyecdysone (20E), 3, 14–15, 18–20,
  25, 31–33, 48–49, 102, 115, 132–134,
  147–149, 156, 158, 162, 177, 183, 197
  biosynthesis of, 34*f*, 35, 38, 41, 52*f*, 53–54
20-hydroxylase, 48

I

IGF-binding proteins, 264
IGR(insect growth regulator), 132, 143
Insect diuretic hormone receptors, 242–243
Insect genomes, 237
  insect physiology and, 259–260
Insectatachykinins. *See* TRP
Insulin receptor (IR), 247, 250. *See also*
  Receptor tyrosine kinases
  insulin receptor substrate (IRS)
    interactions in, 250
Insulin receptor substrate (IRS), 247, 250
Insulin-like growth factor (IGF), 247
Ion transport peptides (ITP), 262

J

JH. *See* Juvenile hormones
Juvenile hormone (JH), 176, 181, 183
  Juvenile hormone (JH),
    synthesis of, 182
  allatostatins, 182

Juvenile hormone (JH) (*continued*)
  allatotropins, 182
  DAR-1, 182
  DAR-2, 182
  allatostatin receptor genes, 182
  JH epoxyhydrolase (JHEH), 182
Juvenile hormone esterase, 182
  apterous (ap), 183
  circklet (clt), 183
  Methoprene tolerant (Met), 183
  ultraspiracle (usp), 183
Juvenile hormone oogenesis
  YP gene1, 185–186
  YP2 gene, 185–186
  YP3 gene, 186
  YP gene expression, 186
  yolk protein synthesis, 185
  vitellogenesis, 185
Juvenile hormone receptor, *Drosophila*
  JH response element (JHRE), 190
  methoprene tolerant (Met) gene effects on, 190
  USP, 190–191
Juvenile hormone receptor, insects, 189
Juvenile hormone, effects on
  Drosophila, 181–182
Juvenile hormone, gene regulated by,
  Aedes aegypti and, 192
  DRNaseZ, 193–194, 194*f*
  Drosophila and, 192
  E75, 196, 196*f*
  20E gene activation in holometabolous insects and, 191–192
  Galleria mellonella and, 192
  JH esterase (Jhe) gene, 197
  JhI-21 and minidisks (mnd), 195
  Locusta migratoria, 193
  Manduca and, 192
  mir-34, 195
  Plodia interpunctella and, 192
Juvenile hormone, hormonal crosstalk by, 197
  Broad-Complex (BR-C), 197–198
  BR-C in *Drosophila, Manduca, Bombyx*, 198–199, 220*f*
  E75A, 199–201, 200*f*
Juvenile hormone, Spermatogenesis, 184–185
Juvenoids, 190. *See* Juvenile hormones

K

K497, 72

L

Larval cuticular protein (LCP-14), 152
Larva–pupa transition, 178
lethal (1) giant (l(1)g), 180
Leucine-rich repeat-containing GPCRs (LGRs), 241–242
LF/SIFa-like peptides, 262
Ligand binding domain (LBD), Ecdysteroid receptors (EcR), 64–65, 64*f*, 67, 68*f*, 70*f*, 71
  dimerization interface, 67, 68*f*, 69
  holo-EcR, 69
  hyperdimerization, 69
  ligand-binding, 65, 67, 68*f*, 69, 77
  ligand-binding pocket, 65, 66*f*, 67
  mechanisms transducing signals, 72–74, 73*f*
Ligand binding pocket, Ecdysteroid receptors (EcR), 114*f*, 118*f*, 121*f*, 123–124. *See also* EcR ligand-binding domains (LBDs)
Ligand-controlled transcription factor (LcTF), 63. *See also* EcR
  effect of ligand on dimerization, 69. *See also* Holo-EcR
  ligand-induced dimerization, 69–70, 76. *See also* Holo-EcR
  spontaneous dimerization, 69. *See also* Holo-EcR
Ligand-receptor interaction, 37, 154. *See also* EcR
Liver X receptor (LXR), 119

M

Manduca sexta, 4–5, 9, 11–12, 17–21, 21*f*, 23–24. *See also* Pre-ecdysis; Eclosion hormone; Brain corazonin cells
Metamorphosis, 176
Methionine, 119*f*
Methoxyfenozides, 132, 133*f*, 137–138, 142–145, 150, 152, 162
Methuselah-like receptor, 243
  mth gene, 243
Molecular chaperone-containing heterocomplex (MCH), 76
Molting hormonal activity, 146–147
Molting hormone. *See* 20-hydroxyecdysone
Molting, 3, 25, 32–33
Mutations, 67, 68*f*, 69, 71, 73
Myoinhibiting peptides (MIP), 235
  Myosuppressins and, 238

N

N-and C-terminal modular domains,
   Ecdysteroid receptors (EcR), 75–76
Neuropeptide family (AstC), 235
Neuropeptide Y (NPY) receptors
   for, 232–233
Nonsteroidal ecdysone agonist, 131–132, 158, 159f, 160–161, 160t. See also DAH; RH-5849; tebufenozides; methoxyfenozides; halofenozide; chromafenozides
NPF
   receptors for, 232–234
N-tert-butyl-N,N0-dibenzoylhydrazine (RH-5849), 132, 136-137, 145, 158, 162
   binding affinity, 156
   effects on pupae, adults and eggs of, 140–142
   enzymes and proteins, 149, 151–152
   larvicidal effects of, 136–137
   in vitro effects of, 146–148
   neurotoxic effects of, 152–153
Nuclear receptors
Nuclear receptors, 60, 71
   ligands, 103f, 104
Nurse cells, 179–180. See also Ecdysteroid biosynthesis

O

Orphan nuclear receptor, bFTZ-F1. See Juvenile hormone; Gene regulation

P

Pancreatic polypeptide (PP), 232
Pars intercerebralis (PI), 185
pbur (partner of bursicon), 13
Peptide hormones, 7f
Peptide receptors, 218–219, 260
   ligand-binding proteins in, 264
   ligand-receptor interactions in, 264
   molecular pharmacology and signaling properties of, 262–263
   receptor–ligand couples, 263
Peptide receptors, myokinin
   receptors, 230–231
   Drosophila orthologue (CG10626) in, 231
   kinins, 230
Peptide receptors, signal-transducing, 219–221, 221f

G protein–coupled receptors [GPCR], 219, 221f, 224t–228t, 233–237, 261f, 262
   neuropeptides, 221–220, 237–238
   transmembrane (TM) receptors, 219
Peptide receptors, sulfakinin receptors, 231
Peptide receptors, tachykinin (or neurokinin) receptors, 222
   endokinins (EK) C and D, 222
   hemokinin-1 (HK-1), 222
   SP, NKA, and NKB, 223
   S. calcitrans (*Stc*-TK), 230
   tachykinin-related peptides (TRP), 223–230
PETH, 9–11, 15, 25
Pheromone biosynthesis–activating neuropeptide (PBAN), 236
phm gene, 42–44, 44f, 50, 51f, 53
   physiology and behavior of, 3, 20, 25
Phytophagous, 33
Pigment dispersing hormones (PDH), 262
Piperonyl butoxide (PB), 138
Pleckstrin homology (PH), 252f
Pleiotropic drug resistance 5 (i.e., PDR5), 143
Plodia interpunctella and, 192
ponA. See Ponasterone A
Ponasterone A, 103, 107, 112, 115, 116f, 120, 122f, 144, 148, 154–156, 154f
Post-ecdysis, 5, 23–24
Pre-ecdysis 1, 4, 5f, 7f, 11, 17, 25
Pre-ecdysis II, 4, 11, 17, 25
Pregnane X receptor (PXR), 120
Proctolin (RYLPT), 237
Proctolin receptor, 237
ProETH, 9–10
Progesterone receptor (PR), 60
Prothoracic gland, 33, 37, 40
Prothoracicotropic hormone (PTTH), 37, 177
PRXA peptides, 235
PTTH. See Prothoracicotropic hormone
Pyrokinin (PK), 235

Q

QSAR, 136, 138–139, 151, 157

R

Radioimmunoassay (RIA), 37
Receptor guanylyl cyclases (rGC), 257–259, 257f, 258t
   receptor families in insects of, 248t–249t
Receptor tyrosine kinases (RTKs), 245–247
   properties of, 245–246

Receptor tyrosine kinases, insulin receptor signaling pathway receptors for, 235
Residues mutation, 123, 125f. *See also* Ligand binding pocket of EcR
Retinoid X receptor (RXR), 102
RH-5849, *See* N-tert-butyl-N,N0-dibenzoylhydrazine
RH-5849. *See* N-tert-butyl-N,N0-dibenzoylhydrazine
Rickets, 241
Ring glands, 40
RNA interference (RNAi), 82, 264. *See also* Gene switches; Medical applications
RTK. *See* Receptor tyrosine kinases
RT-PCR, 38

S
S,S,S-tributylphosphorotrithioate (DEF), 144
sad gene, 42–45, 44f
Secretin receptor (GPCR class B) family. *See* insect diuretic hormone receptors; Corticotrophin-releasing factor (CRF); Diuretic hormones
Sequiterpenoid hormones, 181
Sex peptides (SP), 184, 262
shade (shd), 47–48, 180
shadow (sad), 180
 signaling components.of, 253–254
Single transmembrane receptors. *See* Receptor tyrosine kinases
Sitosterol, 33
Smads (R-Smad) signaling proteins, 253. *See also* TGF receptor families in insects
Smoothened (Smo) experiments, 244
START domain proteins, 179, 180. *See also* Ecdysteroid biosynthesis
Stem cells, 254
Steroidogenic regulatory protein (StAR), 38
STKR, 229. *See also* TRP
Superinducer, 71. *See also* Ultraspiracle protein
 Synthesis, 132

T
Target of rapamycin (TOR), 251, 252f
Tebufenozides, 132, 133f, 137, 140–142, 144–150, 152, 156, 158, 162
Terminal hydroxylations, 41–42
Tetrahydroquinoline (THQ), 104
TGF receptor families in insects, 248t–249t, 251, 253
 TGF signaling pathways of, 254
Therapeutic proteins, 82
Tissue Engineering, 82
Transgenes, 87
Transgenic plants, 90
 triggering hormones (ETH), 7, 7f, 9–10
TRP. *See* Peptide receptors, tachykinin-related peptides, tuberous sclerosis complex proteins in, 251, 252f

U
ultraspiracle (usp), 183
Ultraspiracle protein (USP), 60, 65, 67, 69–70, 70f, 72, 73f, 75–76, 102, 134. *See also* EcR/USP heterodimerization
Unusual floral organs (UFO) gene, 90
UTKR, 230. *See also* TRP

V
Verbutin, 145
Vertebrate cholesterol transporter, MLN64, 179. *See also* Ecdysteroid biosynthesis s
Vitamin D receptor (VDR), 119
Vitellogenesis, 179, 185. *See also* Ecdysteroid biosynthesis

W
Wingless (Wg) signaling, 244
Without children (woc) gene, 181

Z
Zaprinast, 22

GILBERT AND WARREN, FIGURE 1. (*Continued*)

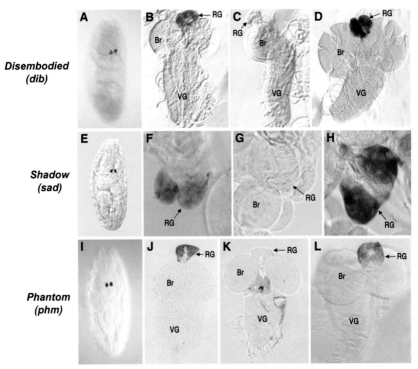

GILBERT AND WARREN, FIGURE 2. *In situ* expression of the Halloween genes *disembodied* (*dib*), *shadow* (*sad*) and *phantom* (*phm*) within the brain–ring gland complex during late embryonic and larval stages. Shown are stage 17 embryos (A, E, and I), late second instar (B, F, and J), and both early (C, G, and K) and late (D, H, and L) third instar brain–ring gland complexes. Note the down regulation of the expression of all three genes between the late second and early third instars and their subsequent up regulation between the early and late third instars. RG, ring gland; Br, brain; VG, ventral ganglion. (Data on embryonic *dib* expression is from Chavez *et al.*, 2000 and larval *dib* and *sad* expression is from Warren *et al.*, 2002. *Phm* expression data is from Warren *et al.*, 2004.)

GILBERT AND WARREN, FIGURE 1. Scheme of 20-hydroxyecdysone (20E) biosynthesis in *Drosophila*. As many as six (or more) cytochrome P450 (CYP) enzymes may be involved, starting from cholesterol (C). Multiple arrows indicate an uncharacterized pathway, perhaps involving more than one biochemical transformation. The Δ^4-diketol is cholesta-4,7-diene-3,6-dione-14α-ol, the diketol is 5β[H]-cholesta-7-ene-3,6-dione-14α-ol, the ketodiol is 5β[H]-cholesta-7-ene-6-one-3β,14α-diol (2,22,25-trideoxyecdysone or 2,22,25dE) and herein the generic term ketotriol represents 5β[H]-cholesta-7-ene-6-one-3β, 14α, 25-triol (2,22-dideoxyecdysone or 2,22dE). The uncharacterized reactions between 7-dehydrocholesterol (7dC) and the Δ^4-diketol are commonly referred to as the "Black Box." Yellow shade delineates the areas of the sterol backbone that are involved in the indicated transformations. The Δ^4-diketol has neither been isolated from insects, nor shown to be a product of radio labeled cholesterol *in vitro* metabolism in insects. It has, however, been shown in crabs (Blais *et al.*, 1996) to be reduced to the diketol, which is then converted into ecdysone.

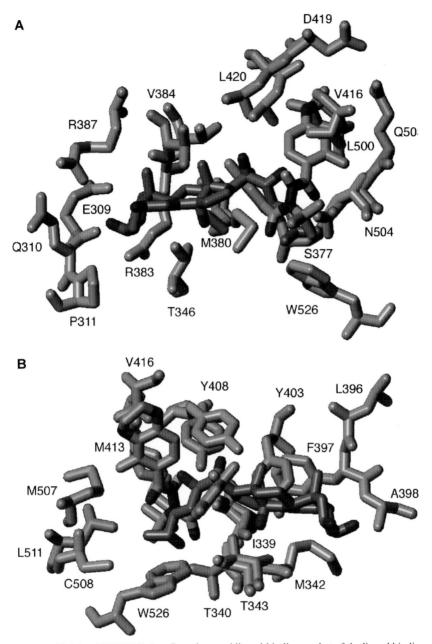

PALLI ET AL., FIGURE 3. Superimposed ligand-binding pocket of the ligand binding domain of *H. virescens* crystallized with ponasterone A (PDB structure 1R1K, residues = yellow, ponasterone A = orange) and crystallized with BY-106830 (PDB structure 1R20, residues, green; BY-106830, cyan). Depicted residues lie within 4 Å of the ligand in each case. (A) Orientation as in Fig. 5 (EcR). (B) 180° rotation of the image (A) along an approximately vertical axis (Billas *et al.*, 2003).

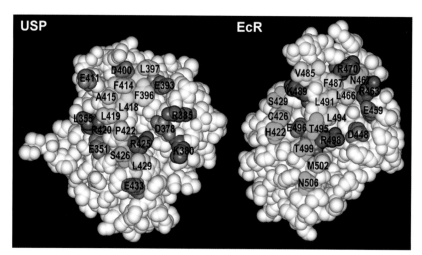

PALLI ET AL., FIGURE 4. Heterodimerization interface of the EcR and USP ligand binding domains. CPK surface representation of the interacting faces of *H. virescens* of the USP LBD (left) and of the muristerone A-bound LBD of EcR (right) (PDB 1R1K). Residues involved in intermolecular contacts at the interfaces are colored (hydrophobic in yellow, polar in green, acidic in red; basic in blue). The model was prepared with WebLab Viewer Pro; contacts were derived with CSU software (Sobolev *et al.*, 1999).

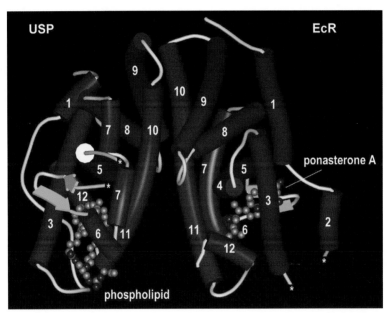

PALLI ET AL., FIGURE 5. Molecular structure of the ligand binding domains (LBDs) of the ecdysteroid receptor (EcR) and ultraspiracle protein (USP). The heterodimer between the EcR LBDs and USP LBDs of *H. virescens* (PDB 1R1K) is shown in a schematic backbone representation with α-helices as red tubes and β-sheets as blue arrows. The α-helices are numbered from 1 to 12. Bound phospholipid and ponasterone A are shown in ball-and-stick representation colored according to atom type (C in grey, O in red, P in orange, N in blue). (*) Loops connecting EcR helices 9 and 10, as well as USP helix 5 and β-sheet ("extra loop") are not depicted (Billas *et al.*, 2003). The model was prepared with WebLab Viewer Pro.

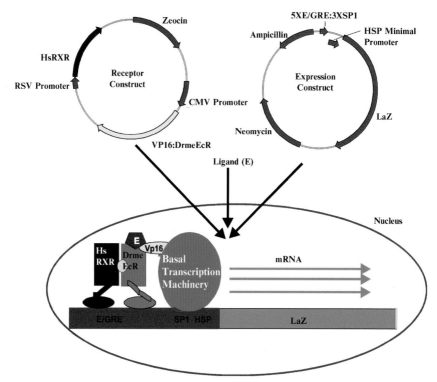

PALLI ET AL., FIGURE 6. First commercial version of EcR-based gene switch marketed by Invitrogen. A fusion proteins of VP16 activation domain and *D. melanogaster* EcR CDEF domains (VP16:DrmeEcR) and complete human RXR are expressed under the control of CMV and RBS promoters respectively. The reporter gene or gene of interest is cloned under the control of heat shock protein (HSP) minimal promoter and 5XE/GRE and 3X SP1 response elements. These two plasmids are co-transfected into mammalian cells and the transfected cells are exposed to ecdysteroids (E), such as ponasterone A or muristerone A.

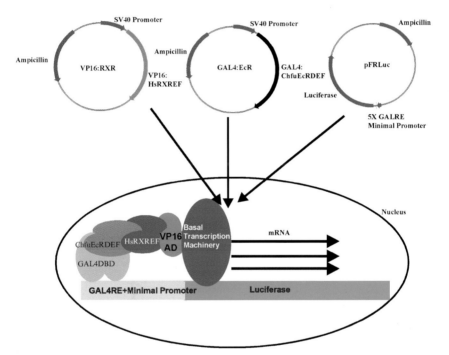

PALLI ET AL., FIGURE 7. Two-hybrid version of EcR-based gene switch. The fusion proteins of GAL4 DNA binding domain and EcR DEF, and VP16 activation domain and RXR EF are expressed under the control of SV40 promoters in two different plasmids. The reporter gene or gene of interest is expressed under the control of synthetic TATAA minimal promoter and 5X GAL4 response elements. These three plasmids are co-transfected into cells, and the transfected cells are exposed to ecdysteroid analog, RG-102240.

```
                H1              H2                         H3                      H4          H5       s1     s2    s3   H6
Bian 107  LTANQQFLIARLVWYQDGYDQPSEEDLKRVTQTWQADTEEIGEASDLPFRQITEMTILTVQLIVEFAKGLPGFAKISQPDQITLLKACSSEVMMLRVSRRYDMSTDSVMFANNQAYTREIYNKAG
Juco 102  LTANQQSLIARLVWYQDGYEQPSDEDLKRVTQTWQEADED.DEESDLPFRQITEMTILTVQLIVEFAKGLPGFAKISQPDQITLLKACSSEVMMLRVARRYDATTDSVLFANNRAYTRENYRKAG
Chfu 284  LTANQQFLIARLIWYQDGYEQPSDEDLKRITQTWQQADDE.NEESDTPFRQITEMTILTVQLIVEFAKGLPGFAKISQPDQITLLKACSSEVMMLRVARRYDAASDSVLFANNQAYTRDNYRKAG
Chsu 290  LTANQQFLIARLVWYQDGYEQPSEEDLKRVTQTWQSNEDE.EEETDLPFRQITEMTILTVQLIVEFAKGLPGFSKISQPDQITLLKACSSEVMMLRVARRYDAASDSVLFANNQAYTRDNYRKAG
Plin 286  LTANQQFLIAGLVWYQDGYEQPSEEDLKRVTQTWQPTDED.DD..DMPFRQITEMTILTVQLIVEFAKGLPGFSKISQPDQITLLKACSSEVMMLRVARRYDAATDSVLFANNQAYTRDNYRKAG
Bomo 349  LSANQKSLIARLVWYQEGYEQPSDEDLKRVTQS....DEE.DEESDLPFRQITEMTILTVQLIVEFAKGLPGFSKISQSDQITLLKASSSEVMMLRVARRYDAASDSVLFANNKAYTRDNYRQGG
Mase 290  LSANQKSLIARLVWYQEGYEQPSEEDLKRVTQTWQLEEEE.EEETDMPFRQITEMTILTVQLIVEFAKGLPGFSKISQSDQITLLKASSSEVMMLRVARRYDAATDSVLFANNQAYTRDNYRKAG
Hevi 290  LTANQKSLIARLVWYQEGYEQPSEEDLKRVTQSTWQ.DED.DEDSDMPFRQITEMTILTVQLIVEFAKGLPGFAKISQSDQITLLKACSSEVMMLRVARRYDAATDSVLFANNQAYTRDNYRKAG
Cavi 457  SSYNQLAVIYKLIWYQDGYEQPSEEDLKRIM.S...SPDENESQHDVSFRHITEITILTVQLIVEFAKGLPAFTKIPQEDQITLLKACSSEVMMLRMARRYDHNSDSIFFANNRSYTRDSYKMAG
Lucu 439  LSYNQLAVIYKLIWYQDGYEQPSEEDLKRIM.S...SPDENESQHDASFRHITEITILTVQLIVEFAKGLPAFTKIPQEDQITLLKACSSEVMMLRMARRYDHNSDSIFFANNRSYTRDSYKMAG
Drme 416  LTYNQLAVIYKLIWYQDGYEQPSEEDLRRIM.S...QPDENESQTDVSFRHITEITILTVQLIVEFAKGLPAFTKIPQEDQITLLKACSSEVMMLRMARRYDHSSDSIFFANNRSYTRDSYKMAG
Ceca 391  LTRNQLAVIYKLIWYQDGYEQPSEEDLKRIM.S...TPDENESPNDISFRHITEITILTVQLIVEFAKGLPAFTKIPQEDQITLLKACSSEVMMLRMARRYDHNSDSIFFANNRSYTRDAYKMAG
Aeae 434  LTANQMAVIYKLIWYQDGYEQPSEEDLKRIMIG...SPNEEEDQHDVHFRHITEITILTVQLIVEFAKGLPAFTKIPQEDQITLLKACSSEVMMLRMARRYDAATDSILFANNRSYTRDSYRMAG
Aeal 333  LTANQMAVIYKLIWYQDGYEQPSEEDLKRIMIG...SPNEEEDQHDVHFRHITEITILTVQLIVEFAKGLPAFTKIPQEDQITLLKACSSEVMMLRMARRYDAATDSILFANNRSYTRDSYRMAG
Anga 280  LTANQMAVIYKLIWYQDGYEQPSEEDLKRIMIN...SPNEEEDPHEIHFRHITEITILTVQLIVEFAKGLPAFTKIPQEDQITLLKACSSEVMMLRMARRYDAETDSILFANNRSYTRDSYKMAG
Chte 275  LTANQVAVIYKLIWYQDGYEQPSEEDLKRITT....LEEEEDQEHEANFRYITEVTILTVQLIVEFAKGLPAFIKIPQEDQITLLKACSSEVMMLRMARRYDHDSDSILFANNTAYTKQTYQLAG
Apmo 578  ISPEQEELIHRLVYFQNEYEQPSEEDLKRITNQ...PS.EGEDISDYKFRHITEITILTVQLIVEFSKRLPGFDELMREDQIALLKACSSEVMMLRMARKYDVQTDSIIFANNQPYTKDSYTVAG
Lomi 305  VSPEQEELIHRLVYFQNEYESPSEEDLRRVTSQ...PT.EGEDQSDVRFRHITEITILTVQLIVEFAKRLPGFDKLLREDQIALLKACSSEVMMFRMARRYDVNSDSILFANNQPYTKDSYNIAG
Temo 260  PEQEELILIHRLVYFQNEYEHPSEEDVKRIINQ...PI.DGEDQCEIRFRHTTEITILTVQLIVEFAKRLPGFDKLLQEDQIALLKACSSEVMMFRMARRYDVQSDSILFVNNQPYPRDSYNLAG
Cama 103  LTREQEELINTLVYYQEQFEQPTEADIKKIRFT....F.DGGDTSDMRFRHITEMTILTVQLIVEFSKQLPGLGTLQREDQITLLKACSSEVMMLRAARFYDAKTDCIVFGNTLPYTQSSYEFAG
Cepu 282  LTREQEELINTLVYYQEEFEQPTEADVKKIRFN....F.DGEDTSDMRFRHITEMTILTVQLIVEFSKQLPGFATLQREDQITLLKACSSEVMMLRAARRYDAKTDSIVFGNNYPYTQASYALAG
Amam 324  LSSSQEDLINKLVYYQQEFESPSEEDMKKTTPF...PLGDSEEDNQRRFQHITEITILTVQLIVEFSKRVPGFDTLAREDQITLLKACSSEVMMLRGARKYDVKTDSIVFANNQPYTRDNYRSAS

            H7           H8                 H9                  H10           H11        H12
Bain  FGYVIENLLHFCRCMYAMSWDNVHYALLTAVVIFSDRPGLENPQLVEEIQRYYLNTIRVYILNQQSGSTRCPVVYGKILSVLSELRSLGMQNSNMCISLKLKNRKLPPFLEEIWD   346
Juco  MSYVIENLLHFCRCMYTMSMDNVHYALLTAIVIFSDRPGLEQPHLVEEIQRYYLNTLRVYIMNQQSASSRCPVVFGKILSILSELRTLGMQNSNMCISLKLKNRKLPPFLEEIWD   339
Chfu  MAYVIEDLLHFCRCMYSMALDNIHYALLTAVVIFSDRPGLEQPQLVEEIQRYYLNTLRIYILNQLSGSARSSVIYGKILSILSELRTLGMQNSNMCISLKLKNRKLPPFLEEIWDV  523
Chsu  MAYVIEDLLHFCRCMYSLSMDNVHYALLTAVVIFSDRPGLEQPQLVEEIQRYYLNTLRVYIMNQHSASPRCAVLYAKILSVLTELRTLGMQNSNMCISLKLKNRKLPPFLEEIWDV  529
Plin  MAYVIEDLLHFCRCMYSMTMDNVHYALLTAIVIFSDRPGLEQPQLVEEIQRYYLNTLRVYIMNQHSDSPRCAVVFGKILSVLTEIRTLGMQNSNMCISLKLKNRKLPPFLEEIWDV  523
Bomo  MAYVIEDLLHFCRCMFAMGMDNVHFALLTAIVIFSDRPGLEQPSLVEEIQRYYLNTLRIYIINQNSASSRCAVIYGRILSVLTELRTLGTQNSNMCISLKLKNRKLPPFLEEIWDV  587
Mase  MSYVIEDLLHFCRCMYSMSMDNVHYALLTAIVIFSDRPGLEQPLLVEEIQRYYLKTLRVYILNQHSASPRCAVLFGKILGVLTELRTLGTQNSNMCISLKLKNRKLPPFLEEIWDV  529
Hevi  MAYVIEDLLHFCRCMYSMMMDNVHYALLTAIVIFSDRPGLEQPLLVEEIQRYYLNTLRVYILNQNSASPRGAVIFGEILGILTEIRTLGMQNSNMCISLKLKNRKLPPFLEEIWDV  528
Cavi  MADNIEDLLHFCRQMYSMKVDNVEYALLTAIVIFSDRPGLEEAELVEAIQSYYIDTLRIYILNRHCGDPMSLVFFAKLLSILTELRTLGNQNAEMCFSLKLKNRKLPKFLEEIWDV  690
Lucu  MADNIEDLLHFCRQMYSMKVDNVEYALLTAIVIFSDRPGLEEAELVEAIQSYYIDTLRIYILNRHCGDPMSLVFFAKLLSILTELRTLGNQNAEMCFSLKLKNRKLPKFLEEIWDV  675
Drme  MADNIEDLLHFCRQMFSMKVDNVEYALLTAIVIFSDRPGLEKAQLVEAIQSYYIDTLRIYILNRHCGDSMSLVFYAKLLSILTELRTLGNQNAEMCFSLKLKNRKLPKFLEEIWDV  652
Ceca  VADNIEDLLHFCRQMYSMKVDNVEYALLTAIVIFSDRPGLEKAQLVEEIQSYYIDTLRVYIINRHCGDSMSLVFFAKLLSILTELRTLGNQNAEMCFSLKLKNRKLPKFLEEIWDV  627
Aeae  MADTIEDLLHFCRQMFSLTVDNVEYALLTAIVIFSDRPGLEQAELVEHIQSYYIDTLRIYILNRHAGDPKCSVIFAKLLSILTELRTLGNQNSEMCFSLKLKNRKLPRFLEEIWDV  570
Aeal  MADTIEDLLHFCRQMFSLTVDNVEYALLTAIVIFSDRPGLEQAELVEHIQSYYIDTLRIYILNRHAGDPKCSVIFAKLLSILTELRTLGNQNSEMCFSLKLKNRKLPRFLEEIWDV  570
Anga  MADTIEDLLHFCRQMYTLTVDNVEYALLTAIVIFSDRPGLEKAELVETIQSYYIDTLRVYILNRHGGDPKCSVTFAKLLSILTELRTLGNQNSEMCFSLKLKNRKLPRFLEEIWDV  517
Chte  MEETIDDLLHFCRQMYALSIDNVEYALLTAIVIFSDRPGLEKAEMVDIIQSYYTETLKVYIVNRHGGESRCSVQFAKLLGILTELRTMGNKNSEMCFSLKLRNRKLPRFLEEVWDV  512
Apmo  MGETIEDLLHFCRQMYAMKVNNAEYALLTAIVIFSERPNLLEGWKVEKIQEIYLEALRAYVDNRR..RPNPGTVFARLLSVLTELRTLGNQNSEMCFSLKFKNKKLPVFLAEIWDV  812
Lomi  MGETIEDMLRFCRQMYAMKVDNAEYALLTAIVIFSERPSLVEGWKVEKIQEIYLEALKAYVDNRR..RPKSGTIFAKLLSVLTELRTLGNQNSEMCFSLKLKNKKLPPFLAEIWDV  539
Temo  MGETIEDLLHFCRTMYSMKVDNAEYALLTAIVIFSERPSLIEGWKVEKIQEIYLEALRAYVDNRR. SPSRGTIFAKLLSVLTELRTLGNQNSEMCISLKLKNKKLPPFLDEIWDV  487
Cama  LGESSQILFRFCRNLCKMKVDNAEYALLSAIIIFSERPNLKELQKVEKLQEIYLDALRAYVCNQR..FPKPGMVFAKLLNILTELRTLGNLNSEVCFSLKLKNKRLPPFLAEIWDV  336
Cepu  LGESAEILFRFCRSLCKMKVDNAEYALLAAIAIFSERPNLKELKKVEKLQEIYLEALKSYVENRR..LPRSNMVFAKLLNILTELRTLGNINSEMCFSLTLKNKRLPPFLAEIWDV  515
Amam  VGDSADALFRFCRKMCQLRVDNAEYALLTAIVIFSERPSLVDPHKVERIQEYYIETLRMYSENHR..PPGKN.YFARLLSILTELRTLGNMNAEMCFSLKVQNKKLPPFLAEIWDI  558
```

PALLI ET AL., FIGURE 9. Alignment of amino acids in the ligand binding domain of ecdysteroid receptors. The amino acid residues present in all EcRs and the amino acids present in only lepidopteran and dipteran EcRs are boxed. The EcR sequences are from *Bicyclus anynana* (Bian, unpublished, gi:6580162), *Junonia coenia* (Juco, unpublished, gi:6580625), *Choristoneura fumiferana* (Chfu [Kothapalli et al., 1995]), *Chilo suppressalis* (Chsu [Minakuchi et al., 2002]), *Plodia interpunctella* (Plin [Siaussat et al., 2004]), *B. mori* (Bomo [Kamimura et al., 1996; Swevers et al., 1995]), *Manduca sexta* (Mase [Fujiwara et al., 1995]), *H. virescens* (Hevi [Billas et al., 2003; Martinez et al., 1999a]), *Calliphora vicina* (Cavi, unpublished, gi:12034940), *Lucilia cuprina* (Lucu [Hannan and Hill, 1997]), *Drosophila melanogaster* (Drme [Koelle et al., 1991]), *Ceratitis capitata* (Ceca, Verras et al., 1999), *A. aegypti* (Aeae [Cho et al., 1995]), *Aedes albopictus* (Aeal [Jayachandran and Fallon, 2000]), *Anopheles gambiae* (Anga, unpublished, gi:55234452), *Chironomus tentans* (Chte [Imhof et al., 1993]), *Apis mellifera* (Apme, unpublished, *XP_394760*), *Locusta migratoria* (Lomi [Saleh et al., 1998]), *Tenebrio molitor* (Temo [Mouillet et al., 1997]), *Carcinus maenas* (Cama unpublished, gi:40748295), *Celuca pugilator* (Cepu [Chung et al., 1998]), and *Amblyomma americanum* (Amam [Guo et al., 1997]).

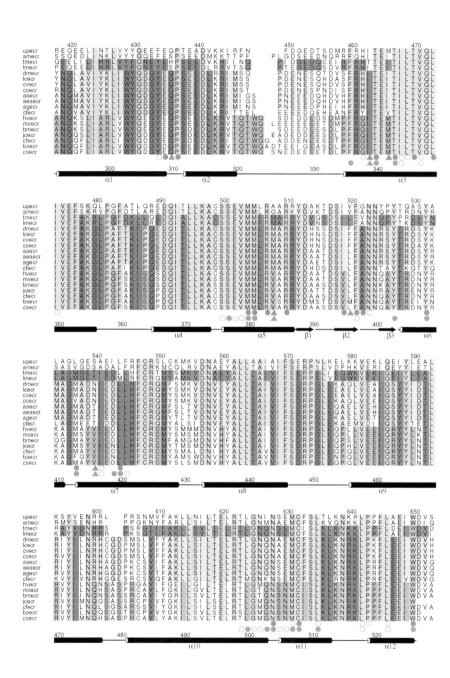

BILLAS AND MORAS, FIGURE 2. Amino acid sequence alignment of the EcR-LBD from different invertebrates. Shown are the EcR-LBD sequences of a crab (up: *Uca pugilator*), a tick (am: *Amblyomma americanu*), a beetle (tm: *Tenebrio molitor*), a grasshopper (lm: *Locusta migratoria*), and dipteran insects (dm: *Drosophila melanogaster*; lc: *Lucilia cuprina*, cv: *Calliphora vicina*, cc: *Ceratitis capitata*, aa: *A. aegypti*, aea: *Aedes albopictus*, ag: *Anopheles gambiae*, ct: *Chironomus tentans*) and lepidopteran insects (ms: *Manduca sexta*, bm: *Bombyx mori*, cf: *C. fumiferana*, ba: *Bicyclus anynana*, jc: *Junonia coenia*, sf: *Spodoptera frugiperda*, hv: *H. virescens*, and cs: *Chilo suppressalis*). Residue numbering corresponding to the sequences of hvEcR and dmEcR are given at the bottom and at the top of the alignment, respectively. The secondary structure elements (α-helices and β-sheets) are indicated at the bottom of the alignment. Color code of the amino acid conservation: light blue: 100% conservation for all sequences; dark blue: 100% conservation for all insects; yellow: 100% conservation for dipteran insects; pink: 100% conservation for lepidopteran insects; grey: 100% conservation for the coleopteran and orthopteran insects. The residues of the ligand binding pocket at less than 4.5 and 6.5 Å from ponA are shown by red full and empty dots, respectively. Residues at less than 4.5 and 6.5 Å from BYI06830 are shown by blue full and empty dots, respectively. Colored arrows indicate residues located at less than 4.5 Å that are differentially conserved in the lepidopteran insect order.

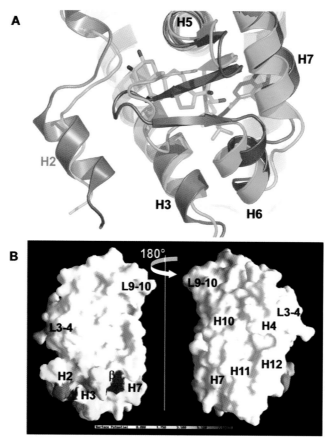

BILLAS AND MORAS, FIGURE 5. Flexible region in the EcR-LBD. (A) Ribbon diagram showing the superimposition of the structures of EcR-LBD in complex with ponA (orange ribbons) and with BYI06830 (green ribbons). The view is restricted to the regions differing the most between the two EcR-LBDs that includes H2, H6, H7, and the β-sheet. PonA and BYI06830 are shown in stick representation with carbon colored in cyan and light grey, respectively, oxygen in red and nitrogen in blue. (B) Surface representations of the ponA-bound EcR-LBD colored according to the root mean square (r.m.s.) deviations calculated residue by residue between the ponA- and the BYI06830-bound EcR-LBDs. The scale ranges from 0 (white) to 7.5 Å (red). The two views are related by a 180° rotation around the vertical axis.

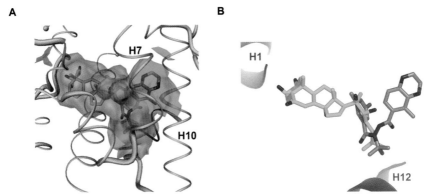

BILLAS AND MORAS, FIGURE 6. The LBDs of EcR complexed to a steroidal and to a nonsteroidal ligands exhibit different, and only partially overlapping ligand-binding cavities. (A) View of the two ligand-binding cavities of EcR-LBD, together with their respective ligand. The ponA-bound EcR cavity is shown in blue and the BYI06830-bound EcR cavity in green. (B) Superimposition of the steroidal and nonsteroidal EcR ligands bound to the EcR-LBD. Atom coloring is red for oxygen, blue for nitrogen, cyan for carbon in ponA and orange for carbon in BYI06830.

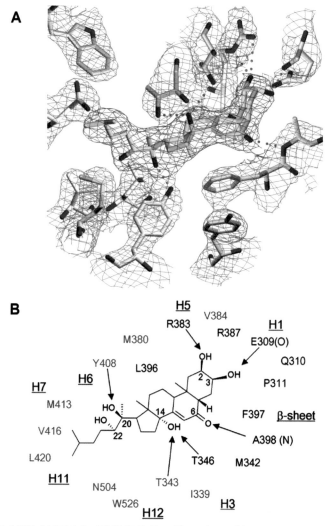

BILLAS AND MORAS, FIGURE 7. The ecdysteroid EcR ligand binding pocket. (A) View of the electron density for the ponA-bound EcR-LBD at 2.9 Å resolution. Two different maps are shown: a sigmaA weighted $2F_{obs} - F_{calc}$ omit map for the ligand (in blue) and a sigma A weighted $2F_{obs} - F_{calc}$ map (in magenta) for selected residues in the ligand-binding pocket. The map is contoured at 1σ and overlaid on the final refined models. Hydrogen bonds between ligand and residues are indicated by green dotted lines. (B) Schematic representation of the interactions of ponA with residues of the binding cavity. Arrows correspond to hydrogen bonds between ligand and amino acid residues. Residues in blue are common to both structures.

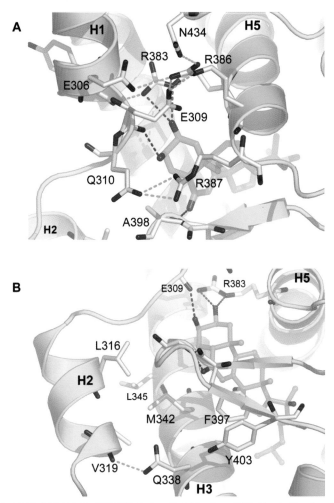

BILLAS AND MORAS, FIGURE 8. Stabilization of H2. PonA directly stabilizes the helical conformation of H2 via an intricate network of hydrogen bonds between the ligand and EcR residues (pink dashed lines) as well as between residues of neighboring structural elements (green dashed lines). Views of (A) the region comprising H1, H5, and the loop H1–H2 and (B) the region downwards encompassing helix H2 and the β-sheet. The amino acid residues and ponA are shown in stick representation with oxygen in red, nitrogen in blue, sulfur in green and carbon in white and cyan for EcR residues and ponA, respectively.

A

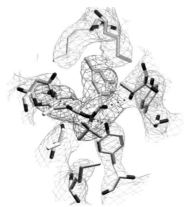

B

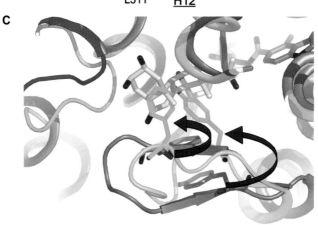

C

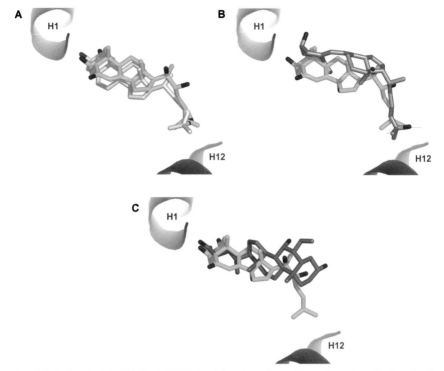

BILLAS AND MORAS, FIGURE 11. Superimposition of ponA with ligands of human NRs. The structures of the LBD of LXR bound to 24(S), 25-epoxycholesterol (eCH), VDR bound to 1α, 25-dihydroxyvitamin D_3 (vitamin D) and FXR bound to the bile acid 6α-ethyl-chenodeoxycholic acid (6ECDCA) were superimposed to the structure of ponA-bound EcR-LBD. Shown are the resulting superimpositions of ponA with (A) eCH, (B) vitamin D, and (C) 6ECDCA. Helices H1 and H12 are indicated. The orientation is identical to that of Fig. 6B. Carbon atoms are shown in cyan for ponA, grey for eCH, pink for vitamin D, and magenta for 6ECDCA. Oxygen atoms are shown in red.

BILLAS AND MORAS, FIGURE 9. The dibenzoylhydrazine EcR ligand binding pocket. (A) View of the electron density for the BYI06830-bound EcR-LBD at 3.0 Å resolution. Two different maps are shown in each figure: a sigmaA weighted $2F_{obs} - F_{calc}$ omit map for the ligand (in blue) and a sigmaA weighted $2F_{obs} - F_{calc}$ map (in magenta) for selected residues in the ligand-binding pocket. The map is contoured at 1σ and overlaid on the final refined models. Hydrogen bonds between ligand and residues are indicated by green dotted lines. (B) Schematic representation of the interactions of BYI06830 with the residues of the binding cavity. Arrows correspond to hydrogen bonds between ligand and amino acid residues. Residues in blue are common to both structures. (C) The structural adaptation of EcR upon binding of synthetic dibenzoylhydrazine compounds involves an inner to outer switch of two aromatic residues F397 and Y403 belonging to the β-sheet. In the BYI06830 EcR complex, these two residues fill the region of the pocket that is occupied by the ponA steroid core and left empty by the dibenzoylhydrazine ligand. The EcR structure and the carbon atom of the corresponding residues are shown in orange and in green for the ponA-bound and the BYI06830-bound complexes, respectively. The carbon atoms of ponA and BYI06830 are colored in grey and cyan, respectively. Oxygen and nitrogen atoms are shown in red and in blue, respectively.

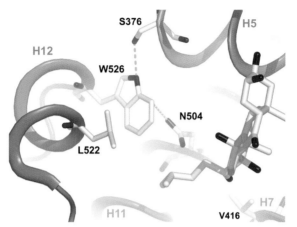

BILLAS AND MORAS, FIGURE 12. Stabilization of H12. The cation–π interaction between Trp526 and N504 (yellow dotted line) helps stabilizing the agonist conformation of the activation helix H12. These two residues are essential for steroidal and nonsteroidal ligand activation of EcR. In addition Trp526 is hydrogen bonded to S376 (green dotted line). The carbon atoms are colored in grey, oxygen atoms in red, and nitrogen atoms in blue.

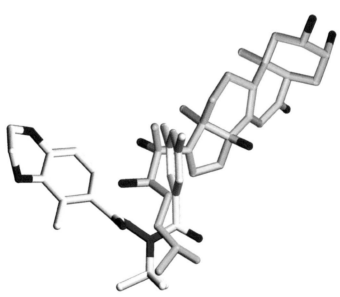

NAKAGAWA, FIGURE 3. Superposition between PonA and BYI06830 (Reproduced with the permission of Nature Publishing Group).

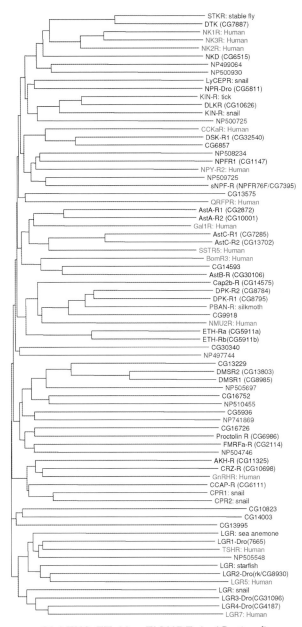

CLAEYS ET AL., FIGURE 1. (*Continued*)

CLAEYS ET AL., FIGURE 1. Dendrogram of a comparison of *Drosophila* GPCRs related to rhodopsin family peptide receptors. Characterized and orphan peptide receptors belonging to the rhodopsin family (GPCR class A) were compared by CLUSTALW analysis. *Drosophila* genome encoded receptors are shown in black, while the names of a number of related GPCRs from human (green), nematode *C. elegans* (violet), other arthropod (blue) and other invertebrate (dark red) origin are shown in color. One of the previously predicted fruit fly GPCRs (CG12610; Hewes and Taghert, 2001) was not included in this comparison, since it is only represented in the genome and protein databases as a partial amino acid sequence. NK1R, NK2R, and NK3R: neurokinin receptors; CCKaR: cholecystokinin A receptor; NPY-R2: Y2 type NPY receptor; QRFPR: receptor for QRFP, an RF-amide peptide; Gal1R: galanin 1 receptor; SSTR5: somatostatin receptor 5; BomR3: bombesin receptor 3; NMU2R: neuromedin 2 receptor; GnRHR: gonadotropin releasing hormone receptor; TSHR: thyroid stimulating hormone receptor; LGR: Leu-rich repeats containing GPCR; CPR1, CPR2: conopressin (vasopressin-like peptide) receptors. The abbreviations of the insect receptors are explained in Table I, as well as in the text. For a list of insect peptides and their sequences, recent review papers can be referred (Gäde *et al.*, 1997; Hewes and Taghert, 2001; Nässel, 2002; Riehle *et al.*, 2002; Vanden Broeck, 2001a).

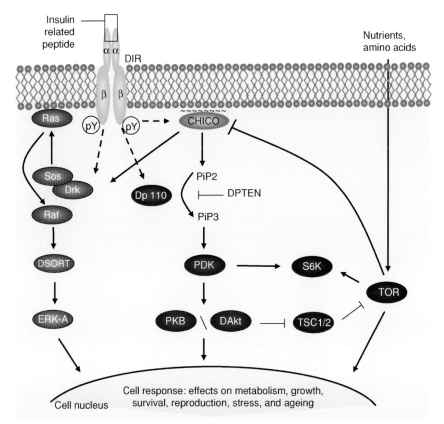

CLAEYS ET AL., FIGURE 2. Model of the insulin pathway in *D. melanogaster*. Upon binding of an insulin-like peptide (dilp), the activated insulin receptor initiates different pathways, namely the Ras/MAPK – and the PI3K/TOR pathway (Blenis, 1993; Shepherd *et al.*, 1998). Directly or indirectly (via CHICO), the activated insulin receptor induces the conversion of the membrane lipid phosphatidylinositol 4,5-bisphosphate (PIP2) into the second messenger phosphatidylinositol 3,4,5-trisphosphate (PIP3) through PI3K (Dp110) (Shepherd *et al.*, 1998). Increased levels of PIP3 cause translocation of the serine/threonine kinases PDK and DAkt to the cell membrane through interactions between PIP3 and the pleckstrin homology (PH) domains of these proteins. PDK then promotes activation of both DAkt (PKB) and S6K through phosphorylation of their activation loops. 40 S ribosomal protein S6 kinase can also be activated through the serine/threonine kinase TOR (Brown *et al.*, 1995). The signaling function of TOR appears to be activated in response to nutrient levels, particularly those of amino acids (Jacinto and Hall, 2003). Recent studies suggest that a link between PI3K and TOR may occur through DAkt-mediated phosphorylation of TSC-2 (tuberous sclerosis complex protein 2), which was found to disrupt and inactivate the TSC1/TSC2 complex (Dan *et al.*, 2002; Inoki *et al.*, 2002, Potter *et al.*, 2002). An additional level of cross-talk between the PI3K and TOR pathways occurs through a negative feedback loop involving TOR-mediated inhibition of IRS (CHICO), an adaptor protein required for the PI3K activation by the insulin receptor. Activation of TOR results in phosphorylation and subsequent proteasomal degradation of CHICO, leading to reduced PI3K signaling (Haruta *et al.*, 2000). The parallel PI3K/TOR pathways are likely to play an important role in coordinating cell growth (and other cell responses) with other metabolic programs. Binding of ILP to the receptor also leads to the formation of the Drk/Sos complex and

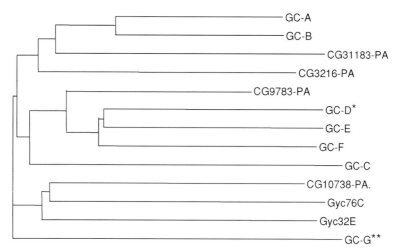

CLAEYS ET AL., FIGURE 3. Dendrogram of a sequence comparison of mammalian and *Drosophila* receptor guanylyl cyclases (rGC). Characterized and orphan receptor guanylyl cyclases (rGC) of mammalian or insect origin were compared by CLUSTALW analysis. The different receptors are discussed in the text and related receptors from other organisms are summarized in Table IV. (*) All mammalian rGC are derived from *H. sapiens*, except for GC-D, which is from *Rattus norvegicus*. (**) Predicted gene (Accession number: XP_497249).

subsequently to the activation of membrane associated Ras-protein. Ras-activation triggers the activation of a kinase cascade that includes Raf, MAPKK (MEK, DSORT), and MAPK (ERK, ERK-A). Once activated MAPK homodimerizes and is imported into the nucleus where it phosphorylates target proteins that regulate transcription, the ultimate goal of the signaling pathway.

Abbreviations: Dilp, "*Drosophila* insulin-like peptide;" DIR, "*Drosophila* insulin receptor;" Drk, "Downstream of RTK" (Grb2); DSORT, orthologue of MEK, "Mitogen-activated ERK-activating kinase"; ERK-A, orthologue of ERK, "Extracellular-signal-regulated kinase"; IRS, "Insulin receptor substrate"; MAPK, "Mitogen-activated protein kinase"; PDK, "Phosphatidylinositol-dependent protein kinase"; PI3K, "Phosphatidylinositol-3-OH kinase"; PKB, "Protein kinase B"; PTEN, phosphatidylinositide phosphatase; S6K, "Ribosomal protein S6 kinase"; Sos, "Son of Sevenless"; TOR, "Target of rapamycin;" TSC1/TSC2, "Tuberous sclerosis complex" protein 1 and 2.